*Palaeozoic Palaeobotany of Great Britain*

# THE GEOLOGICAL CONSERVATION REVIEW SERIES

The comparatively small land area of Great Britain contains an unrivalled sequence of rocks, mineral and fossil deposits, and a variety of landforms that span much of the earth's long history. Well-documented ancient volcanic episodes, famous fossil sites and sedimentary rock sections used internationally as comparative standards, have given these islands an importance out of all proportion to their size. The long sequences of strata and their organic and inorganic contents have been studied by generations of leading geologists, thus giving Britain a unique status in the development of the science. Many of the divisions of geological time used throughout the world are named after British sites or areas, for instance, the Cambrian, Ordovician and Devonian systems, the Ludlow Series and the Kimmeridgian and Portlandian stages.

The Geological Conservation Review (GCR) was initiated by the Nature Conservancy Council in 1977 to assess, document and ultimately publish accounts of the most important parts of this rich heritage. Since 1991, the task of publication has been assumed by the Joint Nature Conservation Committee on behalf of the three country agencies, English Nature, Scottish Natural Heritage and the Countryside Council for Wales. The GCR series of volumes will review the current state of knowledge of the key earth-science sites in Great Britain and provide a firm basis on which site conservation can be founded in years to come. Each GCR volume will describe and assess networks of sites of national or international importance in the context of a portion of the geological column, or a geological, palaeontological or mineralogical topic. The full series of approximately 50 volumes will be published by the year 2000.

Within each individual volume, every GCR locality is described in detail in a self-contained account, consisting of highlights (a précis of the special interest of the site), an introduction (with a concise history of previous work), a description, an interpretation (assessing the fundamentals of the site's scientific interest and importance), and a conclusion (written in simpler terms for the non-specialist). Each site report is a justification of a particular scientific interest at a locality, of its importance in a British or international setting and ultimately of its worthiness for conservation.

The aim of the Geological Conservation Review series is to provide a public record of the features of interest in sites being considered for notification as Sites of Special Scientific Interest (SSSIs). It is written to the highest scientific standards but in such a way that the assessment and conservation value of the site are clear. It is a public statement of the value given to our geological and geomorphological heritage by the earth-science community which has participated in its production, and it will be used by the Joint Nature Conservation Committee, English Nature, the Countryside Council for Wales and Scottish Natural Heritage in carrying out their conservation functions. The three country agencies are also active in helping to establish sites of local and regional importance. Regionally Important Geological/Geomorphological Sites (RIGS) augment the SSSI coverage, with local groups identifying and conserving sites which have educational, historical, research or aesthetic value.

All the sites in this volume have been proposed for notification as SSSIs; the final decision to notify, or renotify, lies with the governing Council of the appropriate country conservation agency.

Information about the GCR publication programme may be obtained from:

Earth Science Branch,
Joint Nature Conservation Committee,
Monkstone House,
City Road,
Peterborough PE1 1JY.

**Titles in the series**

1. **Geological Conservation Review**
   **An Introduction**

2. **Quaternary of Wales**
   S. Campbell and D.Q. Bowen

3. **Caledonian Structures in Britain**
   **South of the Midland Valley**
   Edited by J.E. Treagus

4. **British Tertiary Volcanic Province**
   C.H. Emeleus and M.C. Gyopari

5. **Igneous Rocks of South-West England**
   P.A. Floyd, C.S. Exley and M.T. Styles

6. **Quaternary of Scotland**
   Edited by J.E. Gordon and D.G. Sutherland

7. **Quaternary of the Thames**
   D.R. Bridgland

8. **Marine Permian of England**
   D.B. Smith

9. **Palaeozoic Palaeobotany of Great Britain**
   C.J. Cleal and B.A. Thomas

10. **Fossil Reptiles of Great Britain**
    M. Benton and P.S. Spencer

# *Palaeozoic Palaeobotany of Great Britain*

**C.J. Cleal and B.A. Thomas**

Department of Botany,
National Museum of Wales,
Cardiff, UK.

GCR Editor: W.A. Wimbledon and D. Palmer

**CHAPMAN & HALL**
London · Glasgow · New York · Tokyo · Melbourne · Madras

**Published by Chapman & Hall, 2–6 Boundary Row, London SE1 8HN, UK**

Chapman & Hall, 2-6 Boundary Row, London SE1 8HN, UK

Blackie Academic & Professional, Wester Cleddens Road, Bishopbriggs, Glasgow G64 2NZ, UK

Chapman & Hall GmbH, Pappelallee 3, 69469 Weinheim, Germany

Chapman & Hall USA, One Penn Plaza, 41st Floor, New York NY10119, USA

Chapman & Hall Japan, ITP-Japan, Kyoto Building, 3F, 2-2-1 Hirakawacho, Chiyoda-ku, Tokyo 102, Japan

Chapman & Hall Australia, Thomas Nelson Australia, 102 Dodds Street, South Melbourne, Victoria 3205, Australia

Chapman & Hall India, R. Seshadri, 32 Second Main Road, CIT East, Madras 600 035, India

First edition 1995

© 1995 Joint Nature Conservation Committee

Printed in Great Britain at the University Press, Cambridge

ISBN 0 412 61090 6

Apart from any fair dealing for the purposes of research or private study, or criticism or review, as permitted under the UK Copyright Designs and Patents Act, 1988, this publication may not be reproduced, stored, or transmitted, in any form or by any means, without the prior permission in writing of the publishers, or in the case of reprographic reproduction only in accordance with the terms of the licences issued by the Copyright Licensing Agency in the UK, or in accordance with the terms of licences issued by the appropriate Reproduction Rights Organization outside the UK. Enquiries concerning reproduction outside the terms stated here should be sent to the publishers at the London address printed on this page.

The publisher makes no representation, express or implied, with regard to the accuracy of the information contained in this book and cannot accept any legal responsibility or liability for any errors or omissions that may be made.

A catalogue record for this book is available from the British Library

Library of Congress Catalog Card Number: 94-70931

∞ Printed on permanent acid-free text paper, manufactured in accordance with ANSI/NISO Z39.48-1992 and ANSI/NISO Z39.48-1984 (Permanence of Paper).

# Contents

| | |
|---|---|
| **Acknowledgements** | ix |
| **Access to the countryside** | x |
| **Preface** | xi |

**1 Introduction** — 1
- Palaeozoic vegetational history — 3
- Palaeobotanical problems — 8
- The systematics of the Plant Kingdom — 12
- The choice of GCR sites — 14

**2 History of research on British plant fossils** — 17
- Silurian — 19
- Devonian — 19
- Lower Carboniferous — 20
- Upper Carboniferous — 21
- Permian — 22

**3 Silurian** — 23
- Palaeogeographical setting — 25
- Stratigraphical background — 25
- Evidence of pre-Silurian vegetation — 25
- Silurian vegetation — 26
- Silurian plant fossils in Britain — 29
- Pen-y-Glog Quarry — 31
- Llangammarch Wells Quarry — 35
- Rockhall Quarry — 36
- Cwm Craig Ddu Quarry — 38
- Capel Horeb Quarry — 39
- Perton Lane — 43
- Freshwater East — 45

**4 Devonian** — 51
- Palaeogeographical setting — 53
- Stratigraphical background — 56
- Devonian vegetation — 56
- Devonian plant fossils in Britain — 59
- Targrove Quarry — 61
- Turin Hill — 63
- Llanover Quarry — 67
- Craig-y-Fro Quarry — 71
- Ballanucater Farm (R.J. Rayner) — 75

# Contents

|  |  |
|---|---|
| Auchensail Quarry (R.J. Rayner) | 78 |
| Rhynie | 80 |
| Bay of Skaill | 93 |
| Sloagar | 96 |
| Plaistow Quarry | 100 |

**5 Lower Carboniferous** — **105**

| | |
|---|---|
| Palaeogeographical setting | 107 |
| Stratigraphical background | 107 |
| Early Carboniferous vegetation | 110 |
| Lower Carboniferous plant fossils in Britain | 112 |
| Lennel Braes | 113 |
| Whiteadder | 116 |
| Oxroad Bay (R.M. Bateman, G.W. Rothwell and C.J. Cleal) | 127 |
| Weak Law | 139 |
| Kingwater | 140 |
| Pettycur | 141 |
| Kingswood End | 150 |
| Laggan | 152 |
| Loch Humphrey Burn (R.M. Bateman and C.J. Cleal) | 155 |
| Glenarbuck (R.M. Bateman and C.J. Cleal) | 164 |
| Puddlebrook Quarry | 167 |
| Moel Hiraddug | 172 |
| Teilia Quarry | 175 |
| Wardie Shore | 181 |
| Glencartholm | 184 |
| Victoria Park | 188 |

**6 Upper Carboniferous** — **193**

| | |
|---|---|
| Palaeogeographical setting | 195 |
| Stratigraphical background | 196 |
| Late Carboniferous vegetation | 196 |
| Upper Carboniferous plant fossils in Britain | 203 |
| Nant Llech | 204 |
| Wadsley Fossil Forest | 208 |
| Cattybrook Claypit | 210 |
| Nostell Priory Brickpit | 214 |
| Llanbradach Quarry | 217 |
| Jockie's Syke | 219 |

**7 Permian** — **223**

| | |
|---|---|
| Palaeogeographical setting | 225 |
| Stratigraphical background | 225 |
| Permian vegetation | 225 |
| The Palaeophytic–Mesophytic transition | 229 |
| Permian plant fossils in Britain | 229 |
| Stairhill | 229 |
| Middridge Quarry | 233 |
| Kimberley Railway Cutting | 236 |

| | |
|---|---|
| **References** | **239** |
| **Glossary** | **279** |
| **Index** | **283** |

# *Acknowledgements*

Work on this volume was initiated by the Nature Conservancy Council and has been seen to completion by the Joint Nature Conservation Committee on behalf of the three country agencies, English Nature, Scottish Natural Heritage and the Countryside Council for Wales. Since the Geological Conservation Review was initiated in 1977 by Dr G.P. Black, then Head of the Geology and Physiography Section of the Nature Conservancy Council, many specialists in addition to the authors have been involved in the assessment and selection of sites; this vital work is gratefully acknowledged.

This volume represents the fruits of a 12-year review of Britain's Palaeozoic palaeobotany sites. It was done in collaboration with most of those palaeobotanists who have an interest in these sites, and who were consulted at both the site-selection and writing-up phases of the project. We recognize that without this collaboration, the project would be effectively meaningless, and we thank all those who have given their advice and suggestions. Certain people have provided particularly valuable input, however, and merit specific mention: Dr Dianne Edwards (University of Wales College of Cardiff) and Mr Peter Tarrant (Morville, Shropshire) for help on the Silurian and Devonian; Dr Albert Long (Berwick-on-Tweed), Dr Nick Rowe (Royal Holloway, University of London), the late Dr John Holmes (Université des Sciences et Techniques, Montpellier) and Mr Stan Wood (Edinburgh) for help on the Lower Carboniferous; Mr Cedric Shute (Natural History Museum, London) and Ms Gaynor Boon (Sheffield City Museum) for help on the Upper Carboniferous; and Professor Bob Wagner (Jardin Botanico de Córdoba) and Mr Tim Pettigrew (Tyne & Wear Museums Service) for help on the Permian. Most of the photographic illustrations were prepared by the photographic studios of the National Museum of Wales (Cardiff) and the Natural History Museum (London), to whom we are very grateful. Special thanks must go to Mr Cedric Shute of the Natural History Museum, for taking some of the photomicrographs. Line drawings for the volume were prepared by Ian Foulis Associates (Saltash).

Finally, we would like to thank Dr George Black for inviting us to participate in the project, Dr Bill Wimbledon, for his help and support during site-selection and for his patient editing of the early text, and the GCR publication production team – Dr Des O'Halloran, Neil Ellis (Publications Manager); Valerie Wyld (Text Officer); Nicholas D.W. Davey (Scientific Officer/Editorial Assistant).

# *Access to the countryside*

This volume is not intended for use as a field guide. The description or mention of any site should not be taken as an indication that access to a site is open or that a right of way exists. Most sites described are in private ownership, and their inclusion herein is solely for the purpose of justifying their conservation. Their description or appearance on a map in this work should in no way be construed as an invitation to visit. Prior consent for visits should always be obtained from the landowner and/or occupier.

Information on conservation matters, including site ownership, relating to Sites of Special Scientific Interest (SSSIs) or National Nature Reserves (NNRs) in particular counties or districts may be obtained from the relevant country conservation agency headquarters listed below:

English Nature,
Northminster House,
Peterborough PE1 1UA.

Scottish Natural Heritage,
12 Hope Terrace,
Edinburgh EH9 2AS.

Countryside Council for Wales,
Plas Penrhos,
Ffordd Penrhos,
Bangor,
Gwynedd LL57 2LQ.

# Preface

This volume summarizes the results of a survey of British Palaeozoic palaeobotany sites, undertaken between 1978 and 1990 as part of the Geological Conservation Review (GCR). The GCR was the first attempt to assess the scientific significance of all Britain's geological sites and has proved a landmark in the development of a coherent geological conservation strategy in this country. To ensure that the assessments were based on a firm logical and scientific foundation, the range of scientific interest was divided into ninety-seven discrete blocks, reflecting the natural divisions of stratigraphy, palaeogeography and geological process; Palaeozoic palaeobotany was one of these blocks.

The first stage in the survey was a review of the literature, to establish a comprehensive database of sites. From this, a provisional list of potentially significant sites was made and this was circulated to all relevant specialists in this country and abroad. At the same time, the sites were visited to assess their physical condition and whether the interest was still extant. In some cases, this excavation (so-called 'site-cleaning') was carried out to see if the interest of a site could be resurrected or enhanced. The comments made by the specialists and the field observations were then used to produce a second site list, which again was circulated for comment. This process of consultation continued until a consensus was reached among the specialists about which Palaeozoic sites were of sufficient palaeobotanical interest to justify conservation. The minimum criterion was that it was the best in Britain for yielding a particular assemblage of plant fossils. The resulting GCR sites were thus, at the very least, of national scientific importance, although many, such as Craig-y-Fro Quarry, Rhynie and the various Lower Carboniferous petrifaction sites, were also of international importance.

These GCR sites have been used as building-blocks for establishing a new set of Sites of Special Scientific Interest (SSSIs). If there was no other significant interest at or adjacent to the site, a proposal was made to establish it as an SSSI on the palaeobotanical interest alone. In many cases, however, a site showed other potentially significant features, or it adjoined another site of significance. In these cases, a composite proposed SSSI would be constructed from a set of GCR sites. Despite the heterogeneous nature of such sites, it is important to remember that the palaeobotanical interest is sufficient on its own to justify the conservation of the part of the site yielding the plant fossils. The SSSI proposals that have arisen out of this survey have been sent to the appropriate country conservation agencies (English Nature, Countryside Council for Wales, Scottish Natural Heritage), whose governing Councils are responsible for the final decision to notify them.

This volume is not intended to be a field guide to these sites, nor does it cover the practical problems of their future conservation. Its remit is to put on record the

# Preface

scientific justification for conserving the sites, discussing the interest of the fossils found there, and placing them in a wider palaeobotanical context. Each site is dealt within a self-contained account, consisting of highlights (a précis of its special scientific interest), a general introduction (with a brief historical review of research carried out there), a brief statement as to the stratigraphical context, and a list of all reported plant macrofossil species (including a statement about preservation-types). A detailed interpretation of the significance of the site then follows. This interpretive section has unavoidably had to be couched in technical language, because the conservation value is mostly based on a specialist understanding of the fossils present. The account of each site ends, however, with a brief summary of the interest framed in less technical language, in order to help the non-specialist.

The inclusion of a site in this volume should not be taken as an indication of rights of access, nor should it be taken as an invitation to visit. The majority of the sites are in private ownership and prior permission to visit must always be obtained from the landowner and/or occupier. In many cases the sites are vulnerable to over-exploitation, and it is hoped that those that do visit them will treat them with the respect that should be given to any other part of our unique national heritage.

Finally, it must be emphasized that this volume does not provide a fixed list of the important Palaeozoic palaeobotany sites in Britain. Palaeobotany, like any other science, is an ever-developing pursuit with new discoveries being made continually. During the progress of this very survey, findings at three sites raised them from being of little apparent interest to being of national importance (Targrove Quarry, Kingswood End, Wadsley Fossil Forest). It is inevitable, therefore, that further sites worthy of conservation will be discovered in future years. There is also the problem of potential site loss, with at least one location having come under threat during the time of the survey (Nostell Priory Brickpit). This volume deals with our knowledge of the sites available at the time of the GCR survey (mainly during the 1980s) and must be seen in this context. Nevertheless, the account clearly demonstrates the value of British sites for Palaeozoic palaeobotany, and their important place in Britain's scientific and natural heritage.

# Chapter 1

# Introduction

# Introduction

Palaeobotany is the study of the fragmentary fossilized remains of plants and, as such, can be undertaken for a variety of purposes. Plant fossils can be used for morphological studies and as evidence for whole-plant reconstructions, as well as providing the basis for evolutionary theory, palaeoecological reconstructions and interpretations, and biostratigraphy. This volume deals with the palaeobotany of the Palaeozoic Era, which represents the first 200 million years of land plant history. It was a time of major evolution and diversification of plants, both in a taxonomic and structural sense, as they adapted to terrestrial habitats. This, coupled with regional climatic and edaphic factors, led to geographical isolation of plant groups, regional floras and the complex ecological interrelationships that existed during the latter part of the Palaeozoic. A brief introduction to these points is necessary to appreciate the choice of sites listed in this volume and the species lists included in them. It will also be useful to explain briefly how plant fossils are formed, and how palaeobotanists study and name them. The reader in search of a more detailed understanding of plant fossils, explanations of the various systematic groups of plants and for interpretations of the fossil plants themselves may start by consulting one of the several available general texts on palaeobotany such as Taylor (1981), Thomas (1981a), Bassett and Edwards (1982), Stewart (1983), Meyen (1987), Thomas and Spicer (1987), Stewart and Rothwell (1992) and Taylor and Taylor (1992).

## Palaeozoic vegetational history

The history of terrestrial vegetation can be divided into three broad phases, known as the Palaeophytic, Mesophytic and Cenophytic (Gothan and Weyland, 1954) and corresponding in time approximately to the Palaeozoic, Mesozoic and Cenozoic eras. The phases represent times of radiation of land vegetation, separated by major extinction events - the Permian-Triassic, and the K-T (Cretaceous-Tertiary) boundary events (Erwin, 1990; Halstead, 1990). These events had a major disruptive effect on vegetation, causing the extinction of many plant families (e.g. see Figure 7.1 for the Permian-Triassic event). However, the effective emptying of many habitats also had an accelerating effect on plant evolution, allowing the diversification of new groups that had previously found it difficult to compete with the well-established communities of more primitive forms. For instance, the conifers and cycads first appeared in the Palaeozoic (Clement-Westerhof, 1988; Gao and Thomas, 1989), but it was not until the Mesozoic, when the well established pteridophyte-dominated forests had disappeared, that they underwent a major proliferation. The extinction events in the end proved to be a major driving-force in land plant evolution, by clearing the competitive 'log-jams' presented by the climax communities of the more primitive groups.

Being the first of these great phases in the evolutionary history of plants, the Palaeophytic is in many ways the most fundamental. It can be viewed from two perspectives - structural and taxonomic. Most of the major morphological and anatomical strategies that allowed plants to take advantage of the terrestrial habitats appeared at this time (Chaloner, 1970); only angiospermous flowers (probably) did not appear until the Mesozoic. The progressive appearance of these adaptations is summarized in Figure 1.1. They fall into three broad categories: reproduction-dispersal, architecture and water relations.

### *Reproduction and dispersal*

Clearly, plants would only have been able to occupy terrestrial habitats when their propagules were able to reach and remain viable on land. The spores of most marine algae cannot survive out of water for any length of time. However, some freshwater green algae produce spores with a resistant wall, allowing them to disperse between isolated areas of freshwater, which would be a significant pre-adaptation to a more fully terrestrial life. This pre-adaptive hypothesis is supported by the fact that the earliest known resistant spores with a trilete mark are from the lower Silurian (upper Llandovery; Burgess, 1991), significantly pre-dating the other evidence for the existence of land plants.

In pteridophytic reproduction, an asexual sporophytic plant or generation produces spores which disperse and germinate to produce a gametophyte plant or generation. The gametophytes produce eggs and male gametes, which fuse to form a zygote, from which the next sporophyte generation grows. There is increasing evidence that some early land plants had sporophyte and gametophyte generations that looked essentially similar (Remy, 1991). However, this symmetry soon broke down and in vascular plants the sporophyte became much larger and the

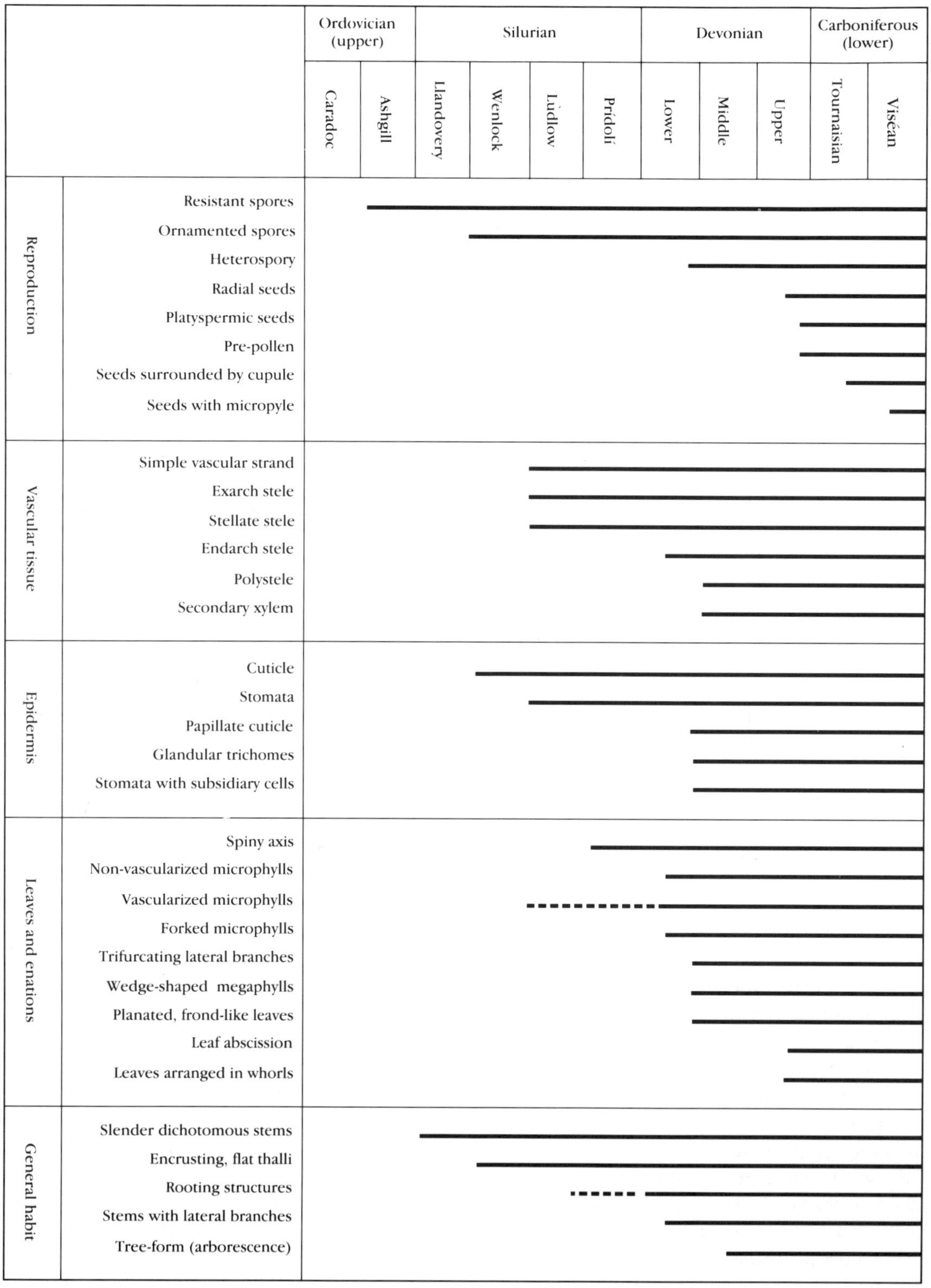

| | | Ordovician (upper) | | Silurian | | | | Devonian | | | Carboniferous (lower) | |
|---|---|---|---|---|---|---|---|---|---|---|---|---|
| | | Caradoc | Ashgill | Llandovery | Wenlock | Ludlow | Přídolí | Lower | Middle | Upper | Tournaisian | Viséan |
| **Reproduction** | Resistant spores | | —————————————————————————————— | | | | | | | | | |
| | Ornamented spores | | | ———————————————————————— | | | | | | | | |
| | Heterospory | | | | | | | | ——————————————— | | | |
| | Radial seeds | | | | | | | | | | ———————— | |
| | Platyspermic seeds | | | | | | | | | | ———————— | |
| | Pre-pollen | | | | | | | | | | ———————— | |
| | Seeds surrounded by cupule | | | | | | | | | | ——————— | |
| | Seeds with micropyle | | | | | | | | | | | ——— |
| **Vascular tissue** | Simple vascular strand | | | | ——————————————————————— | | | | | | | |
| | Exarch stele | | | | ——————————————————————— | | | | | | | |
| | Stellate stele | | | | ——————————————————————— | | | | | | | |
| | Endarch stele | | | | | | | | ——————————————— | | | |
| | Polystele | | | | | | | | | ——————————— | | |
| | Secondary xylem | | | | | | | | | ——————————— | | |
| **Epidermis** | Cuticle | | | | ———————————————————————— | | | | | | | |
| | Stomata | | | | | ——————————————————— | | | | | | |
| | Papillate cuticle | | | | | | | | ——————————————— | | | |
| | Glandular trichomes | | | | | | | | ——————————————— | | | |
| | Stomata with subsidiary cells | | | | | | | | ——————————————— | | | |
| **Leaves and enations** | Spiny axis | | | | | | ———— | ———————————————— | | | | |
| | Non-vascularized microphylls | | | | | | | | ——————————————— | | | |
| | Vascularized microphylls | | | | - - - - - - - - | | | ——————————————— | | | | |
| | Forked microphylls | | | | | | | | ——————————————— | | | |
| | Trifurcating lateral branches | | | | | | | | ——————————————— | | | |
| | Wedge-shaped megaphylls | | | | | | | | ——————————————— | | | |
| | Planated, frond-like leaves | | | | | | | | ——————————————— | | | |
| | Leaf abscission | | | | | | | | | ——————————— | | |
| | Leaves arranged in whorls | | | | | | | | | | ———————— | |
| **General habit** | Slender dichotomous stems | | | | ———————————————————————— | | | | | | | |
| | Encrusting, flat thalli | | | | ———————————————————————— | | | | | | | |
| | Rooting structures | | | | | | - - - - - | ——————————————— | | | | |
| | Stems with lateral branches | | | | | | | | ——————————————— | | | |
| | Tree-form (arborescence) | | | | | | | | | ——————————— | | |

**Figure 1.1** The progressive appearance of adaptations for life on land in plants.

# Introduction

gametophyte an ephemeral, small object, known as a prothallus; in most pteridophytes, the 'plant that you see' is the sporophyte (the situation is reversed in mosses and liverworts). The problem with this reproductive strategy is that it restricts the plants (or at least the gametophyte generation) to damp conditions, as the motile male gametes have to swim to the egg. However, within this constraint the strategy has proved successful and still exists in a number of extant plant groups, most notably the ferns.

In the Devonian, a number of modifications to this basic pteridophytic reproductive strategy occurred. Instead of all gametophytes producing both eggs and male gametes, some plants produced separate male and female gametophytes, helping to reduce inbreeding. With only a few exceptions, the spores that produced the male and female gametophytes were produced in separate sporangia. Also, in many cases it proved advantageous for the gametophyte to be further reduced in size and retained within the spore wall, providing added protection to this vulnerable stage in the plant's life cycle. Thirdly, it became advantageous for the female spore and gametophyte to be larger than the male, as this allowed it to provide some food reserves to help in the early growth of the sporophyte; this asymmetry in spore size is known as heterospory.

Heterospory still occurs in a number of extant plant groups, such as the lycopsid *Selaginella* and the water-ferns *Marsilea* and *Pilularia*. However, it was in essence a transitory phase, eventually culminating in the situation where the female sporangium would contain just a single functional spore (megaspore); this is essentially what we now call the seed habit. There is, however, more to being a seed than just having a single megaspore, the most significant being that the spore (in a seed known as an embryo sac) is not released from the sporangium (or nucellus); rather, the whole structure is shed, usually after fertilization. This has obvious advantages, as the sexual phase of the reproductive cycle could take place in a more protected environment, on the parent plant. However, it also presents a problem, as a male gamete produced by a gametophyte on the ground is no longer able to find its own way to the egg. This was overcome by transporting the spore to the seed, either by wind or an animal vector. It was then 'captured' by a specialized structure in the apical part of the ovule (in primitive forms this was a lagenostome, in more advanced species a micropyle), which provided a protected environment for the germination of the spore and the production of the gametes. Another feature of the early ovules and seeds was that they were enclosed by protective sheaves, an inner one known as an integument and sometimes an outer one known as a cupule.

The seed habit was one of the most significant developments for land vegetation, as it freed plants from needing wet conditions for their reproduction. It seems to have happened first in the Late Devonian and, although pteridophytic plants remained important for the rest of the Palaeozoic, seed plants (gymnosperms) soon came to dominate much of the Earth's vegetation, especially after the Permian–Triassic extinction event.

## *Plant architecture*

When living in an aqueous environment, architecture is not a major problem for plants due to the support supplied by the water. On land, however, gravity causes clear problems. One solution, adopted by some early land plants, was to remain small, encrusting the land surface (e.g. *Parka*, *Nematothallus*). However, this is clearly restricting and any plant that could remain upright would have an adaptive advantage. Some plants seem to have done this by developing a zone of thickening, known as a sterome, around a terete stem; some mosses still use this strategy. More commonly, however, it was achieved by the development of a central strand of woody tissue (xylem) in the stem (it also functioned as a vascular structure – see below). The earliest example of a woody stele that we know of is from the Upper Silurian.

Various morphologies developed in the Devonian that helped maximize the strength of the xylem, while keeping its mass to a minimum. These included steles with a stellate cross-section (actinostele), and divided steles, consisting of several discrete strands (polystele). Nevertheless, there was a limit to the size of plant that could be supported by such structures. Some plant groups, most notably the lycopsids, overcame this by having additional support provided by other strengthened tissue in the stem, such as secondary cortex. This allowed the plants to grow to a considerable size (40 metres or more), and was a very efficient means of growth. However, the resulting trees were not particularly strong and would have been vulnerable to traumatic (e.g. storm) damage.

# Introduction

Much stronger and, in the long run more successful, was the development of secondary wood, which first appears in the Middle Devonian.

Another restriction on size was the photosynthetic efficiency of the plant. If a plant consisted merely of terete axes, an increase in size would result in a reduction in the surface area to volume ratio. An early strategy to increase the surface area was the growth on the stem of emergences or spines, which had already appeared in the Silurian. These emergences progressively increased in size and became vascularized. These are what are called microphyllous leaves, such as seen today in clubmosses. However, there is a limit to the efficiency of this strategy, and an alternative approach appeared in the Middle Devonian – the megaphyllous leaf. The early phase in the evolution of megaphylls was where terete axes were clustered at the extremities of the plant, such as in the progymnosperm *Protopteridium*. In itself, this only marginally increased the available photosynthetic area, but improved when the axes formed into a flattened configuration, and then by filling the gaps with additional photosynthetic tissue (mesophyll).

Roots are another fundamental feature necessary for increasing plant size, both in providing anchorage, and as a means of obtaining water and nutrients. Unfortunately, roots are rarely preserved in the fossil record, and our understanding of the evolution of this organ is poor. Some of the Early Devonian land plants, such as *Rhynia*, had horizontal rhizome-like axes, from which slender rhizoids extended into the ground. By the Late Devonian, the progymnosperm *Archaeopteris* had roots very much like modern trees. How the latter developed is not known, but it was clearly a successful strategy.

## Water relations

Plants living out of water are also faced with problems of water supply and control. The problem of supply was overcome by the most successful plants in tandem with that of mechanical support, discussed above. The stele consists, partially or exclusively, of xylem (tissue used mainly for the movement of water) and phloem (used for the movement of organic matter). Although tracheid-like tubes occurred in a number of algae that were becoming adapted to the land environment in the Silurian and Devonian (e.g. *Prototaxites*), and probably fulfilled a vascular function, only in the so-called vascular plants did this tissue form in clusters, sufficient for it also to fulfil a mechanical function.

The problem of water loss was overcome by covering most of the exposed parts of the plant with a cutinized 'skin' known as a cuticle. Cuticles developed in most of the plants becoming adapted to the land in the Silurian and Devonian (e.g. *Nematothallus*, *Parka*). However, the vascular plants developed a strategy for controlling the passage of water between the plant and the atmosphere, which clearly gave them an adaptive advantage. This was achieved by having small pores in the cuticle. Most significantly, these pores (known as stomata) were surrounded by specialized cells called guard cells that could control the size of the pore.

## Taxonomic radiation

The vascular plants were only one of a number of plant groups that were becoming adapted to terrestrial habitats in the Late Silurian and Early Devonian. However, the combination of resistant spores, a stele combining mechanical and vascular functions, and a cuticle with controllable stomata, clearly gave them an advantage over these other groups, and they rapidly came to dominate land vegetation.

The origin of the vascular plants is still not known for certain, although it is widely thought that the Chlorophycophyta ('green algae') may represent the ancestral stock. The Charales ('stoneworts'), in particular, have been advanced as possible ancestors. However, as their stratigraphical range only extends down to the Upper Silurian (Ishchenko, 1975), it is more likely that they share a common ancestor with the vascular plants, rather than being their ancestors.

Vascular plants first appeared in the Late Silurian (c. 420 Ma). The broad pattern of their subsequent evolution in the Palaeozoic is summarized in Figure 1.2. The earliest vascular plants form a plexus of phylogenetic lineages, that are difficult to place in a coherent taxonomic framework, but the scheme most widely used recognizes a series of classes, including the Rhyniopsida, Zosterophyllopsida, Trimerophytopsida and Horneophytopsida. These were all already present in the Early Devonian and presumably reflect the initial radiation of vascular plants into the previously empty terrestrial habitats.

These primitive classes had a relatively short life, all becoming extinct by the end of the

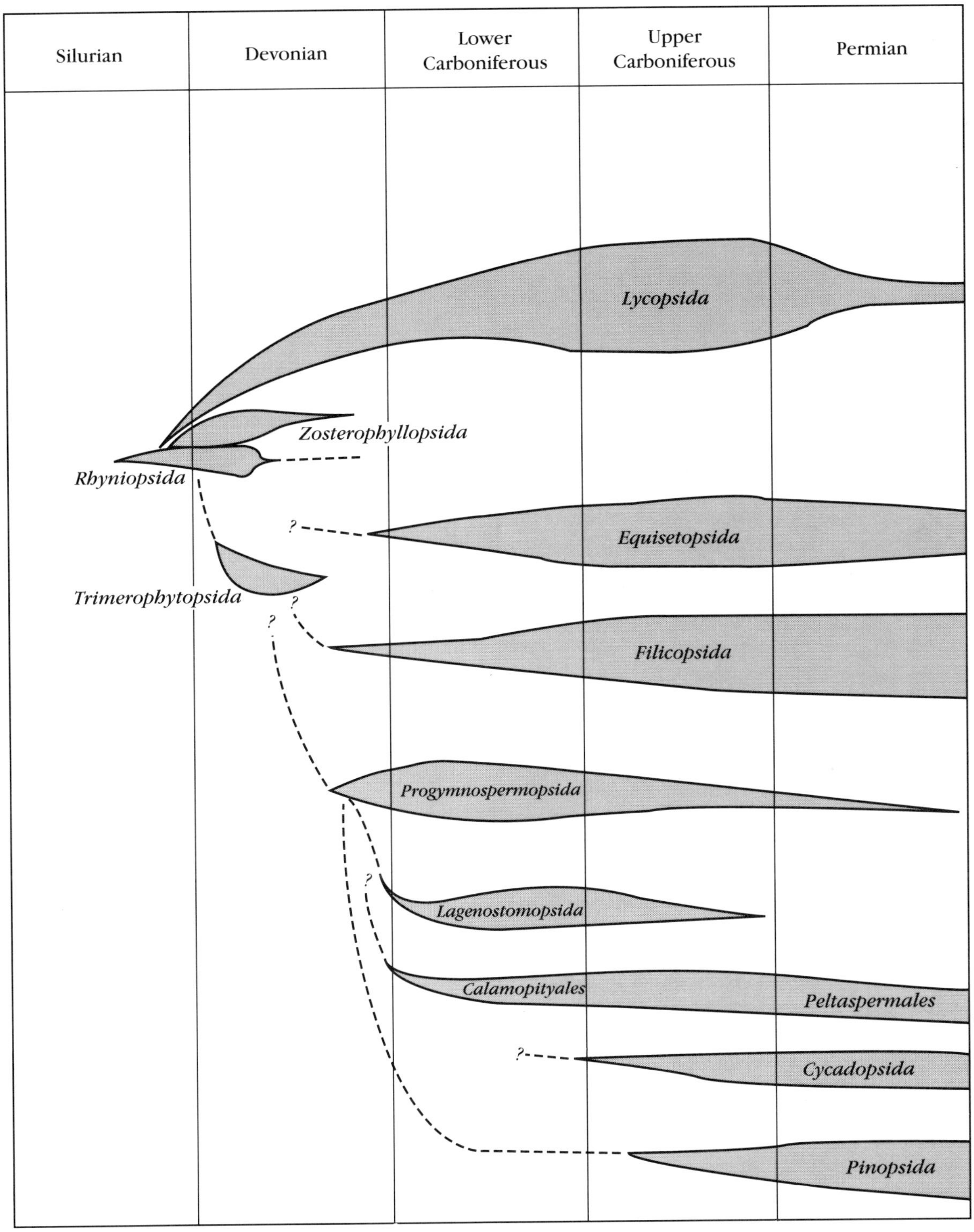

# Introduction

**Figure 1.2** The ranges of the main plant classes through the Palaeozoic.

# Introduction

Devonian (355 Ma). Although successful at times of relatively low competition, they proved unable to survive when more advanced forms appeared. They were nevertheless of great significance in providing the origins of the more advanced groups. For instance, the Lycopsida ('club mosses') originated from the Zosterophyllopsida, and the Filicopsida ('ferns') and Progymnospermopsida originated from the Trimerophytopsida; the Progymnospermopsida in turn gave rise to the Gymnospermophyta ('seed plants'). The origin of the other major class of Palaeozoic plants, the Equisetopsida ('horsetails'), is not known, but was probably from either the Rhyniopsida or Trimerophytopsida.

By the start of the Carboniferous, all of the classes of Pteridophyta had appeared, and the origins of the seed plants can be seen. The subsequent Carboniferous Period saw a further diversification within the pteridophyte classes, that can be seen at the family level (Figures 5.1, 6.2 and 7.1). Also, the main classes of seed plants start to become recognizable, including the Lagenostomopsida, Cycadopsida and Pinopsida. The Late Carboniferous (320–290 Ma) saw the culmination of Palaeophytic vegetation, primarily in the palaeoequatorial belt (including Britain), where dense forests consisting mainly of pteridophytic plants dominated the landscape.

Towards the end of the Carboniferous and in the Early Permian, the tropical forests disappeared (the reasons for this are discussed in more detail in the introduction to Chapter 7). Forests instead developed in higher latitudes, both north and south. This shift inevitably had an impact on plant evolution, causing a number of extinctions and originations, especially at the rank of family. However, at the rank of class, nothing really significant changed, except for the extinction of the Progymnospermopsida and Lagenostomopsida; all of the other classes persisted through to the end of the Permian.

The Palaeophytic vegetation came to an end at the close of the Permian (250 Ma). Although no classes disappeared, there was a major turn-over of orders and families (Figure 7.1). Also, two of the leading Palaeophytic pteridophyte classes, the Lycopsida and Equisetopsida, underwent a dramatic decline; they have managed to persist up to the present day, but only in very reduced numbers of species and genera.

## Palaeobotanical problems

Palaeobotany differs from other branches of palaeontology in a number of ways; these differences relate to how plants easily fragment and to the fact that the pieces can be fossilized in several ways. A fossil is 'any specimen that demonstrates physical evidence of occurrence of ancient life (i.e. Holocene or older)' (Schopf, 1975). However, as the majority of plant remains are either eaten or decayed through microbial action, fossilization is the exception rather than the rule (Figure 1.3). The fossils that are discovered, therefore, represent only a fraction of organisms that lived in the past. The study of fossilization processes is often referred to as taphonomy (Bateman *in* Cleal, 1991).

### *Fossilization processes*

Only occasionally are plants preserved in the place where they grew. Perhaps the most famous exception is the chert at Rhynie, where silicified peat deposits provide information on a whole community of plants and animals (see Chapter 4). Other notable examples are where the bases of trees are preserved *in situ* as internal sedimentary moulds. The most commonly found examples are of the stigmarian bases and the lowermost parts of the trunks of arborescent lycopsids such as the large specimen described by Williamson (1887) now in the Manchester Museum, those in the Fossil Grove, Glasgow (MacGregor and Walton, 1972; Gastaldo, 1986), and those at Wadsley Fossil Forest, Sheffield (Sorby, 1875). Petrified examples of *in situ* lepidodendroid stems have been found on Arran (Walton, 1935).

The norm, however, was for fragments to be detached from the plant and transported away from where they grew. The detachment may have been part of the normal life-process of the plant – the shedding of leaves or twigs, or the dispersal of seeds, pollen or spores as part of the reproductive cycle. Alternatively, the detachment may have been traumatic, perhaps through storm or flood damage. It may even just have been the result of the post-mortem breakdown of the plant. The subsequent transportation usually involved a combination of air (wind) and water (river-flow, current, tide) vectors, taking the fragment to its eventual site of burial in sediment. This site will normally have been subaqueous, usually in a lake or other form of non-marine standing-body of water, or occasionally in the sea; plant fragments

# Introduction

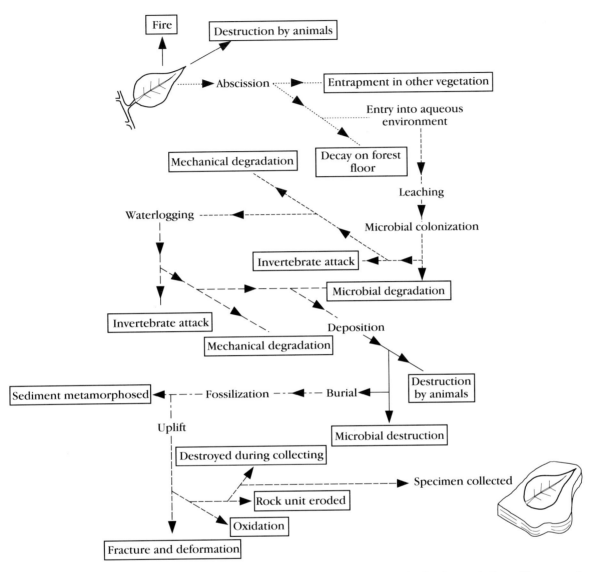

**Figure 1.3** The potential processes involved in a plant fragment passing into the fossil record. From Thomas and Spicer (1987, figure 4.1).

coming to rest on dry land usually decay before they can become buried.

The fossil record provides a very biased view of past vegetation, being controlled largely by the availability of the plant fragments that can enter the fossilization process (Spicer, 1980). Only plants growing in lowland habitats, and exposed to winds or growing adjacent to flowing water, will normally be sampled. Wind will selectively carry fragments from the more exposed parts of the vegetation, so understorey plants will be under-represented in fossilferous deposits. Water transport, similarly, will selectively carry away those plants which grow near rivers or lakes. The action of water is, however, more complex than that of wind, because dispersed organs are more susceptible to mechanical fragmentation and attack by micro-organisms and re-sorting by water turbulence. Patterns of deposition may be extremely complex and only detectable by quantitative sampling and statistical analysis (Spicer and Hill, 1979).

Substantial selection and sorting of plant remains takes place during their transportation to

# Introduction

places of active deposition. The more durable parts, such as wood, may stand the greater chance of not being eaten or of decaying, but it is the more frequent, lighter and less rigid parts, such as leaves, pollen and spores, that are carried over greater distances to reach sites of deposition and potential fossilization (Hughes, 1976). It was on the whole the availability, rather than the durability, of organs that controlled the chances of their being fossilized.

In order to obtain a better idea of how plant fragments enter the fossil record, and how this distorts our view of the vegetation growing at the time, a number of studies have been made on modern analogues for fossil depositional environments. For the Late Carboniferous deltaic environment seen in the British sequences, for example, studies on the Orinoco Delta by Scheihing and Pfefferkorn (1984) and on the Mobile Delta in Alabama by Gastaldo *et al.* (1987) provide instructive points of comparison.

## Specimen interpretation

It is only small plants that are preserved whole, and then only in exceptional circumstances of fossilization. Large plants will certainly not be preserved whole. As the association of dispersed organs is not sufficient evidence for assuming that they were once parts of a single plant, we rely on the chance finding of organs which are still organically connected or the recognition of very distinctive anatomical features. A great deal of work has been undertaken on attempting to reconstruct whole plants, but as yet very few have been successfully completed. The problems and many of the successes are reviewed by Chaloner (1986).

Plant remains may be preserved in a number of ways, depending on the burial and the subsequent sedimentary and geological processes. Sometimes the remains are so durable that they survive virtually unchanged (cuticles, spore and pollen walls, charcoal). The majority are, however, altered in some way. Soluble compounds are quickly lost and microbial activity commences soon after death or abscission, so the degree of preservation depends to a large extent on the speed of fossilization. Many fossils contain organic residues that can be extracted and biochemically identified (Thomas, 1986). Some of these residues are very easily extractable. For example, lignin derivatives have been obtained from a wide range of sub-fossil woods and plant fossil compressions (Logan and Thomas, 1987).

The various modes of preservation in which Palaeozoic plant fossils are normally found are summarized in Figure 1.4. The commonest mode is as an 'adpression', which may either be a 'compression', where there are compressed coalified remains of the plant, or an 'impression', where there are no actual plant remains, but only an impression on the matrix. The majority of the sites discussed in this volume yield adpression plant fossils. Three-dimensionally preserved remains may be formed, sometimes through sediment infiltration of organs, producing 'casts' and/or 'moulds'. If the plant remains are infiltrated with mineral-rich water, then fossilization can result in a petrifaction where the internal cellular detail is preserved, e.g. calcium carbonate in Upper Carboniferous coal-balls and in the Lower Carboniferous plant fossils at Pettycur and Whiteadder. Sometimes compressions may be partially pyritized or secondarily limonitized such as those found, for example, at the Targrove, Llanover, Craig-y-Fro and Ballanucater Farm sites (see Chapter 4).

Plant fossils can be studied in a number of ways, depending on their modes of preservation, but there are a number of techniques that can yield a surprising amount of detail (see Lacey, 1963; Schopf, 1975, or the general textbooks listed on p. 3 for details of the techniques). For instance, by extracting the cuticle from a compression, it is often possible to see microscopic details of the epidermal cells, including stomata and tiny hairs (e.g. Edwards *et al.*, 1982; Cleal and Zodrow, 1989; Cleal and Shute, 1991, 1992). Thin sections through mineralized petrifactions can show extremely fine detail of the cell structure of the plant fragment (e.g. the work on Rhynie Chert by Kidston and Lang, summarized in Chapter 4). It is only by careful and often painstakingly detailed studies that full interpretations of the fossilized plants are possible. By this means our knowledge of the once-living plants themselves and of the floral assemblages in which they grew has expanded rapidly in the last few decades.

## Naming plant fossils

A consistent nomenclature is clearly vital for the interpretation and communication of information on plant fossils, especially when so much reliance and importance is placed on species lists; as indeed it is here in the site descriptions. However, the problem confronted in studying fossil plants is how to name and classify the very different plant

# Introduction

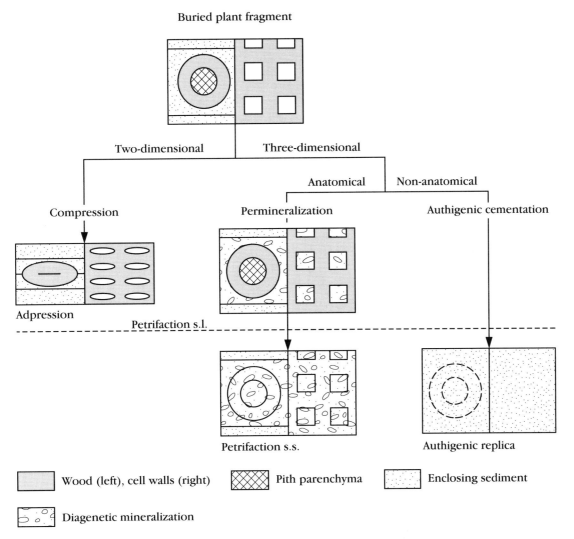

**Figure 1.4** Summary of modes and nomenclature of plant fossil preservation. Each of the major modes of preservation is represented by a rectangle. The left-hand side of each rectangle shows a transverse-section through a hypothetical stem, and the right-hand side a close-up section showing cells. Based on Bateman *in* Cleal (1991, figure 2.2).

organs, or pieces of organs that were naturally shed, broken off, or traumatically detached from the parent plant prior to fossilization. To overcome this problem, palaeobotanists have developed a system of nomenclature for isolated organs, whereby different parts of plants are given different generic and specific names. Cleal (1986b) and Thomas (1990) have given general discussions on this, and Collinson (1986) has reviewed the problems relating to the use of modern generic names for plant fossils. The taxonomy of plant fossils, therefore, differs from that of living plants in many respects and has its own 'rules' (Greuter *et al.*, 1988). Species lists must, therefore, be treated with caution when they are being used for an estimate of species diversity or a comparison of assemblages. The number of genera or species of fossils will not necessarily equate to the number of different plant species that gave rise to the remains making up the assemblage and certainly does not equate to the number of different plant species that grew in the original vegetation.

Much debate has centred on the concept of the genus as applied to the isolated fossil organs of plants and to their inclusion in any taxonomic

# Introduction

hierarchy. Some genera based on organs may be assigned to families, while others (form-genera) can only be referred to a taxon of higher rank than a family. The application and use of these terms can even vary from one author to another. For example, Thomas and Brack-Hanes (1984) have suggested that families should be based on female or bisexual reproductive organs and that other organs, thought to belong to the same parent plants, can only be included there as satellite taxa.

Occasionally, as mentioned above, two genera of different organs may be shown to represent parts of one plant, but it is normal to retain the use of both names. There is no certainty that all species of these two genera were connected in this way as either one or both may be very broadly defined. It is also highly likely that different organs evolved at different rates, thereby having different stratigraphical ranges. Although it may seem unnecessarily complicated, this system of taxonomy works well for plant fossils and it is not one that is meant to equate to the taxonomy of living plants. Thomas (1990) and various authors in Spicer and Thomas (1986) summarize the many problems relating to the taxonomy and systematics of plant fossils.

## *Information from plant fossils*

Plant fossils are used for a variety of purposes, either as individuals or as whole assemblages. One of their main roles has been to provide a documentation of the progressive evolution of land vegetation. This has been particularly important during the early phases of this process, in the Silurian and Devonian (see Chapters 3 and 4). Its value diminishes for later parts of this history; as plants made greater inroads into the hinterland, far away from areas of sediment deposition, the proportion of the total land vegetation that could find its way into the fossil record became less and less. Nevertheless, it is the only direct means that we have of looking at this evolutionary history and, provided its message is interpreted in the appropriate geological context, it can provide a means of testing indirect phylogenetic models, developed using methods such as cladistic and DNA analyses.

Plant fossils can also be used to interpret the living communities from which they were derived i.e. palaeoenvironmental analyses. Here the interrelationships between the fossil assemblages and the lithology of the sediments may provide evidence of different living assemblages (e.g. Scott, 1979; Gastaldo, 1985, 1987; Cleal and Thomas, 1988). This can be extended to an interpretation of regional floras (Raymond, 1985; Raymond *et al.*, 1985; Raymond and Parrish, 1985; Vakhrameev *et al.*, 1978). Plant fossils can also be used as indicators of palaeoclimates (Chaloner and Creber, 1990) and the former positions of continental plates (Chaloner and Creber, 1988)

## The systematics of the Plant Kingdom

There have been many proposed classifications for plants. The one adopted here essentially follows Cleal (1993) for the Pteridophyta and Gymnospermophyta. Due to the frequent absence of details of reproductive structures, the divisions of algae and bryophytes have not been further subdivided, and effectively follow the classification of Taylor (1981). To help clarify the taxonomic lists given with the various site descriptions in this volume, the classification (down to the rank of family) will be summarized below, together with some of the common names of the higher taxa (only those taxa relevant to the volume are given).

Division Cyanochloronta ('blue-green algae')
Division Chlorophycophyta ('green algae')
Division Rhodophycophyta ('red algae')
Division Phaeophycophyta ('brown algae')
Division Bryophyta (mosses and liverworts)
Division Pteridophyta
    Class Rhyniopsida
        'rhyniophytoid fossils'
        Order Rhyniales
            Family Rhyniaceae
    Class Zosterophyllopsida
        Order Zosterophyllales
            Family Zosterophyllaceae
    Class Horneophytopsida
        Order Horneophytales
            Family Horneophytaceae
    Class Trimerophytopsida
        Order Trimerophytales
            Family Trimerophytaceae
    Class uncertain
        Order Barinophytales
            Family Barinophytaceae
    Class Lycopsida ('club-mosses')
        Order Drepanophycales
            Family Drepanophycaceae

# Introduction

Order Protolepidodendrales
    Family Protolepidodendraceae
    Family Eleutherophyllaceae
Order Lycopodiales
    Family Lycopodiaceae
Order Selaginellales
    Family Selaginellaceae
Order Lepidocarpales
    Family Cyclostigmaceae
    Family Flemingitaceae
    Family Sigillariostrobaceae
    Family Lepidocarpaceae
    Family Spenceritaceae
    Family Caudatocarpaceae
    Family Pinakodendraceae
    Family Sporangiostrobaceae
    Family Pleuromeiaceae
Order Miadesmiales
    Family Miadesmiaceae
Order Isoetales ('quillworts')
    Family Isoetaceae
    Family Chaloneriaceae
Class Equisetopsida ('horsetails')
    Order Pseudoborniales
        Family Pseudoborniaceae
    Order Bowmanitales ('sphenophylls')
        Family Bowmanitaceae
        Family Eviostachyaceae
        Family Cheirostrobaceae
    Order Equisetales
        Family Archaeocalamitaceae
        Family Calamostachyaceae
        Family Tchernoviaceae
        Family Gondwanostachyaceae
        Family Equisetaceae
        Family Echinostachyaceae
Class Filicopsida ('ferns')
    Order Cladoxylales
        Family Cladoxylaceae
    Order Ibykales
        Family Ibykaceae
    Order Coenopteridales
        Family Rhacophytaceae
        Family Zygopteridaceae
        Family Stauropteridaceae
        Family Corynepteridaceae
        Family Biscalithecaceae
    Order Botryopteridales
        Family Psalixochlaenaceae
        Family Tedeleaceae
        Family Botryopteridaceae
        Family Sermeyaceae
    Order Urnatopteridales
        Family Urnatopteridaceae
    Order Crossothecales
        Family Crossothecaceae
    Order Marattiales
        Family Asterothecaceae
        Family Marattiaceae
    Order Osmundales
        Family Osmundaceae
    Order Filicales
        Family Gleicheniaceae
        Family Cynepteridaceae
        Family Matoniaceae
        Family Dipteridaceae
        Family Polypodiaceae
        Family Dicksoniaceae
Class Progymnospermopsida
    Order Aneurophytales
        Family Aneurophytaceae
        Family Protokalonaceae
        Family Protopityaceae
    Order Archaeopteridales
        Family Archaeopteridaceae
    Order Noeggerathiales
        Family Noeggerathiaceae
        Family Tingiostachyaceae
    Order Cecropsidales
        Family Cecropsidaceae
Division Gymnospermophyta ('seed plants')
    Class Lagenostomopsida
        Order Lagenostomales
            Family Elkinsiaceae
            Family Genomospermaceae
            Family Eospermaceae
            Family Lagenostomaceae
            Family Physostomaceae
    Unnamed Class
        Order Calamopityales
            Family Calamopityaceae
        Order Callistophytales
            Family Callistophytaceae
        Order Peltaspermales
            Family Peltaspermaceae
            Family Cardiolepidaceae
            Family Umkomasiaceae
        Order Leptostrobales
            Family Leptostrobaceae
        Order Arberiales ('glossopterids')
            Family Arberiaceae
            Family Caytoniaceae
        Order Gigantonomiales ('gigantopterids')
            Family Emplectopteridaceae
    Class Cycadopsida
        Order Trigonocarpales ('medullosans')
            Family Trigonocarpaceae
            Family Potonieaceae

# Introduction

        Order Cycadales ('cycads')
           Family Cycadaceae
    Class Pinopsida
        Order Cordaitanthales ('cordaites')
           Family Cordaitanthaceae
           Family Rufloriaceae
           Family Vojnovskyaceae
        Order Dicranophyllales
           Family Dicranophyllaceae
           Family Trichopityaceae
        Order Pinales ('conifers')
           Family Emporiaceae
           Family Utrechtiaceae
           Family Majonicaceae
           Family Ullmanniaceae
           Family Voltziaceae

## The choice of GCR sites

The British Isles has an outstanding number of sites that yield Upper Palaeozoic plant fossils, especially from the Upper Silurian, Lower Devonian, and Carboniferous. Many of these, such as the Rhynie Chert locality, are of national and international importance and have been known to the scientific community for a great many years. If palaeobotany is to survive as a viable research subject in Britain then these important sites must be conserved and maintained for further studies to be possible. A full rationale for conserving sites of palaeobotanical importance, the methods for selecting them and the problems in managing them have been discussed in detail by Cleal (1988).

Based on the presence of certain broad guidelines, only those sites which could be regarded as nationally or internationally significant have been chosen for inclusion in this volume. They include the following:

1. Sites yielding a unique assembly of species which have contributed significantly to our understanding of plant fossils (e.g. Perton Lane, Whiteadder).
2. Sites where species are exceptionally well-preserved, showing structural features not seen elsewhere (e.g. Pettycur).
3. The best available sites in Britain for showing the major plant fossil assemblages (e.g. Teilia Quarry).

The sites selected have been chosen to include an adequate cover of the range of Britain's plant fossil assemblages. The list is the result of selection from the vast number of sites known at present but, of course, the collection of new and exciting species at existing sites, the reinterpretation of existing specimens, or the discovery of new sites may bring about its modification in the future.

The survey resulted in the selection of 42 sites, as summarized in Table 1.1. The distribution of the sites in Britain is shown in Figure 1.5, using the site codes given in Table 1.1.

**Table 1.1** The Palaeozoic palaeobotany GCR sites

| Geological System | Site Name | Site Code |
|---|---|---|
| Permian | Kimberley Railway Cutting | P3 |
| | Middridge Quarry | P2 |
| | Stairhill | P1 |
| U. Carboniferous | Jockie's Syke | U6 |
| | Llanbradach Quarry | U5 |
| | Nostell Priory Brickpit | U4 |
| | Cattybrook Claypit | U3 |
| | Wadsley Fossil Forest | U2 |
| | Nant Llech | U1 |
| L. Carboniferous | Victoria Park | L16 |
| | Glencartholm | L15 |
| | Wardie Shore | L14 |
| | Teilia Quarry | L13 |
| | Moel Hirradug | L12 |
| | Puddlebrook Quarry | L11 |
| | Glenarbuck | L10 |
| | Loch Humphrey Burn | L9 |
| | Laggan | L8 |
| | Kingswood | L7 |
| | Pettycur | L6 |
| | Kingwater | L5 |
| | Weak Law | L4 |
| | Oxroad Bay | L3 |
| | Whiteadder | L2 |
| | Lennel Braes | L1 |
| Devonian | Plaistow Quarry | D10 |
| | Sloagar | D9 |
| | Bay of Skaill | D8 |
| | Rhynie | D7 |
| | Auchensail Quarry | D6 |
| | Ballanucater Farm | D5 |
| | Craig-y-Fro Quarry | D4 |
| | Llanover Quarry | D3 |
| | Turin Hill | D2 |
| | Targrove Quarry | D1 |
| Silurian | Freshwater East | S7 |
| | Perton Lane | S6 |
| | Capel Horeb Quarry | S5 |
| | Cwm Craig Ddu Quarry | S4 |
| | Rockhall Quarry | S3 |
| | Llangammarch Wells Quarry | S2 |
| | Pen-y-Glog Quarry | S1 |

# Introduction

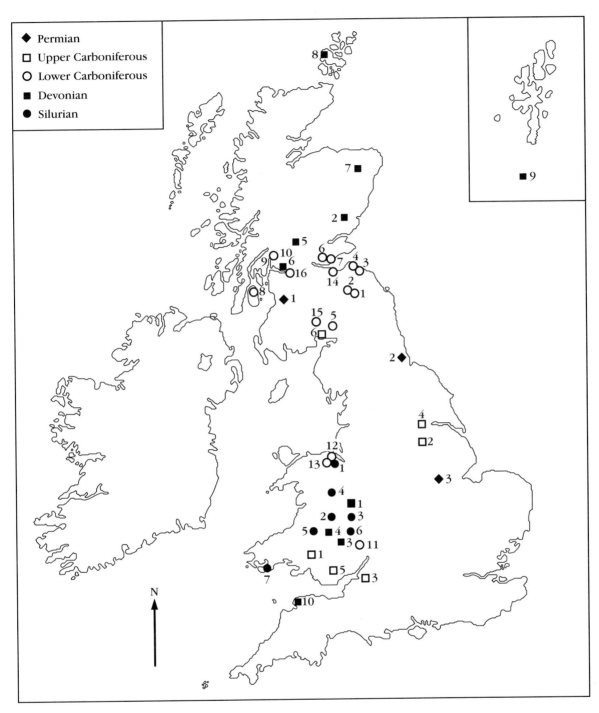

**Figure 1.5** The distribution of the GCR Palaeozoic palaeobotany sites in Britain. See Table 1.1 for site names.

# Chapter 2

# *History of research on British plant fossils*

## SILURIAN

British sites, particularly in Wales and the Welsh Borders, have played a key role in developing ideas about the evolution of land plants during the Silurian (440–410 Ma). This may be due, at least in part, to it being the type area for much of the Silurian and having been subject to more intensive investigation than other outcrops of this age (Bassett, 1984; Holland and Bassett, 1989). It has meant that there is a long history of records of plant fragments from these strata, dating back to the mid-nineteenth century (e.g. Phillips, 1848; Brodie, 1869, 1871; see Lang, 1937 for a more complete account of the early records). However, no serious attempt was made to investigate the form and structure of these difficult fossils until Lang's (1937) classic monograph. For many years, Lang's paper was regarded as the definitive statement on Silurian vegetation, and not until the 1970s was any attempt made to re-evaluate his observations, principally by Dianne Edwards (e.g. Edwards, 1979a, 1982; Edwards and Davies, 1976; Edwards and Rogerson, 1979; Edwards et al., 1979, 1986; Fanning et al., 1988, 1990, 1991). Edwards and her co-workers are continuing to investigate the fossil evidence for Silurian terrestrial vegetation.

## DEVONIAN

Plant fossils from the Devonian (410–355 Ma) have been investigated in Britain for over 150 years, but this long history can be broken down into three main phases. The first effectively started with the work of Hugh Miller (Rosie, 1981), who collected from a number of Scottish localities. Although there had been earlier records (Fleming, 1811, 1831; Williams, 1838; De la Beche, 1839; Murchison, 1839), Miller was the first to describe and discuss such fossils in detail (Miller, 1841, 1849, 1855, 1857). His contributions to palaeobotany have been largely overshadowed by his work on vertebrate palaeozoology. However, he was probably the first palaeontologist in the world to treat Devonian plant fossils seriously, and he laid the foundations for the studies of later nineteenth century workers (e.g. Dawson, 1888).

Considering the poor preservation of most Devonian plant fossils, it is not surprising that many of the nineteenth century palaeobotanists had difficulty interpreting them. For a time, it became widely believed that the 'fossil plants of the Devonian rocks in Europe resemble generically, with very few exceptions, those of the coal-measures' (Lyell, 1865, p. 542). The discovery of 'conifer' wood (now known in fact to be from a progymnosperm) in the Middle Devonian of Scotland tended to reinforce this misconception (Miller, 1841; M'Nab, 1871). Some geologists such as Dawson (1859, 1870, 1871) argued that at least some of the Devonian fossils represented truly simple and primitive plants, but others just regarded them as poorly preserved fragments of more advanced plants. Despite this confusion, the nineteenth and early twentieth centuries saw a gradual accumulation of a pool of information on the Devonian plant fossils, which is admirably summarized by Arber (1921).

The second phase of research on British Devonian plant fossils commenced with the description of the Rhynie Chert assemblage by Kidston and Lang (1917b, 1920a, b, 1921a, b). This was the first unequivocal evidence that Devonian plants really were simple and primitive, and provided a significant impetus to the further investigation of the less well-preserved adpression assemblages from elsewhere. In Britain, the field was dominated at this time by Lang. Following the work at Rhynie, he continued to collaborate with Kidston on the Scottish assemblages (Kidston and Lang, 1923a, b, 1924). After Kidston's death in 1924, Lang continued to work on the Scottish sites, but later moved on to material from Wales and the Welsh Borders. During this second phase, Devonian palaeobotany became an established discipline and many of our presently held views on early land plant evolution were developed.

From about 1945, there was a lull in the investigation of British Devonian plant fossils. During the mid-1960s, however, the third and most recent phase in their investigation began, mainly as a result of the work of Dianne Edwards. She has instigated new work on many of the classic Devonian plant adpression localities, collecting fresh material and applying new and improved techniques to its study. She has also searched out new localities, such as Auchensail Quarry (see also Edwards and Richardson, 1974; Edwards and Rose, 1984; Edwards et al., 1986). This has all provided significant new information on the morphological variation and anatomy of many of these plants. Recent attempts at biostratigraphical (Banks, 1980), palaeoecological (Edwards, 1980b) and palaeogeographical analyses (Raymond et al., 1985; Edwards, 1990) have leaned heavily on this new data. There has also been significant recent progress on the Rhynie fossils, mainly through the work of Lyon (1957, 1962, 1964), Bhutta (1972, 1973a, b), El-Saadawy and Lacey (1979a, b), Remy (1978, 1980a, b, 1991), Remy and Remy (1980a, b), Remy et al. (1980b), D.S. Edwards (1980, 1986), Edwards and Lyon (1983), Lyon and Edwards (1991) and Remy and Hass (1991a, b). This was mainly as a result of using new techniques, particularly serial sectioning using acetate peels, which allowed far more refined reconstructions of the plants to be achieved.

Despite the recent resurgence of interest, there are many aspects of British Devonian palaeobotany that remain neglected. For instance, other than papers by Chaloner (1972), Edwards (1976) and Allen and Marshall (1986), there has been no recent work on the Middle Devonian plant fossils from northern Scotland. This is despite there being numerous sites available for further collecting. The British Upper Devonian assemblages have been even more neglected, the only recent references to them being by Long (1973) and Fairon-Demaret and Scheckler (1987). There is evidently considerable potential for further work in this country.

## LOWER CARBONIFEROUS

Plant adpressions were reported from the Lower Carboniferous (355-320 Ma) of Britain in the early to mid-nineteenth century, in particular from the Burdiehouse Limestone near Edinburgh (e.g. Lindley and Hutton, 1831-1837; Miller, 1857). However, intensive work on fossils from these strata did not start until the late nineteenth century, principally by Kidston (Edwards, 1984). Kidston's immense contribution to the subject culminated in his classic monographs published between 1923 and 1925, which remain the definitive publications on the Lower Carboniferous adpression palaeobotany of Britain (those parts of this monograph not published before his death were subsequently updated and published by Crookall, 1955-1975). Probably the most significant aspect of Kidston's work, other than the fact that he placed on record so many fossils of this age, was his analysis of the frond architecture of the early pteridosperms and his consequent refinement of their classification.

The next major phase of work on the British fossils of this age was by Walton (1926, 1931, 1941), Benson (1904, 1933, 1935a, b) and later Lacey (1962). They mainly concentrated on sites in Wales and south-west England, although Walton also reported on adpressions from the Clyde Plateau Volcanic Formation near Glasgow (Walton et al., 1938). Walton's work was particularly important for helping establish reconstructions of the plants, especially for determining the connection between foliage and fructifications. Lacey's work is also of considerable significance, if only because it is one of the few attempts to prepare cuticles from plant fossils of this age.

Most recently, the only significant contribution to British Lower Carboniferous adpression palaeobotany has been Rowe's (1988a, b, c) work on material from the Drybrook Sandstone in the Forest of Dean. Although previously studied by Lele and Walton (1962b), Rowe has applied new techniques to provide a fresh insight into the fossils.

The first records of British Lower Carboniferous petrifactions were in the early nineteenth century, by Witham (1831, 1833), who described material from a number of sites in the Cementstone Group of southern Scotland. Witham's work is mainly significant in having developed the method of thin sectioning rocks, which was not only important for the future development of palaeobotany, but was perhaps more significant for petrology and mineralogy. The first major advance in this field for palaeobotany, however, was the discovery in the 1870s of the petrifactions from the Pettycur Limestone. The work of Williamson and Scott at this site established it as of international significance (Williamson, 1872, 1873, 1874a, b, 1877, 1880, 1883, 1895; Williamson and Scott, 1894, 1895; Scott, 1897, 1901); until the discoveries 40 years later at Rhynie (see Chapter 4), it provided some of the oldest evidence of cell structure in plants and was central to ideas about early plant evolution.

Various other petrifaction sites were discovered in Scotland during the late nineteenth and early twentieth centuries, but these were mainly found by chance. The first concerted effort to search for new sites was by Gordon (1935a, b, 1938, 1941), who discovered petrifactions at various localities in the volcanogenic deposits exposed along the south-east coast of Scotland. Gordon described a variety of new taxa from both these new and some of the old localities, his work on the ferns being particularly important. Other major contributions to Lower Carboniferous petrifaction palaeobotany during the mid-twentieth century were by Calder (1934, 1935, 1938), Walton (1935,

1949a, b, c, 1957, 1969), Beck (1958), Chaphekar (1963) and Chaphekar and Alvin (1972).

In more recent years, by far the most important work on Lower Carboniferous petrifactions has been by Long (1959-1987) on the Cementstone Group sites. By utilizing the peel method, Long has revolutionized our view of the Early Carboniferous plants, particularly of the seed plants. His views on the evolution of these plants have not always met with universal acceptance, but the quality of his observations and the degree to which other palaeobotanists have had to use them in developing their ideas concerning seed plant evolution, in particular, are indisputable.

Another approach to the study of the British Lower Carboniferous petrifaction sites has been by Scott, who has integrated the analysis of species distribution and sedimentology to develop ideas about the plant ecology of the time (Scott *et al.*, 1984, 1985, 1986; Scott and Rex, 1987; Rex and Scott, 1987). Scott's work has produced particularly interesting results from exposures of the volcanogenic strata in southern Scotland, such as the Pettycur Limestone and the Oxroad Bay tuffs. He has also instigated the collection of additional new material from many of these sites, which have been worked on mainly by other palaeobotanists (Rothwell and Wight, 1989; Galtier and Scott, 1986a, b; Meyer-Berthaud, 1986; Meyer-Berthaud and Galtier, 1986a, b; Bateman, 1988; Bateman and Rothwell, 1990). The review paper by Scott *et al.* (1984) provides one of the best sources of information on the distribution of Lower Carboniferous petrifactions in Britain.

## UPPER CARBONIFEROUS

Upper Carboniferous (320-290 Ma) plant fossils have been known from Britain since at least the seventeenth century, the first published account usually being credited to Lhuyd (1699). A number of subsequent publications by British naturalists during the eighteenth and early nineteenth centuries, described Upper Carboniferous plant fossils, the most widely quoted being Artis (1825) and Lindley and Hutton (1831-1837) (for a fuller account of early British work on these fossils, see Kidston, 1923a and Andrews, 1980). Unlike continental Europe, however, Britain failed to produce any significant contributors to the subject until the mid-nineteenth century. The first major British palaeobotanist to deal with the adpression plant fossils was Kidston, whose impressive list of publications started in the 1880s (Crookall, 1938; Edwards, 1984) and culminated in his classic 1923-1925 monographs. During the first half of the twentieth century, notable contributions were also made by Arber (1904b, 1912, 1914, 1916).

Most of this work was of a floristic nature, documenting assemblages and species distributions; Kidston's work on fern fructifications and pteridosperm frond architectures being the most significant exceptions. More botanically orientated, morphological/anatomical study was instead concentrated on the Langsettian coal-balls from Yorkshire and Lancashire. Petrified plant fossils were first noted in coal-balls in the 1850s by Binney, but their true significance was established first by Williamson in a series of papers starting in the 1860s, and later by Scott. The results of this main phase of British coal-ball work is admirably summarized by Scott (1920-1923).

After this 'golden period' of British Upper Carboniferous palaeobotany, ending in the mid-1920s, interest in the subject suffered a decline. Crookall attempted to continue Kidston's work on the adpressions, although this amounted mainly to further documenting the distribution of species in Britain. His main achievement was the 1955-1975 monographs, which described those plant groups not covered in the Kidston volumes. However, they were published some time after Crookall's main phase of work, in the 1930s, and despite some attempts to update them, they have a rather archaic feel. It is also worth mentioning here the work of Dix (1933, 1934, 1935) on the biostratigraphy and Davies (1929) on the palaeoecology of the Upper Carboniferous plant fossils, although in both cases their publications are weakened by a failure to document fully the taxa they were recording.

The most significant recent contributions to Upper Carboniferous palaeobotany in Britain have been on the lycopsids, by Chaloner (much of whose work is summarized by Chaloner *in* Boureau *et al.*, 1967), Thomas (1967a, b, 1970, 1977, 1978a, b, 1981b) and Boulter (1968). There has also been some recent interest in the ferns and pteridosperms (Thomas and Crampton, 1971; Cleal and Laveine, 1988; Shute and Cleal, 1989; Cleal and Shute, 1991, 1992). Finally, there has also been interest in the use of plant fossils for Late Carboniferous palaeoecology (e.g. Scott, 1977, 1978, 1979) and biostratigraphy (Wagner and Spinner, 1972; Cleal, 1978, 1984b, 1986c, 1987b; Cleal and Thomas, 1988).

Coal-ball work also underwent somewhat of a decline after the 1930s. Contributions were made by Holden and Long (reviewed briefly by Andrews, 1980). The most significant work has, however, been that of Holmes on some of the herbaceous ferns found in the coal-balls (reviewed by Holmes, 1989).

## PERMIAN

Most work on the Permian (290–250 Ma) palaeobotany of Britain has been on the Marl Slate and its equivalents. The best historical account of studies on these fossils is by Stoneley (1958), who notes records dating back to the mid-nineteenth century (Sedgwick, 1829; Lindley and Hutton, 1937; King, 1850; Kirkby, 1862, 1864, 1867). Stoneley provides the only attempt at a monographic analysis of these fossils, although a useful review is also provided by Schweitzer (1986). Individual taxa have also been dealt with by Townrow (1960) and Poort and Kerp (1990).

*Chapter 3*

*Silurian*

# Evidence of pre-Silurian vegetation

The Silurian was a period of great significance for the evolution of the Plant Kingdom (Figure 3.1). Plants had been in existence for over three thousand million years, but had been largely restricted to aqueous, mainly marine habitats: the terrestrial habitats presented a number of environmental barriers, which the pre-Silurian algal plants had been unable to overcome fully. Ultraviolet radiation was previously thought to be one such barrier (Lowry *et al.*, 1980), but this now seems unlikely as it is thought that the ozone layer had almost attained its present-day thickness by the Cambrian. A more significant obstacle was desiccation, and it was not until the Late Ordovician and more especially the Silurian that plants developed strategies to overcome this. These included: (a) an outer cuticle with stomata to control water loss, (b) a vascular system to transport water and nutrients around the plant, and incidentally to provide upright support, and (c) spores impregnated with sporopollenin, that could survive in a non-aqueous environment. These developments facilitated the sudden radiation of plant taxa in the Lower Devonian fossil record (see Chapter 4). Much of this very early history of land vegetation has been demonstrated using fossils found in Britain.

## PALAEOGEOGRAPHICAL SETTING

The best available review of Silurian plant fossil distribution is by Edwards (1990). She has shown that most fossils represent equatorial to subequatorial vegetation, particularly of Laurussia (Figure 3.2). In addition to the British localities, there are records from the USA (Schopf *et al.*, 1966; Banks, 1972, 1973; Pratt *et al.*, 1978), Ireland (Edwards *et al.*, 1983), the former Czechoslovakia (Obrhel, 1962, 1968), Ukraine (Ishchenko, 1969, 1975), Kazakhstan (Senkevich, 1975) and China (Edwards, 1990).

Laurussia would have provided an ideal setting for the first migration of plants onto land. The climate was warm and moist, and the extensive delta-systems generated by Caledonian earth movements provided extensive intermediate areas between marine and the drier terrestrial habitats.

An assemblage from Raudfjorden in Spitsbergen was originally considered to be of Late Silurian age (Høeg, 1942), but is now thought to be Early Devonian (Banks, 1972).

Some Silurian plant fossils from Gondwana indicate a quite different and apparently advanced vegetation. The best documented belong to the lower of the *Baragwanathia* assemblages of Australia (e.g. Tims and Chambers, 1984). The stratigraphical position of these fossils as upper Ludlow is now generally accepted (see below), but they represent a level of evolutionary development not otherwise seen below the Lower Devonian (Siegenian) in Laurussia. A comparable assemblage has also been reported from Libya (Klitzsh *et al.*, 1973; Boureau *et al.*, 1978; Douglas and Lejal-Nicol, 1981), but there are problems with its stratigraphical assignment (Edwards *et al.*, 1979); currently available evidence indicates that they are in fact Devonian (Edwards pers. comm.). The specimens described by Daber (1971) from indisputably Přídolí strata of Libya are more compatible with the type of plant fossils found in Laurussia.

## STRATIGRAPHICAL BACKGROUND

The chronostratigraphical divisions of the Silurian are summarized in Figure 3.5. The stage boundaries are defined at stratotypes in mainly basinal marine facies, and located on the basis of graptolite zones. However, palynology has allowed detailed correlations to be made between the basinal sequences and the near-shore facies, in which most of the Silurian plant fossils occur (Richardson and Edwards *in* Holland and Bassett, 1989).

The lithostratigraphy of the Silurian strata in Britain is summarized by Cocks *et al.* (1971) and Holland and Bassett (1989).

All of the Silurian plant fossils belong to the *Cooksonia* Zone in the Banks (1980) classification.

## EVIDENCE OF PRE-SILURIAN VEGETATION

Microscopic algae inhabited Early Palaeozoic soils (Retallack, 1986). The oldest evidence of macroscopic terrestrial plants is cuticles and spores from the Ordovician (Gray, 1985), while the oldest macrofossils are from the Silurian. The only records of macrofossils from below the Silurian in Britain are from the Ordovician of Pembrokeshire (Hicks, 1869) and Cumbria (Sedgwick, 1848; Nicholson, 1869), but these are poorly documented. From outside this country, the most widely quoted pre-Silurian land plant is *Aldanophyton*, from a Cambrian marine limestone in Siberia (Kryshtofovich, 1953). It was

# Silurian

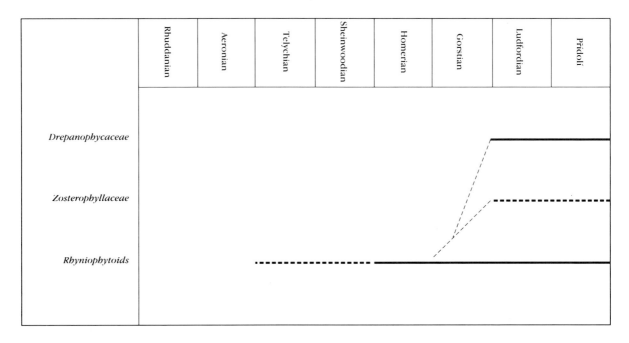

**Figure 3.1** The distribution of families of vascular plants in the Silurian. Based on data from Cleal (1993).

initially described as a lycopsid, but this is now generally discounted (e.g. Stewart, 1960, 1983), although its affinities remain a mystery. Records from the Ordovician of Poland (Greguss, 1959; Koslowski and Greguss, 1959) and the former Czechoslovakia (Obrhel, 1959) have been queried by Chaloner (1960). The poorly preserved 'fossils' from an Ordovician sandstone in Kazakhstan (Senkevich, 1963) may not be plant remains.

## SILURIAN VEGETATION

Characters that were adaptive to life in the terrestrial habitats appeared in several groups of plants. Most significant were small plants with naked, dichotomous axes, known as rhyniophytoids, and including *Cooksonia* and *Steganotheca* (Figure 3.3). Britain has yielded the most abundant and well-preserved fossils of these small plants, and most of what we know about them is based on British work. Our knowledge of them is still incomplete, though; even their size is a matter of conjecture, although they were probably little more than a few millimetres high. It is generally assumed that they were the earliest known vascular plants. They show undoubted evidence of a cuticle with stomata, and of cutinized spores (Edwards *et al.*, 1986), but no evidence of vascular tissue has been found in the Silurian. They are thus provisionally designated as 'rhyniophytoid', implying that they have the appearance of a primitive vascular plant, but without yielding direct evidence of vascular tissue (Edwards and Edwards, 1986). The oldest examples of well-preserved rhyniophytoid remains are from the Wenlock Series.

Unequivocal evidence of vascular plants in the Silurian of Britain is relatively limited. Slender axes with xylem tissue have been reported from the Ludlow Series (Edwards and Davies, 1976). No fructifications were attached, but it is generally assumed that they are fragments of rhyniophytes, the archetypal simple vascular plants, with naked, branching axes bearing single, terminal sporangia. Outside of Britain, the evidence is more convincing. The most widely discussed example is the lycopsid *Baragwanathia*, which is best known from Australia. Its stratigraphical position has been the subject of much controversy, but a Ludlow age is now widely accepted (Garratt, 1979, 1981; Edwards *et al.*, 1979; Douglas and

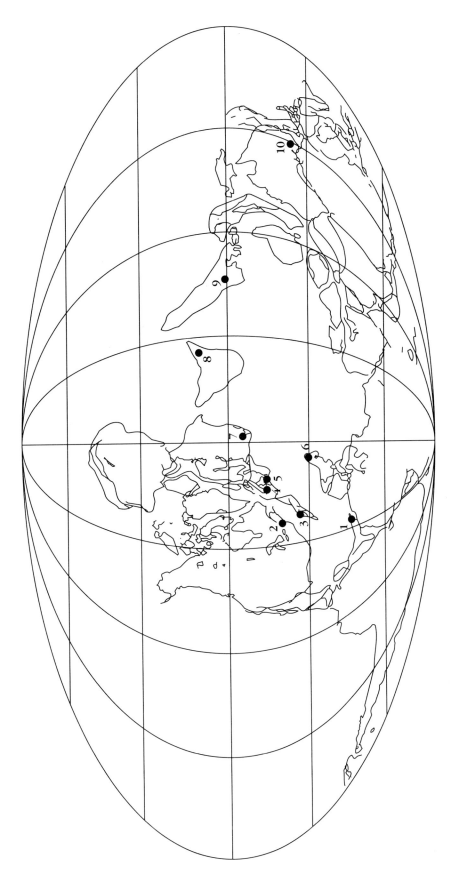

**Figure 3.2** The palaeogeography of the Silurian, showing the location of the major fossil floras of this age. 1 – Virginia; 2 – Maine; 3 – New York State; 4 – Tipperary; 5 – Wales; 6 – Bohemia; 7 – Podolia; 8 – Kazakhstan; 9 – Xinjiang; 10 – Victoria. Based on Scotese and McKerrow (1990).

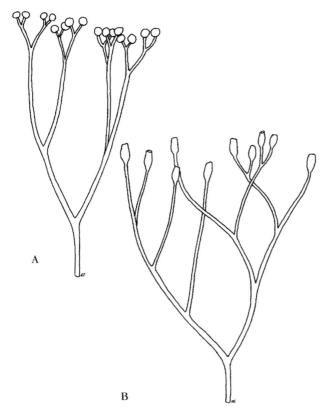

**Figure 3.3** Reconstructions of two typical Silurian rhyniophytoid plants. (A) *Cooksonia*. (B) *Steganotheca*. From Thomas and Spicer (1987, figure 3.1; after D. Edwards).

1979). The general consensus now seems to be that they originated from the green algae. Charophytes have been thought to be one possible ancestral group, although their lowest fossil occurrence, in the Přídolí (Ishchenko, 1975), rather weakens this argument. Despite the extensive fossil record of early vascular and vascular-like plants now available, the origin of the group remains obscure.

Of the other plant groups which appear to have been developing strategies to adapt to terrestrial habitats, the most spectacular was *Prototaxites* and allied form-genera. They had thick axes or 'trunks', up to one metre thick, which appear to have lain prostrate on the land surface. The axes were constructed of a mass of tubes, some of which may have functioned in a similar way to vascular tissue in vascular plants. However, little is known of the overall form of these plants or of their mode of reproduction.

Another enigmatic plant with features apparently adapting it to terrestrial conditions, is represented by cutinized sheets known as *Nematothallus*. At one time, these were thought to be the leaf-like organs of *Prototaxites* but, following work on some well-preserved British material, they are now considered to be thalli that encrusted the land surface (Edwards and Rose, 1984).

Upper Silurian terrestrial sedimentary rocks in Britain frequently contain spherical objects known as *Pachytheca*. It is assumed that they are the remains of another type of plant becoming adapted to a terrestrial mode of life, although whether they are individual organisms or part of a larger plant is not certain.

These enigmatic non-vascular land plants have been traditionally united under the umbrella-term 'nematophytes', but it is unlikely that they were closely related. They appear to represent evolutionary dead-ends, since none are known above the Devonian. Their origins and affinities are unclear. Possible affinities for *Prototaxites* with the brown algae and *Pachytheca* with the green algae have been suggested (e.g. Taylor, 1981), but the evidence is far from clear.

In contrast to the terrestrial vegetation, our knowledge of Silurian marine algae remains poor (Johnson and Konischi, 1959). It is known that thalloid algae occurred as far back as the Cambrian Period, mainly through the fossilized remains of forms that secreted a calcareous skeleton, such as the dasyclads. As lime secretion was presumably a relatively advanced specialization, it

Lejal-Nicol, 1981; Gould, 1981; Hueber, 1983; Garratt *et al.*, 1984). Its occurrence in strata of about the same age as the oldest known ?rhyniophytes is disconcerting. Similar, but poorly preserved material has also been reported from the Ludlow of Saxony in Germany (Roselt, 1962). The discovery of rhyniophytoids in the underlying Wenlock Series (Edwards and Feehan, 1980; Edwards *et al.*, 1983) has partially resolved the problem, although the time-gap is still uncomfortably narrow. To some, this is evidence that vascular plants are polyphyletic (e.g. Banks, 1968), but there are so many characters which appear to unify the group that this seems most unlikely (Delevoryas, 1962; Stewart, 1983). A more likely explanation is that it simply reflects the incompleteness of the fossil record of plants.

It was once thought that vascular plants originated from among the bryophytes (e.g. Campbell, 1895), but there is no support for this thesis from the fossil record; liverworts do not appear before the Middle Devonian (Ishchenko and Shylakov,

# Silurian plant fossils in Britain

is likely that macroscopic branching algae occurred well back into the Precambrian, but fossil evidence is generally poor due to problems of preservation. There are some records of Silurian non-calcareous branching forms, the best documented being from Britain (e.g. *Powysia*, *Inopinatella*), but they tell us relatively little about the diversity and evolutionary history of such algal plants.

## SILURIAN PLANT FOSSILS IN BRITAIN

Britain has the most complete record of Silurian land-plant fossils in the world, and has played a central role in developing a phylogenetic and palaeoecological model for the terrestrialization of vegetation. Most of the sites occur in Wales and the Welsh Borders, including all those described in this chapter (Figures 3.4 and 3.5). The oldest are transported fragments in deep-water, Wenlock turbidites from north Wales (records from the Llandovery of Pembrokeshire are now regarded as doubtful – Keeping, 1882, 1883; Nathorst, 1883a, 1883b). More complete evidence occurs in the Ludlow to Přídolí interval, which has yielded diverse assemblages of rhyniophytoid plants.

Elsewhere in Britain, Silurian fossils have also been reported from southern Scotland. Etheridge (1874) described specimens from the Llandovery of Scotland as *Parka decipiens* Fleming, but he failed to illustrate them. Since this species is normally restricted to the Lower Devonian, the identification must be regarded as suspect. A

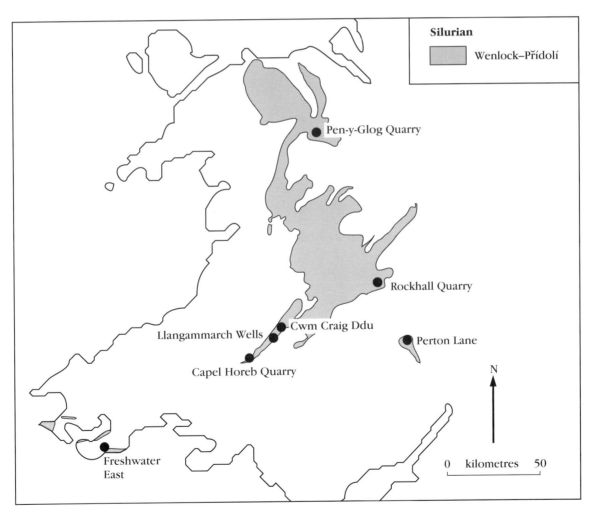

**Figure 3.4** Outcrop of Wenlock to Přídolí strata in Wales and the Welsh Borderland, showing the locations of the Silurian GCR palaeobotany sites.

| Series | Stages | GCR Palaeobotany sites | Main areas outside Britain ||||
|---|---|---|---|---|---|---|
| | | | Western Europe | Eastern Europe | Asia/Australasia | North America |
| Přídolí | | Freshwater East<br>Perton Lane<br>Capel Horeb Quarry | | Bohemia<br>Podolia | Kazakhstan<br>Xinjiang | New York State |
| Ludlow | Ludfordian | Capel Horeb Quarry | | | | |
| | Gorstian | Cwm Craig Ddu Quarry<br>Rockhall Quarry<br>Llangammarch Wells Quarry | | | Victoria | |
| Wenlock | Homerian | | Tipperary | | | |
| | Sheinwoodian | Pen-y-Glog Quarry | | | | |
| Llandovery | Telychian | | | | | Maine |
| | Aeronian | | | | | |
| | Rhuddanian | | | | | Virginia |

**Figure 3.5** Chronostratigraphical classification of the Silurian, and the positions of the GCR and other major palaeobotanical sites in this system.

variety of other problematic plant fossils have been recorded from the Hagshaw Hills and Lesmahagow area of Scotland (Crookall, 1930; Ritchie, 1963), which are probably late Llandovery or early Wenlock in age (Cocks et al., 1971). Crookall's record of *Taitia* from here was the only one to be adequately described and illustrated, and even this is difficult to interpret, and may not even be a plant.

## PEN-Y-GLOG QUARRY

### Highlights

Pen-y-Glog Quarry has yielded the oldest, well-preserved *Prototaxites–Pachytheca* assemblage in Great Britain, and the oldest preserved as petrifactions from anywhere in the world (Figure 3.6). It has also yielded a number of other enigmatic plant fossils, including *Berwynia carruthersii* Hicks. The assemblage provides a valuable insight into mid-Silurian floras, and into the nature of early land vegetation.

### Introduction

This quarry (SJ 107422), which lies on the north side of the River Dee valley, near Corwen, Clwyd, has yielded some of the oldest plant fossils in Great Britain. The fossils were described by Hicks (1881, 1882) and Dawson (1882). More recently, some poorly preserved material has been discussed by Burgess and Edwards (1988).

### Description

#### Stratigraphy

There is no detailed stratigraphical section published for this locality. Approximately 30 metres of the Pen-y-Glog Slate Formation is overlain by 15 metres of the Pen-y-Glog Grit Formation (together, the Pen-y-Glog Group). The geological distribution of the two facies-associations is shown in Figure 3.7. The Pen-y-Glog Slate consists of uniform dark grey shales, and has yielded a typical off-shore marine graptolite fauna of the *Cyrtograptus murchisoni* Zone (Elles, 1900). The Pen-y-Glog Grit consists of alternating coarse sandstones (with plant fossils), siltstones and shales, which have been interpreted as turbidites infilling the Denbigh trough (Cummins, 1957), and which contain a *Monograptus riccartonensis* Zone fauna (Elles, 1900). The biostratigraphical evidence suggested is clearly indicative of a lower Sheinwoodian (early Wenlock) age.

#### Palaeobotany

The best preserved specimens are petrifactions from the sandstones, and include *Prototaxites hicksii* (Etheridge) Dawson and *Pachytheca* sp., together with some enigmatic spherical bodies. The underlying shales have yielded *Berwynia carruthersii* Hicks.

### Interpretation

Other than some spores and cuticle fragments (Burgess, 1991; Burgess and Richardson, 1991), Pen-y-Glog has yielded the oldest known evidence of land plants. The best evidence is in the form of small fragments of *Prototaxites hicksii*, no more than 50 mm long. They show little of the gross morphology, but internal structure can be clearly seen in thin section (Hicks, 1881, pl. 25). It conforms with that normally associated with *Prototaxites*, consisting of wide and narrow sets of tubes, except the former are rather smaller and denser than in most other species (12–22 μm in diameter and *c.* 2500 tubes per cm$^2$ in cross-section). Barber (1892) suggested that it might be the same as *Prototaxites storrei* (Barber) Dawson, found in South Wales, but the smaller, denser tubes in the Pen-y-Glog specimens may indicate that they are different. According to Burgess and Edwards (1988), the thicker tubes may be internally thickened, in which case they would belong to their new form-genus *Nematasketum*. However, they were unable to confirm this in freshly collected material, and were unable to examine the original type specimens, and so made no formal proposal of transference. They are probably the oldest *Prototaxites*-type specimens found in Great Britain to date. Arber (1904a) makes passing comments to other occurrences in North Wales but, without further information about the localities, their age cannot be determined. The previously mentioned *P. storrei* specimens described from Rumney Quarry near Cardiff (Barber, 1892) are from the upper Wenlock (*Cyrtograptus lundgreni* Zone) and are thus younger. From outside of Great Britain, there is only one reliable record from older strata, from

# *Silurian*

**Figure 3.6** Pen-y-Glog Slate Quarry. Cleaved Pen-y-Glog Slates in the lower part of the quarry face, that have yielded *Berwynia*. These are overlain by turbidites of the Pen-y-Glog Grits, that contain *Prototaxites*. (Photo: C.J. Cleal.)

# Pen-y-Glog Quarry

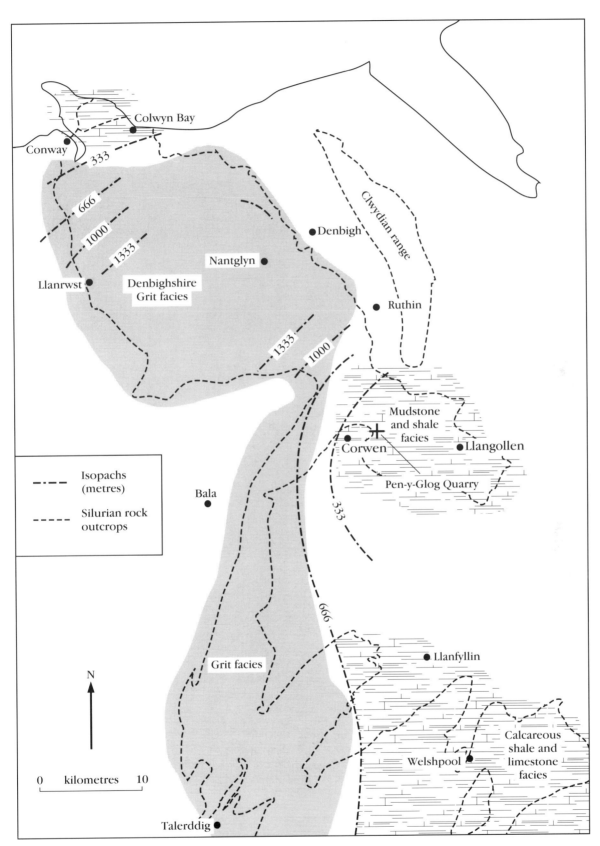

**Figure 3.7** Distribution of grit and shale facies in the Wenlock of North Wales, showing the position of Pen-y-Glog Slate Quarry. Based on Smith and George (1961, figure 20).

the lower Llandovery of Virginia, USA (Pratt *et al.*, 1978), but this was based on tubes macerated from compression fossils. The records from the Ordovician of the Sahara (Arbey, 1973; Koeniguer, 1975) were based on compressions showing no internal structure to confirm the identification.

Associated with the *Prototaxites* are small spheres of *Pachytheca*. Hicks (1881) was able to describe some of their internal structure, but his illustrations are rather diagrammatic and the microscope slides are now lost. Barber (1889) reported examining the slides, however, and stated that the preservation was poor. He confirmed that they were *Pachytheca*, but could not place them in a particular species.

These are amongst the oldest specimens of *Pachytheca* found in Great Britain. The only slightly older specimen is from the lower Wenlock Buildwas Beds of Shropshire and is poorly preserved (Andrew, 1925). There are, however, several records of *Pachytheca* from slightly younger strata in the Wenlock Series of Wales and the Welsh Borders (Harris, 1884; Hooker, 1889; Barber, 1889, 1891; Thiselton-Dyer, 1891; Storrie, 1892; Seward, 1895; Strahan and Cantrill, 1912). Ritchie (1963) mentioned *Pachytheca* from possible Wenlock strata in Scotland, but did not illustrate the specimens. The exact age of these specimens is open to question.

A larger spherical body has come from the Pen-y-Glog sandstones (British Geological Survey collections, specimen no. zl 363); it is not well-preserved, but is larger (*c.* 8 mm in diameter) than the *Pachytheca* spheres found here, and does not show the characteristic two-layered structure of that genus. In a letter (dated 19th February 1946) kept with the specimen, W.H. Lang wrote that there was little doubt that it was a plant, but that it could not be identified beyond 'sphaerical body *incerta sedis*'.

The species from the sandstones are all of uncertain affinity, but both *Prototaxites* and *Pachytheca* are widely believed to be land plants. Their presence in mid-basinal marine sediments may be due to the sandstones being turbidites, the plant fragments having been transported into the deeper parts of the Welsh Basin by turbidity currents from a landmass, probably somewhere to the south.

*Berwynia carruthersii* Hicks (Figure 3.8) represents parallel-sided, sometimes branching axes, preserved as anthracitic coal (Hicks, 1882). Many have a rugose surface, which Hicks interpreted as spirally arranged leaf bases, but it is too irregular

**Figure 3.8** *Berwynia carruthersii* Hicks. Enigmatic, possibly algal plant; Natural History Museum, London, specimen V.5887. Pen-y-Glog Slate Formation (Wenlock), Pen-y-Glog Slate Quarry. × 0.5. (Photo: Photographic Studio, Natural History Museum, London.)

for this to be likely. Also visible are zones along the margins of the axes, which show a rather finer patterning. However, this and the irregular rugose patterning are probably just a taphonomic effect. In the absence of any internal structure being preserved, it is impossible to be certain as to what group of plants *Berwynia* belongs.

Hicks (1882) described some other enigmatic fragments from the Pen-y-Glog shales as *Parka*, but they are too poorly preserved for this to be confirmed.

## Conclusion

Other than some spores, Pen-y-Glog has yielded the oldest evidence of land vegetation in Britain, about 427 million years old. The fossils are also the oldest-known land plant fossils in the world that show details of cell structure, but are not closely related to anything living today.

## LLANGAMMARCH WELLS QUARRY

### Highlights

Llangammarch Wells Quarry is the only known locality to yield undoubted specimens of *Powysia bassettii* Edwards, one of only two non-calcareous algae known from the Silurian.

### Introduction

This site, a small quarry in lower Ludlow marine siltstones near Llangammarch Wells, Powys (SN 937472), has yielded unusually well-preserved compressions of algae. They were first figured by Bassett and Edwards (1973, p. 4) and were subsequently named and described in full by Edwards (1977).

### Description

#### Stratigraphy

The geology of the quarry has been discussed by Bassett (*in* Baker and Hughes, 1979). The sequence consists of dark grey, graptolitic siltstones and shales, representing a marginal marine setting. The part of the quarry containing the plant fossils has yielded graptolites of the lower *Neodiversograptus nilssoni* Zone, indicating a Gorstian (early Ludlow) age. This is supported by acritarch evidence (K.J. Dorning, pers. comm., 1979). The general geology of the area is shown in Figure 3.9.

#### Palaeobotany

Just one species of plant fossil has been found: *Powysia bassettii* Edwards. They are mostly brown or yellow stained impressions, although some areas of carbonaceous residue are also present.

### Interpretation

The fossils show a complex, branching thallus with a basal hold-fast (Figure 3.10). The branches appear to consist of a mass of intertwined, longitudinal tubes, with no parenchymatous tissue preserved. Animal fossils, particularly dendroid graptolites, can sometimes develop similar branching structures, but Rickards (*in* Edwards, 1977) believed that these were not graptolites. Geochemical analysis of one of the specimens by Niklas (*in* Edwards, 1977) gave added support to its algal affinities. It was probably a marine benthic alga, although Edwards (1977) suggested that it might have been a freshwater plant that had been swept into the sea before burial.

This is the only locality known to yield well-preserved *Powysia*. Specimens in the Department of Palaeontology, The Natural History Museum, London, collected from Wenlock Series beds at Dudley and in the Pentland Hills, show similar branching structures. None, however, show any structural details to confirm that they belong to *Powysia*. It is one of only two fully described non-calcareous algae known from the Silurian, the other being *Inopinatella* (see Rockhall Quarry, below).

### Conclusion

Llangammarch Wells Quarry has yielded one of the very few examples of a non-calcareous marine alga (*Powysia*) known from the Silurian (423 Ma). Although a few other sites have yielded fossils of this type, the Llangammarch Wells material is the best preserved, yielding some details of internal structure.

# Silurian

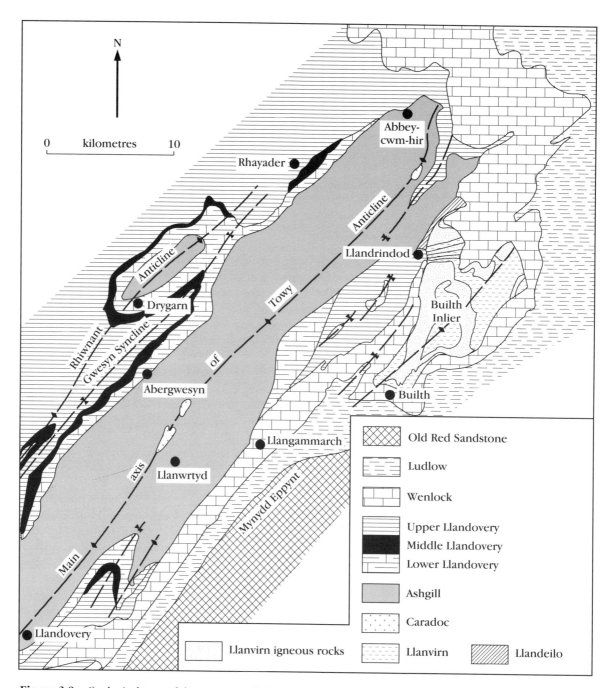

**Figure 3.9** Geological map of the area around Llangammarch Wells. The quarry yielding the plant fossils lies just to the east of the village as marked on the map. Based on George (1970, figure 14).

## ROCKHALL QUARRY

### Highlights

Rockhall Quarry is the only known locality for *Inopinatella lawsonii* Elliott, which is the only Palaeozoic example of what might be a non-calcified dasyclad alga. It is thus potentially significant for helping us to understand the early evolutionary development of this important family of marine plants.

# Rockhall Quarry

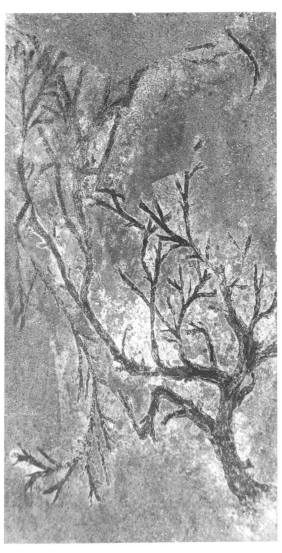

**Figure 3.10** *Powysia bassettii* D. Edwards. Enigmatic branched alga; National Museum of Wales, specimen 72.39G1a (holotype). Graptolitic shales (lower *Neodiversograptus nilssoni* Zone – Gorstian), Llangammarch Wells Quarry. × 1.6. (Photo: Photographic Studio, National Museum of Wales.)

## Introduction

This Silurian limestone quarry lies just north of the village of Aymestry, in the county of Hereford and Worcester (SO 423655). Plant fossils were described from here by Elliott (1971), who interpreted them as probably algal in origin.

## Description

### Stratigraphy

This is the type locality for the Aymestry Limestone Formation, which is a lower Gorstian (lower Ludlow) shallow marine deposit (Holland *et al.*, 1963).

### Palaeobotany

Elliott (1971) described the plant fossils as *Inopinatella lawsonii* Elliott; they are mostly preserved as coalified compressions.

## Interpretation

This is the only locality to yield *I. lawsonii* (Figure 3.11), which probably grew on the edge of a shallow marine shelf. It has a main stem *c.* 0.3 mm wide and more than 30 mm long, and with branches attached in whorls of four. No reproductive structures have been found, but Elliott noted a similarity to the juvenile stages of the extant dasyclad *Neomaris*, and suggested that it may have been a primitive non-calcified example of that family. If correct, then *Inopinatella* is the only known non-calcified dasyclad to have been found in the Palaeozoic. The Dasycladales has a fossil record that extends back to the Cambrian (Meyen, 1987), but the preservation potential of non-calcified forms would be very low, which could explain their absence in the pre-Silurian fossil record.

## Conclusion

Rockhall Quarry has yielded the only known examples of a marine alga, *Inopinatella*, which is about 420 million years old. It is thought to belong to the group known as the dasyclads, which have been important components of benthic vegetation for over 500 million years. Most members of the group have a calcified body, and at one time in the geological past (*c.* 200 million years ago) they were major reef-building organisms. *Inopinatella* was not calcified, however, and is thought to have been a primitive, early representative of the group.

# *Silurian*

**Figure 3.11** *Inopinatella lawsonii* Elliott. Non-calcified, possibly dasyclad alga; Natural History Museum, London, specimen V.31287. Aymestry Limestone Formation (upper Gorstian), Rockhall Quarry. × 2. (Photo: Photographic Studio, Natural History Museum, London.)

## CWM CRAIG DDU QUARRY

### Highlights

Cwm Craig Ddu has yielded the oldest known specimens of *Cooksonia* in Great Britain, and is only marginally pre-dated by the oldest known specimens from anywhere in the world. They represent what is probably the most primitive type of vascular plant, and are thus of key importance for understanding the early phases in the development of vegetation on land.

### Introduction

This locality comprises a small roadside quarry in Ludlow-aged siltstones on the north-eastern slopes of Mynydd Eppynt, between Llanwrtyd Wells and Builth Wells, Powys (SN 962475). Plant fragments from here were first reported by Straw (1953), and more were discovered in 1973 during a Geologists' Association field meeting (Baker and Hughes, 1979). The most comprehensive account of the flora is given by Edwards *et al.* (1979), who described a number of sterile and fertile fragments of rhyniophytoid plants (*sensu* Edwards and Edwards, 1986).

### Description

#### Stratigraphy

The geology of this quarry has been discussed by Straw (1930, 1937), Bassett (*in* Baker and Hughes, 1979) and Edwards *et al.* (1979). The exposed sequence belongs to the Pterinea Beds, the basal member of the Wilsonia Shales Formation (Straw, 1937). The beds represent deposits formed on the palaeoslope on the eastern margin of the Welsh basin. It contains a fauna, including rhynchonellid brachiopods, bryozoans, nautiloids, bivalves and rare graptolites, belonging to the *Saetograptus leintwardinesis incipiens* Zone (Straw, 1937; Bassett *in* Baker and Hughes, 1979), indicating a late Gorstian (early Ludlow) age, although acritarchs collected by K.J. Dorning (pers. comm., 1979) have given a mid-Gorstian age.

#### Palaeobotany

Edwards *et al.* (1979) recorded *Hostinella* sp., *Cooksonia pertoni* Lang, *Cooksonia* sp. and cf. *C. cambrensis* Edwards from here. They are mostly preserved as coalified compressions. Specimens are rare and generally fragmentary, rarely exceeding a few millimetres in size.

# Capel Horeb Quarry

## Interpretation

The specimens found here have slender, occasionally dichotomous, branching axes. Mostly, they are sterile and thus placed in the form-genus *Hostinella*. A dark central line, suggestive of vascular tissue, occurs along some of them. Even using film pulls, Edwards *et al.* (1979) were unable to obtain microscopic evidence of tracheids to confirm that this is vascular tissue, but they did reveal some evidence of epidermal and/or cortical cells.

Three of the specimens described by Edwards *et al.* showed thin axes terminated by single, small sporangia (one of these specimens is represented by both the part and counterpart). The sporangia varied in shape and were assigned to different 'species', although Edwards *et al.* were clearly uneasy at identifying more than one species from here based on such limited material. There can be little doubt, however, that they belong to the form-genus *Cooksonia* (Figure 3.12).

*Cooksonia* is widely regarded as the most primitive vascular land plant (Taylor, 1981; Thomas and Spicer, 1987; Edwards *et al.*, 1992). For a time, the Cwm Craig Ddu specimens were the oldest examples of *Cooksonia* known from anywhere in the world. Although older specimens are now known from the Wenlock of Ireland (Edwards and Feehan, 1980; Edwards *et al.*, 1983), they are still the earliest found in Great Britain. They are thus of considerable interest for charting the early history of the British land floras and demonstrating the morphological simplicity of the first vascular plants to appear on land in this part of the world. They contrast with the more advanced plants found in approximately coeval strata in Gondwana (the *Baragwanathia* flora) and suggest that even at this early time there was a clear difference between the land vegetation of the southern low palaeolatitudes and the rest of the world.

## Conclusion

Cwm Craig Ddu Quarry has yielded specimens of *Cooksonia* that are some 420 million years old, and which are the oldest examples of this primitive land plant to have been found in Great Britain. This is widely thought to be the ancestral form of most if not all land vascular plants. It is only marginally pre-dated by similar fossils in Ireland, which are at present the oldest examples of such plants from anywhere in the world.

**Figure 3.12** *Cooksonia pertoni* Lang. Earliest examples of rhyniophytoid land plants known from Great Britain; National Museum of Wales, specimen 79.17G3. Wilsonia Shales Formation (lower Ludlow), Cwm Craig Ddu Quarry. × 5. (Photo: Photographic Studio, National Museum of Wales.)

## CAPEL HOREB QUARRY

### Highlights

Capel Horeb Quarry has yielded the earliest examples of plant axes with vascular tissue found anywhere in the world (Figure 3.13). This is thus the oldest unequivocal evidence of a land vascular plant.

### Introduction

This site is a disused quarry near Llandovery, Powys (SN 844323), and has yielded plant fossils of Ludlow and also possibly Přídolí age. The fossils are usually very fragmentary, and some early descriptions of the site make little or no mention of them (Straw, 1930; Potter and Price, 1965).

# Silurian

**Figure 3.13** Capel Horeb Quarry. The bedding plane on the left of the picture is of upper Ludlow beds of the Upper Roman Camp Formation. These are overlain by upper Ludlow or lower Přídolí beds of the Long Quarry Formation. (Photo: C.J. Cleal.)

# Capel Horeb Quarry

Better preserved material has been found, as described by Heard (1939), Edwards (1970b, 1982), Edwards and Davies (1976) and Edwards and Rogerson (1979).

## Description

### Stratigraphy

The geology of this site has been described by Straw (1930), Potter and Price (1965), Edwards and Richardson (in Friend and Williams, 1978) and Edwards and Rogerson (1979). The lower part of the sequence consists of shallow marine siltstones and sandstones with a restricted shelly fauna, and belongs to the Upper Roman Camp Formation. Faunal evidence discussed by Potter and Price (1965) suggests a late Ludlow (mid-Ludfordian) age for these strata, a conclusion supported by the microfossils (Dorning in Edwards and Davies, 1976). Lying unconformably above these beds is the Long Quarry Formation (Figure 3.14), which has been interpreted as either upper Ludlow (Richardson and Lister, 1969) or lower Přídolí (Potter and Price, 1965). They are probably littoral siltstones and sandstones. The topmost part of the sequence belongs to the Red Marls Formation.

### Palaeobotany

The following plant fossils have been reported from the Upper Roman Camp Formation here:

Phaeophycophyta(?):
 *Nematothallus* sp.
 *Nematoplexus* sp.

Rhyniophytoids:
 *Cooksonia hemisphaerica* Lang
 cf. *C. caledonica* Edwards
 *Steganotheca striata* Edwards

The Long Quarry Formation has yielded the following rhyniophytoids:

 *Cooksonia* cf. *pertoni* Lang
 *S. striata* Edwards
 cf. *Renalia* sp.

The thalloid-like structure described by Heard and Jones (1931a) as *Eohepatica dyfriensis*, and by Heard and Jones (1931b) as *Thallomia llandy-*

**Figure 3.14** Close-up of part of the face of Capel Horeb Quarry, showing the relationship between the two plant-bearing intervals. (Photo: B.A. Thomas.)

*friensis*, is now believed to be part of a dictyocarid arthropod (Rolfe, 1969).

## Interpretation

The Upper Roman Camp flora appears to be dominated by what are probably non-vascular land plants, such as *Nematothallus*. Bulk maceration has yielded cuticles showing a reticulate pattern on their inner surface, characteristic of the genus (Edwards, 1982). There is considerable variation in this patterning, but it is not yet clear whether more than one species is present. The evidence from Capel Horeb suggests that *Nematothallus* was a thalloid, encrusting plant, rather than a leaf-like structure of a *Prototaxites*, as suggested by Lang (1937) and Jonker (1979).

Edwards (1982) also found a variety of fine tubes in her bulk maceration samples. The majority were isolated tubes of uncertain affinity. There were, however, some clusters of tubes that resemble those found in *Prototaxites* axes. These were named *Nematoplexus* by Edwards, following the nomenclature of Lyon (1962).

Rhyniophytoid plants are represented here by a number of different taxa recognized on the basis of their reproductive structures. The most abundant belong to *Cooksonia*. The specimens from the Upper Roman Camp Formation have mostly globose sporangia, which can be identified as *C. hemisphaerica* (Edwards and Rogerson, 1979), although one was assigned to cf. *C. caledonica* (Edwards and Rogerson, 1979, pl. 1 fig. d). Other specimens from these strata had more elongate sporangia, which lie outside the circumscription of *Cooksonia* as defined by Lang (1937), and were not identified by Edwards and Rogerson. *C. hemisphaerica* is known to have a variety of sporangial shapes (Edwards, 1979a) and the possibility that these slightly more elongate forms belong there cannot be excluded.

Only one fertile specimen of *Cooksonia* has been reported from the Long Quarry Formation here (Edwards and Rogerson, 1979, pl. 1 fig. h). This showed squatter sporangia than those from the Upper Roman Camp Formation and was identified as *C.* cf. *pertoni* Lang.

This is the type locality for another type of rhyniophytoid plant: *Steganotheca striata* Edwards, 1970b (fig. 8b and Figure 3.15). It occurs in both the Upper Roman Camp and Long Quarry formations. Like *Cooksonia*, it has slender, dichotomous axes with terminal sporangia. The sporangia are, however, elongate and less well individualized, and usually show a heavily carbonized, lenticular structure at the apex. It is at present unclear whether this structure is simply due to compression of the sporangial tip, or is evidence of a dehiscence structure. Because of the elongate shape of the sporangia and isotomous branching, Edwards (1970b) initially placed *Steganotheca* in the Rhyniaceae, but it has subsequently been described as rhyniophytoid (Pratt *et al.*, 1978; Edwards and Edwards, 1986).

The holotype of *Cooksonia downtonensis* Heard, which originated from Capel Horeb (Heard, 1939), was transferred to *S. striata* by Edwards (1970b). It is arguable that Heard's name for this species should take priority, but this is not the place to propose a new combination.

A single specimen from the Long Quarry Formation (Edwards and Rogerson, 1979, pl. 1 fig. i) shows a more complex branching pattern than

**Figure 3.15** *Steganotheca striata* D. Edwards. One of the earliest known land plants; National Museum of Wales, specimen 69.64G32a (holotype). Upper Roman Camp Formation (upper Ludlow), Capel Horeb Quarry. × 1.5. (Photo: Photographic Studio, National Museum of Wales.)

the other rhyniophytoid species found at Capel Horeb. Edwards and Rogerson compared it with *Cooksonia hemisphaerica* Ananiev and Stepanov, 1969, *non* Lang, which Gensel (1976) has in turn compared with *Renalia*. The latter shows characters intermediate between the Rhyniaceae and Zosterophyllaceae and there has been some disagreement as to its taxonomic position. However, Edwards and Edwards (1986) have recently argued that it belongs to the Rhyniaceae, and that its zosterophyll-like characters are due to evolutionary convergence.

Far commoner than the fertile rhyniophytoid specimens discussed above are unbranched and dichotomous axes without sporangia. Being sterile, it is impossible to identify them beyond *Hostinella* sp. However, examples from the Upper Roman Camp Beds have been shown to have *in situ* tracheids (Edwards and Davies, 1976), the earliest known examples of axes with such tissue. Prior to Edwards and Davies' (1976) study, the oldest known axes with *in situ* tracheids were from the basal Devonian at Targrove

Quarry (Lang, 1937). As with the Capel Horeb specimens, none of the Targrove axes with tracheids bore sporangia and so it has been impossible to confirm their identification. However, they were found in association with fertile specimens of *Cooksonia hemisphaerica*, and it has been widely asserted that they belonged to the same plant (e.g. Taylor, 1981). Dispersed tracheids have been reported from strata as old as the Cambrian (Gosh and Bose, 1952; Jacob *et al.*, 1953) but Banks (1975a) has argued that these cannot be relied on as evidence of the presence of the Tracheophyta, because of possible reworking or contamination. The Capel Horeb specimens are thus the oldest indisputable evidence of land vascular plants.

## Conclusion

Capel Horeb has yielded the oldest unequivocal evidence of plants with vascular conducting tissue (xylem) from anywhere in the world; the fossils are *c.* 415 million years old. The development of this tissue was one of the key steps that helped plants overcome the hydraulic problems inherent with living on the land, and thus paved the way for the evolution of land vegetation as we see it today.

## PERTON LANE

## Highlights

Perton Lane is a classic site for Přídolí Series plant fossils, being the first to be subject to a detailed palaeobotanical investigation. It is the type locality for *Cooksonia*, widely regarded as the most primitive known vascular plant, and the only known locality for *Caia* and the enigmatic *Actinophyllum*.

## Introduction

This small roadside exposure of Ludlow and Přídolí shales in the village of Stoke Edith, Hereford and Worcester (SO 598403) is one of the classic sites for British Silurian palaeobotany. There has been some confusion over the name of the site, since it has been referred to as Perton Quarry (e.g. by Lang, 1937). As pointed out by Edwards *et al.* (1979), however, Perton Quarry is a large exposure of middle Ludlow limestones, *c.* 300 metres south of the Perton Lane Section, and the latter name is now generally used for the fossil-bearing outcrop.

Plant fossils from here were first recorded by Phillips (1848) and Phillips and Salter (1848), who described what may be algal fertile structures. Slender, branching axes were described by Brodie (1869, 1871) and were compared by Carruthers (*in* Brodie, 1871) with *Psilophyton*. Plant fossils from here have also been briefly discussed by Barber (1889), Richardson (1907), Stamp (1923) and Straw (1926). Until recently, the most complete account of the assemblage was by Lang (1937). Subsequently, however, there have been significant contributions by Fanning (1987), Fanning *et al.* (1990, 1991) and Burgess and Edwards (1988).

## Description

### Stratigraphy

The geology of this site is covered by Brodie (1871), Straw (1926), Squirrell and Tucker (1960, 1967) and Edwards *et al.* (1979). The sequence consists of Ludfordian (upper Ludlow) Upper Perton Formation, overlain by Přídolí Rushall Formation. The Rushall Formation, which yields the plant fossils, belongs to interval I.1 in King's (1925) lithostratigraphy. It consists mainly of buff to light grey mudstones with thin sandstone bands, and probably represents littoral deposits.

### Palaeobotany

The plant fossils from here are preserved as coalified compressions. The following species have been found to date:

Phaeophycophyta(?):
 *Nematothallus pseudovasculosa* Lang
 *Nematasketum diversiforme* Burgess and
  Edwards

Chlorophycophyta(?):
 *Pachytheca* sp.

Rhyniophytoids:
 *Cooksonia pertoni* Lang
 *Pertonella dactylethra* Fanning, Edwards and
  Richardson

# Silurian

*Caia langii* Fanning, Edwards and Richardson
*Salopella* sp.
*Hostinella* sp.

Uncertain affinities:
  *Actinophyllum* sp.

## Interpretation

This is the type locality for *Cooksonia pertoni*, which is the type species for the form-genus (Figure 3.16). It is widely believed to be the most primitive known vascular plant. It is difficult to envisage an upright land plant with a simpler morphology, with its thin, isotomously forked aerial shoots, no leaves or other macroscopic emergence, and terminally-borne sporangia lacking a dehiscence structure. Specimens from elsewhere (locality details not yet published) have shown evidence of stomata and peripheral supporting tissue around the axes (Edwards *et al.*, 1986) and, most recently, a vascular strand (Edwards *et al.*, 1992).

Evidence from *in situ* spores suggests that at least three species of plant bore *C. pertoni*-type sporangia (Edwards *et al.*, 1986). However, the shape of the sporangia is morphologically indistinguishable in all three and so the name *C. pertoni* may be retained as a form-species for such structures.

Two further species of rhyniophytoid have been described recently by Fanning *et al.* (1990, 1991), both of which are characterized by prominent spines. *Pertonella dactylethra* is morphologically very similar to *C. pertoni*, except for the spinose sporangia. *Caia langii*, on the other hand, has significantly more elongate sporongia, rather resembling *Horneophyton* from Rhynie. Fanning *et al.* gave various suggestions as to possible functions for the spines: (a) they increased the photosynthetic area near the sporangia, where there would be considerable energy-demands; (b) they protected the sporangia from predation; and (c) they trapped moisture, helping to protect the developing sporangia from desiccation.

Fossils of uncertain affinity but given the coral-related name *Actinophyllum* are known only from this exposure. Phillips (1848) compared the genus with the fertile structures of the extant dasyclad alga *Acetabularia* but, as it is only known from isolated specimens, the point is difficult to confirm. Straw (1926), who has provided the best photographic record of these fossils, discussed the possibility of it being a coral, but finally came to the conclusion that Phillips' suggestion was more likely to be correct.

Lang (1937) described some poorly preserved specimens from this locality as *Prototaxites*, but better material has since been obtained by Burgess and Edwards (1988). These new specimens differ from *Prototaxites* principally in the presence of internally differentially-thickened tubes. The functional significance of this feature is still unclear. However, it was regarded as sufficiently different from *Prototaxites* to justify the establishment of a new genus and species, *Nematasketum diversiforme* (holotype from Lye Stream near Morville, Shropshire).

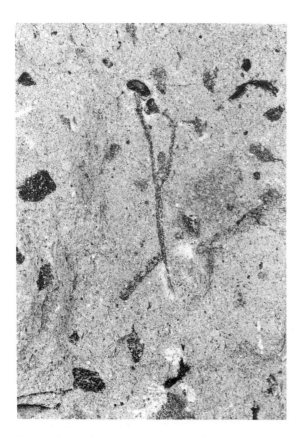

**Figure 3.16** *Cooksonia pertoni* Lang. A fertile specimen from the type locality of this important early land plant; Natural History Museum, London, specimen V.58009. Rushall Formation (Přídolí), Perton Lane. × 3. (Photo: Photographic Studio, Natural History Museum, London.)

## Conclusion

Perton Lane is the classic locality for the study of the earliest land vegetation, living *c.* 410–420 Ma,

# Freshwater East

and is where W.H. Lang made many of his pioneering discoveries in the 1930s. It is the type locality for *Cooksonia*, which is widely regarded as the archetypal primitive land plant, and is central to any discussion on the earliest evolutionary history of the vascular plants. It is also the only known locality for *Caia*, which has very unusual spiny reproductive organs.

## FRESHWATER EAST

### Highlights

Freshwater East has yielded the most diverse Silurian flora from anywhere in the world, and is particularly rich in rhyniophytoid species (Figure 3.17). This has proved of great importance for determining the ranges of morphological variation of these primitive land plants, and particularly of their reproductive structures. It is the type locality for *Cooksonia cambrensis* Edwards and *Tortilicaulis transwalliensis* Edwards, and has yielded the oldest known examples of spiny axes.

### Introduction

Sandstone exposures on the north side of Freshwater East Bay, Dyfed (SS 024981) have yielded a variety of upper Silurian plant fossils. The fossils were first noted here by Dixon (1921), and a number were included in Lang's (1937) classic study on English and Welsh Přídolí Series floras. A more recent account of part of the flora has been provided by Edwards (1979a).

### Description

#### Stratigraphy

The geology of this site has been described by Dixon (1921). The plant fossils were found in a

**Figure 3.17** Freshwater East. View across bay towards the Late Silurian plant-bearing exposures in the Freshwater East Formation. (Photo: C.J. Cleal.)

0.3 m-thick grey sandstone within the Milford Haven Group (Figures 3.18 and 3.19). King (1934) correlated these beds with the Ledbury Marl Formation of the Welsh Borderland, and a Přídolí age has been confirmed by spores described by Richardson and Lister (1969). Allen and Williams (*in* Edwards, 1979a) have interpreted this part of the Freshwater East sequence as having accumulated on coastal mudflats.

### Palaeobotany

The following species have been reported from here:

Phaeophycophyta(?):
    *Nematothallus pseudovasculosa* Lang
    *Prototaxites* sp.

Chlorophycophyta(?):
    *Pachytheca* sp.

Rhyniophytoids:
    *Cooksonia hemisphaerica* Lang *emend.*
        Edwards
    *C. cambrensis* Edwards
    cf. *C. caledonica* Edwards
    cf. *C. pertoni* Lang
    *Cooksonia* sp.
    *Tortilicaulis transwalliensis* Edwards
    cf. *Salopella* sp.
    *Hostinella* sp.
    *Psilophytites* sp.

The specimens are preserved as adpressions. Occasionally, some iron oxide and iron sulphide occurs on them, but no evidence of permineralization has been found (Edwards, 1979a).

## Interpretation

This flora is characterized by an abundance of fertile rhyniophytoid specimens. By far the commonest were identified as *Cooksonia* by Edwards (1979a), who used 83 of the most complete examples to analyse the variation in sporangial shape and attachment. This allowed her to identify five form-species from here, one of which (*C. cambrensis*) was new. It also allowed her to emend Lang's (1937) diagnosis for *C. hemisphaerica*. She recognized the problem of identifying species on what are relatively minor differences, but argued that it was the only practical way of analysing such morphologically simple and fragmentary plant fossils. This is by far the most diverse *Cooksonia* assemblage reported from anywhere in the world.

*Tortilicaulis transwalliensis* was described from a number of unbranched axes bearing terminal, elongate bodies (Figure 3.20). The latter were presumed to be sporangia, although no spores were obtained from them. Neither was any evidence of tracheids found in the axes. A distinctive feature of the species is that the axes have a non-rigid appearance, often seeming to be twisted. It has elongate sporangia, similar to those of *Sporogonites*, but lacks some of the well-defined characters of the latter, such as the longitudinal folds and the sterile basal region. Consequently, Edwards placed it in a separate form-genus. Edwards also found some features shared with certain extant bryophytes, particularly the twisted axes, but again the preservation precluded an assignment.

A small number of specimens were axes, this time showing no evidence of twisting, which bore rather smaller, elongate sporangia. Edwards (1979a) compared these with *Salopella* of Edwards and Richardson (1974), but was unable to place them in any particular species.

As is normal in Přídolí age floras, fragmentary sterile axes are far commoner than the fertile specimens. The most common are smooth, either simple or dichotomous axes, which were identified as *Hostinella* by Edwards (1979a). No evidence of tracheids has been found in these specimens. Edwards also described three specimens showing a cluster of branches, which she compared with the 'K-branching' thought to be from the base of the *Zosterophyllum* plant (Walton, 1964a). However, she emphasized that there was no unequivocal evidence that the Freshwater East specimens were basal structures, and that they could equally be from the aerial part of a plant.

Of considerable interest was Edwards' discovery of four examples of axes, one of which shows a dichotomy, with spines attached (Figure 3.21). In the absence of attached reproductive structures, it was impossible to say what group of plants they belonged to, and so they were placed in the generalized form-genus *Psilophytites*. Their interest is in being the oldest examples of spiny axes recorded from anywhere in the world.

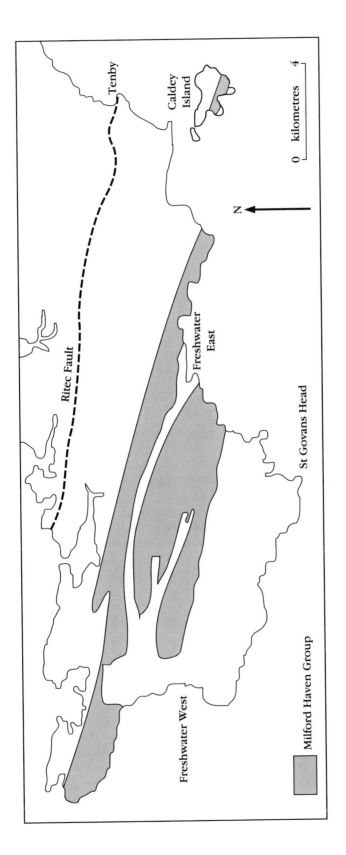

**Figure 3.18** The outcrop of the Milford Haven Group (Upper Silurian and Lower Devonian) in Pembrokeshire, showing position of Freshwater East and other localities that yield plant fossils. Based on Williams *et al.* (1982, figure 2).

## Silurian

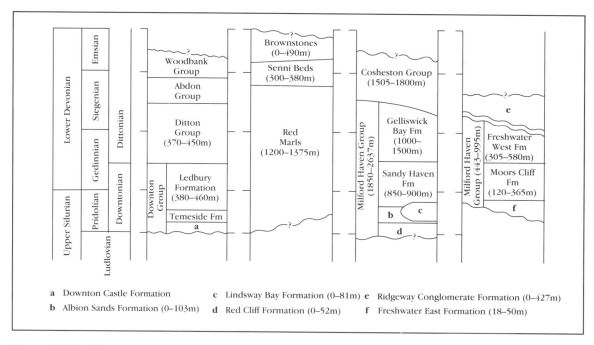

**Figure 3.19** Lithostratigraphical divisions of the Přídolí and Lower Devonian of South Wales and the Welsh Borderland. From left to right, the four columns represent the sequences in (1) the Welsh Borderland, (2) central South Wales, (3) Dyfed north of the Ritec Fault, and (4) Dyfed south of the Ritec Fault. Based on Friend and Williams (1978, figure 31).

**Figure 3.20** *Tortilicaulis transwalliensis* D. Edwards. A fertile specimen of an early rhyniophytoid land plant; National Museum of Wales, specimen 77.6G2. Freshwater East Formation (lower Milford Haven Group – Přídolí), Freshwater East. × 4. (Photo: Photographic Studio, National Museum of Wales.)

# *Freshwater East*

## Conclusion

The Freshwater East plant fossils, which are 410–420 million years old, represent the most diverse Silurian flora from anywhere in the world. Small, upright, leafless plants, known as rhyniophytoids, are particularly abundant and diverse here, and the site has been of considerable importance in the study of these archetypal, primitive land plants. It has also yielded the earliest evidence of plant stems with spines, which may represent the evolutionary precursors of leaves. The flora provides a highly significant insight into land vegetation just before the major radiation that occurred about 10 million years later in the Early Devonian, and which is the subject of the next chapter.

**Figure 3.21** *Psilophytites* sp. The oldest known examples of a plant axis with spines; National Museum of Wales, specimen 77.6G56a. Freshwater East Formation (lower Milford Haven Group – Přídolí), Freshwater East. × 10. (Photo: Photographic Studio, National Museum of Wales.)

# Chapter 4

# *Devonian*

# Palaeogeographical setting

During the Devonian, there was a dramatic change in terrestrial vegetation. From the humble stock of rhyniophytes, rhyniophytoids and primitive lycopsids in the Silurian, there was a rapid radiation of plant groups, as vegetation evolved to fill the newly-available, terrestrial ecological niches (Figure 4.1). This new plant cover included the first appearance of equisetes, 'advanced' lycopsids (e.g. protolepidodendrids), trimerophytes, fern-like plants (e.g. rhacophytids, cladoxylids), progymnosperms and, eventually, fully gymnospermous plants. By the end of the period, all of the major groups of vascular plants except the angiosperms were present. During the period, there was also a number of major morphological innovations in the Plant Kingdom, including laminate foliage, the seed as a means of propagation, and the arborescent habit (Figure 1.1). Together, these phylogenetic and morphological developments provided the spring-board from which the luxuriant Carboniferous floras could develop.

The Devonian plant fossil record is particularly good in Britain, especially in the lower part of the system. The flora found at Rhynie stands pre-eminent, but there are also a number of other key sites in southern Scotland, Wales and the Welsh Borders. As a consequence, British Devonian palaeobotany has played a central role in developing our ideas on early land plant evolution.

## PALAEOGEOGRAPHICAL SETTING

During the Devonian, most of Britain was on the south-east margins of Laurussia and thus very near the equator (Figure 4.2). Apart from the marine environments of south-west England, sedimentation was mainly non-marine, in fluvio-deltaic or lacustrine environments (Allen, 1979; Trewin, 1985), and resulted in red-beds of the Old Red Sandstone 'magnafacies' (Erben, 1964). The proximity of Britain to the equator indicates tropical temperatures but the evidence for precipitation rates is equivocal. Clearly, in the upland regions there must have been considerable precipitation, in order to produce large enough rivers to generate the extensive deltas. On the deltas themselves, however, there is little direct sedimentological evidence of precipitation, and there is indication that, at least periodically, the climate was comparatively dry (e.g. the '*Psammosteus* Limestone' in the Welsh Borders – Allen, 1985).

The geographical distribution of Devonian plant fossils has been discussed by Edwards (1973, 1990), Raymond *et al.* (1985), Raymond (1987), Allen and Dineley (1988), and Edwards and Berry (*in* Cleal, 1991). Through much of the Devonian, the general pattern seen in the Silurian continues to be recognizable. Devonian plant fossils have been most widely found in the low palaeolatitudes, particularly in Laurussia (which includes the British localities) and thus represent tropical and subtropical vegetation. Within these low palaeolatitudes, some provincialism can be recognized, such as in northern Gondwana (e.g. Malone, 1968; Tims and Chambers, 1984), Kazakhstania (Yurina, 1969; Senkevich, 1975) and Cathaysia (Li and Cai, 1978; Li and Edwards, 1992).

Northern high palaeolatitudes, as represented in present-day Siberia, had a quite distinctive vegetation throughout the Devonian (Petrosyan, 1968; Stepanov, 1975). However, the plant fossils from here have not been studied to the same extent as those of the low palaeolatitudinal Laurussia, and a detailed comparative analysis will be needed to establish the exact extent of the differences.

Fossils representing the southern high palaeolatitude vegetation occur in South Africa, South America and Antarctica. Their stratigraphically lowest occurrence is in the Middle Devonian (Anderson and Anderson, 1983). In contrast to the low palaeolatitude vegetation, especially of Laurussia, lycopsids such as *Archaeosigillaria* and *Leptophloeum* predominated, and this remained the situation through into the Early Carboniferous.

Edwards (1973) argued that, in the Upper Devonian, plant fossil provincialism is much less extreme, and that most assemblages are dominated by taxa such as *Archaeopteris*, *Cyclostigma* and *Rhacophyton* (see also Edwards and Berry *in* Cleal, 1991). However, the Upper Devonian palaeobotanical record (especially in the Frasnian) is much poorer than that of the Lower and Middle Devonian (Banks, 1980), and this may be the real reason for the apparent reduction in provincialism. That the reduction is artificial is supported by the similarity of the palaeophytogeography of the Middle Devonian and the Lower Carboniferous, suggesting that there was an essential floristic continuity from the Middle Devonian to Early Carboniferous.

There has been disagreement as to the degree of global provincialism in Devonian plant fossil assemblages. Raymond (1987), using statistical techniques, such as polar ordination analysis, claims to recognize considerable differentiation,

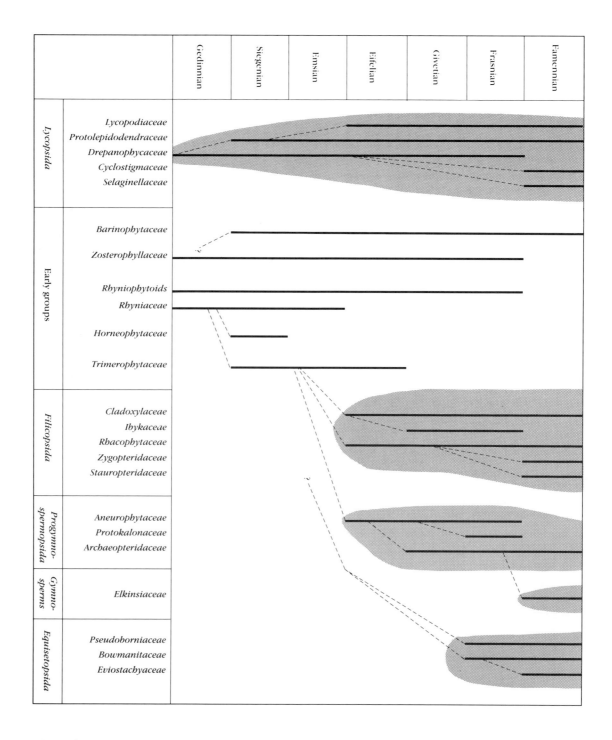

**Figure 4.1** The distribution of families of vascular plants in the Devonian. Based on data from Cleal (1993).

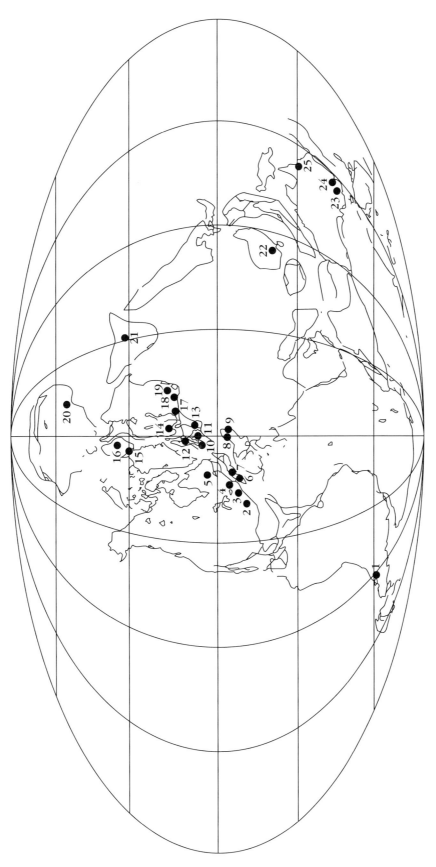

**Figure 4.2** The palaeogeography of the Devonian, showing the location of the major fossil floras of this age. 1 – North Argentina; 2 – West Virginia; 3 – Pennsylvania; 4 – Maine; 5 – Quebec; 6 – New York State; 7 – New Brunswick; 8 – Rhineland; 9 – Bohemia; 10 – Ireland; 11 – Wales and the Welsh Borderland; 12 – Scotland; 13 – Belgium; 14 – Norway; 15 – Bear Island; 16 – Spitsbergen; 17 – Poland; 18 – Moldavia; 19 – Ukraine; 20 – Kuzbass; 21 – Kazakhstan; 22 – Yunnan; 23 – Victoria; 24 – New South Wales; 25 – Queensland. Based on Scotese and McKerrow (1990).

especially in the upper Lower Devonian. In contrast, Allen and Dineley (1988) argue that, other than in Kazakhstan (the Kazakhstania Palaeocontinent) little provincialism can be recognized until the Lower Carboniferous. An important factor may be that the Laurussian assemblages have been subject to considerably more study than those of the other continents. It cannot be ruled out, therefore, that at least some of the apparent provincialism may be an artifact introduced by palaeobotanists working in partial isolation in distant parts of the globe.

## STRATIGRAPHICAL BACKGROUND

Figure 4.9 (p. 60) summarizes the chronostratigraphy used in this chapter, which essentially follows House *et al.* (1977). There have been proposed changes in the Lower Devonian, with the introduction of the Lochkovian and Pragian as the two lower stages of the system (Ziegler and Klapper, 1985). However, there are practical difficulties with identifying accurately the boundaries of these two stages in non-marine sequences and so, following Edwards (1990), the traditional stage divisions (Gedinnian and Siegenian) have been retained. Although all of the stages shown in Figure 4.9 are defined in marine sequences, it is now possible to use palynology to establish reasonably detailed correlations with the continental sequences where most Devonian plant fossils are found in Britain (Streel *et al.*, 1987).

The lithostratigraphy of the British Devonian sequences is reviewed by House *et al.* (1977), where details will be found of the formational intervals described herein (see also Simpson, 1959).

The most recently published biostratigraphical scheme for Devonian plant macrofossils is given by Edwards and Berry (*in* Cleal, 1991), which essentially follows that of Banks (1980); this is summarized in Figure 4.9. It provides a sequence of assemblage zones, whose boundaries are identified by the tops and bottoms of the ranges of particular species. However, the zones are also defined in terms of significant morphological developments, such as the appearance of stems with centrarch primary xylem in the *Psilophyton* Zone and of the arborescent habit in the *Svalbardia* Zone. This departs from traditional biostratigraphical methodology and makes it perhaps less robust for establishing detailed stratigraphical correlations. Nevertheless, it makes it a particularly useful framework around which the evolutionary history of terrestrial vegetation during the Devonian can be fitted.

## DEVONIAN VEGETATION

The Early Devonian vegetation of Britain appears to have been dominated by small, morphologically primitive plants. In Gedinnian times, rhyniophyte and rhyniophytoid plants (e.g. *Cooksonia*, *Salopella*) continue to be dominant (see Chapter 3). Perhaps the single most important locality for these plants is in the Siegenian of Britain (Rhynie), where petrified fossils show details of their cell structure. However, this represents a rather atypical, volcanically-influenced habitat and elsewhere in the Siegenian the group was in serious decline; by the Emsian, it had become virtually extinct.

One of the first groups of plants to replace the rhyniophytes and rhyniophytoids in the British floras was the zosterophylls (Figure 4.3). Evidence of 'H'- and 'K'-style branching, widely regarded as characteristic of the zosterophylls, is known from the uppermost Silurian, but fertile specimens are not known below the Gedinnian. They rapidly diversified during the Early Devonian and reached their acme in the late Siegenian and early Emsian (Niklas and Banks, 1990); however, soon after they underwent a rapid decline and eventually became extinct in the Late Devonian. The early

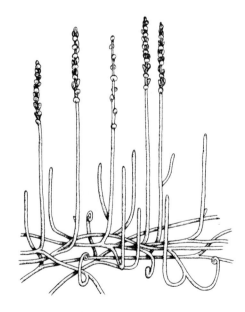

**Figure 4.3** Reconstruction of *Zosterophyllum*. From Gensel and Andrews (1984).

forms (e.g. *Zosterophyllum*) had naked axes with sporangia aggregated into terminal spikes. During the Siegenian, however, some developed leaves (e.g. *Sawdonia*), while in others changes can be seen in the reproductive structures (e.g. *Gosslingia*, where the sporangia are not clustered into terminal spikes). Although Britain is not the only place where fossils of these plants occur, some of the best preserved examples have been found here and have been the basis of much of what we know about them, especially of their reproductive and vegetative anatomy (e.g. Lele and Walton, 1962a; Edwards, 1969a, b, 1970a, 1972, 1975).

From an evolutionary standpoint, the zosterophylls are important as they were probably closely related to the early lycopsids (Hueber, 1992), which became one of the most important components of land vegetation during the Late Devonian and Carboniferous (Thomas, 1978a, 1992). This phylogenetic model is strongly supported by the sequence of intermediaries *Sawdonia-Asteroxylon-Drepanophycus* that can be clearly demonstrated in the British Devonian fossil record (Figure 4.4). There is the problem of the presence of the lycopsid *Baragwanathia* in the Silurian of Australia, which pre-dates the earliest known zosterophylls, but this probably just reflects the incompleteness of the fossil record. In Britain, the earliest lycopsids are found in the Siegenian.

Another characteristic plant group of Early Devonian Britain was the trimerophytes (Figure 4.5). The taxa varied considerably in general morphology, and could have naked (*Psilophyton dawsonii* Banks *et al.*) or spinose (*P. princeps*

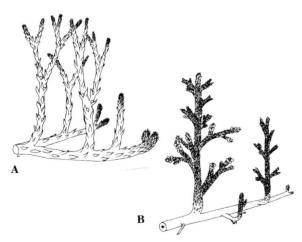

**Figure 4.4** Reconstructions of Early Devonian lycopsids. (A) *Asteroxylon*. (B) *Drepanophycus*. From Thomas and Spicer (1987, figure 3.6; after Kräusel and Weyland, and Kidston and Lang).

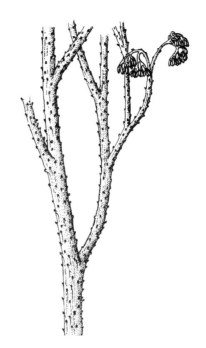

**Figure 4.5** Reconstruction of the Devonian trimerophyte *Psilophyton*. From Thomas and Spicer (1987, figure 3.7).

Dawson) axes, although most appear to have been herbaceous. However, they are all characterized by their fructifications, which consisted of loose trusses of sporangia (e.g. Andrews *et al.*, 1977). The group seems to have been relatively short-lived, appearing first in the Siegenian, reaching its acme in the Emsian, and becoming extinct during the Eifelian. However, it was of critical importance for the evolution of vascular plants, since it was probably ancestral to both the ferns and progymnosperms (and thus the gymnosperms).

In addition to the above, the Early Devonian vegetation included a number of enigmatic plants. Among the vascular plants, there is the Barinophytopsida, whose earliest occurrence is in the Siegenian of Britain (*Krithodeophyton* from Craig-y-Fro Quarry). Although they are clearly vascular, and even show some points of similarity with the zosterophylls and lycopsids, many have distinctive fertile structures with pinnately arranged sporangia, sometimes separated by sterile appendages. In some cases they are heterosporous, with mega- and microspores borne in the same sporangia. The barinophytopsids do not readily fit into any known group of pteridophytes, either living or extinct.

Among the non-vascular land plants, *Prototaxites*, *Nematothallus* and *Pachytheca* continued to be abundant in the Early Devonian, although all but the first became extinct before the mid-Devonian. In addition, *Parka decipiens* Fleming flourished for a short time, in the Gedinnian. This was a cutinized, encrusting alga, whose thallus was covered in large sporangia. Hemsley (1989, 1990a) has suggested that it might represent an immediate evolutionary position between the algae and the bryophytes.

The mid-Devonian saw a major change in vegetation. Most of the dominant groups of the Early Devonian became extinct or were declining in importance. Only the lycopsids continued to play a major role in land vegetation, particularly in the higher latitudes. Most lycopsids retained a herbaceous habit in the mid-Devonian, although some larger forms (*Lepidodendropsis*) made their first appearance here, providing a foretaste of some of the arborescent forms that came to dominate much of Carboniferous vegetation.

The characteristic rhyniophyte/zosterophyll/trimerophyte complex of the Early Devonian was gradually replaced in mid-Devonian times by early, fern-like plants and progymnosperms (Figure 4.1). Both groups probably evolved from the trimerophytes. The fern-like plants include representatives of the Cladoxylales, Iridopteridales and Coenopteridales, and can be recognized to be allied with the ferns mainly on the basis of their stem anatomies (Scheckler, 1974). However, they were quite different in outward appearance from recent ferns, most noticeably in not having laminate fronds. The foliage instead consisted of three-dimensionally branching systems of terete axes, which would often bear terminal trusses of sporangia. All known forms were herbaceous, although some could be up to one metre or more high (e.g. *Pseudosporochnus*, Figure 4.6).

The earliest and most primitive progymnosperms are found in Britain (*Protopteridium*, Figure 4.7). Like their presumed ancestors, the trimerophytes, these early progymnosperms did not have fully laminate foliage, and are distinguished mainly by the presence of secondary wood in the stems, and the greater complexity of the protoxylem (Beck, 1976). The first of these features is of great significance, as it allowed the evolution of the arboreal habit and thus of forest habitats. The size of these earliest progymnosperms is still a matter of conjecture, although they were probably at least several metres high.

The Late Devonian saw the extinction of the typical Early Devonian forms, while the lycopsids,

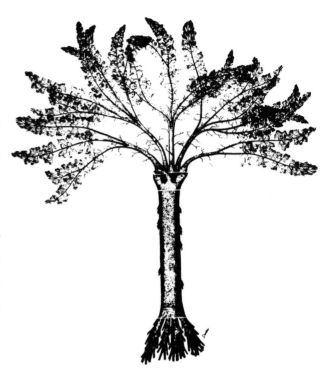

**Figure 4.6** Reconstruction of the Devonian fern-like plant *Pseudosporochnus*. Based on Leclercq and Banks (1962).

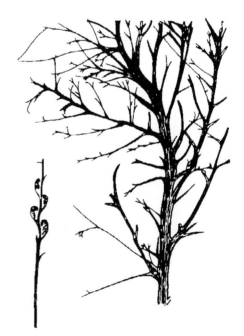

**Figure 4.7** Partial reconstruction of the Devonian progymnosperm *Protopteridium*. Based on Meyen (1987, figure 38M).

progymnosperms and fern-like plants continued to diversify. The arborescent habit became fully developed, with trees growing to over 20 metres high (Beck, 1962).

Equisetes, in the form of the Bowmanitales and Pseudoborniales, also made their first appearances in the Late Devonian (Figure 4.1). It was once thought that equisetes first appeared in the mid-Devonian, represented by fossils such as *Hyenia*. However, the latter are now generally thought to belong to the cladoxylaleans.

The earliest British examples of seed plants are found in the Famennian of Devon (Arber and Goode, 1915), and are only just pre-dated by examples from North America (Rothwell *et al.*, 1989). Little is still known about the Devonian gymnosperms, although they seem mostly to have been herbaceous and probably belonged to the lagenostomalean pteridosperms. The development of seeds was of critical importance for the evolution of terrestrial vegetation, as it finally freed plants from the constraint of requiring moist conditions to facilitate fertilization. From the end of the Devonian, plants became adapted to take advantage of the so-called 'upland' or extra-basinal habitats. Thereafter, the fossil record provides an increasingly selective record of the Earth's total vegetation.

## DEVONIAN PLANT FOSSILS IN BRITAIN

The record of Devonian plant fossils in this country is most complete in the Lower Devonian, with an extensive range of sites from the Gedinnian to the Emsian in Wales, the Welsh Borders (Figure 4.8) and in Scotland. From these, it is possible to document the early phases in the Devonian radiation of land vegetation, mentioned at the beginning of this chapter. Gedinnian plant fossils are found widely in the Arbuthnott Group of Tayside, Scotland, although there are also a number of sites in the Ditton Group of Shropshire and

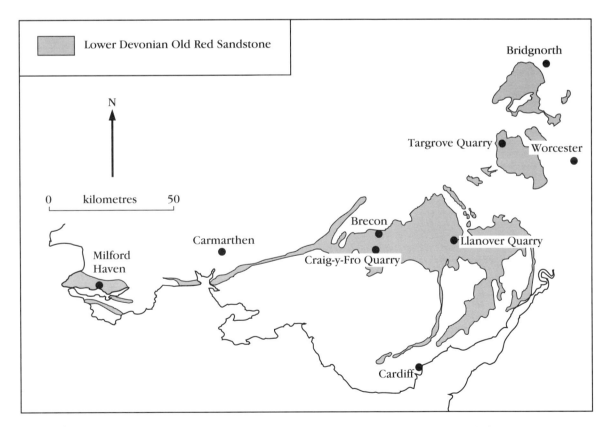

**Figure 4.8** Distribution of Lower Devonian plant fossil-bearing beds in South Wales and the Welsh Borderland showing position of GCR sites. Based on Kenrick and Edwards (1988b, figure 25).

| Chronostratigraphy | | | Plant Fossil Biostratigraphy | GCR Palaeobotany sites | Main areas outside Britain | | | | |
|---|---|---|---|---|---|---|---|---|---|
| Series | Stages | | | | Western Europe | Eastern Europe | Asia | Southern Continents | North America |
| Upper Devonian | Famennian | | *Rhacophyton* | Plaistow Quarry | Bear Island Belgium Ireland | Ukraine | Kazakhstan Kuzbass | New South Wales | Pennsylvania New York St. West Virginia |
| | Frasnian | | *Archaeopteris* | | Bear Island | Ukraine | Kazakhstan Kuzbass | New South Wales | New York St. Quebec |
| Middle Devonian | Givetian | | *Svalbardia* | Sloagar | Spitsbergen Belgium Rhineland | Bohemia Ukraine | Kazakhstan Kuzbass Yunnan | Queensland | New York State |
| | Eifelian | | *Hyenia* | Bay of Skaill | Spitsbergen Belgium Rhineland | Bohemia Ukraine | Kazakhstan Kuzbass | | New Brunswick |
| Lower Devonian | Emsian | | *Psilophyton* | Auchensail Ballanucater Farm Rhynie Chert Craig-y-Fro Quarry Llanover Quarry | Spitsbergen Norway Belgium Rhineland | Poland Ukraine Moldavia | Kuzbass Yunnan | Victoria N. Argentina | New Brunswick Maine |
| | Siegenian | | *Zosterophyllum* | Turin Hill Targrove Quarry | Spitsbergen Belgium Rhineland | Bohemia | Kazakhstan Kuzbass Yunnan | Victoria | |
| | Gedinnian | | | | | | | | |

**Figure 4.9** Chronostratigraphical classification of the Devonian, and the positions of the GCR and other major palaeobotanical sites in this system.

the St Maughan's Group of South Wales (Edwards, 1990). Siegenian plant fossils are best represented in the Senni Beds and its equivalents in South Wales and the Welsh Borders (Croft and Lang, 1942). There is also the world-famous Rhynie Chert in the Grampians of Scotland, which is now thought to be Siegenian in age. The Strathmore Group of central Scotland, provides the most abundant Emsian plant fossils in this country (Henderson, 1932; Rayner, 1983, 1984).

Middle Devonian plant fossils are mainly found in the fossil fish beds of northern Scotland, ranging from Cromarty north to the Shetlands. There are also records of poorly-preserved fossils from Cornwall (Fox, 1900, 1901, 1904; Lang, 1929; Hendriks, 1935, 1966; Hendriks *et al.*, 1971).

For reasons which are still far from clear, Upper Devonian plant fossil occurrences are rare throughout the world, and Britain is no exception to this. The best British examples are found in the Baggy Formation and its lateral equivalents in north Devon (Arber and Goode, 1915). In Ireland, the Kiltorcan assemblage is of about the same age, and has yielded well-preserved examples of *Archaeopteris* and *Cyclostigma* (Chaloner, 1968). Comparable assemblages have also been found in Scotland and the Forest of Dean (Miller, 1857; Crookall, 1939; Long, 1973). However, the Forest of Dean site no longer yields fossils and, despite an extensive search being made as part of the Geological Conservation Review, the Scottish site could not be located. A reference to plant fossils of this age from southern Scotland by Smith (1862) remains to be properly documented.

From this range of Devonian palaeobotanical sites in Britain, ten have been selected as GCR sites. The stratigraphical positions of these sites are shown in Figure 4.9.

## TARGROVE QUARRY

## Highlights

Targrove Quarry has yielded the most diverse assemblage of fertile rhyniophytoid specimens known from anywhere in the world, and is the type locality for *Cooksonia hemisphaerica* Lang. It has yielded exceptionally preserved examples of *Nematasketum diversiforme* Burgess and Edwards, and one of the earliest known examples of *in situ* tracheids in slender axes.

## Introduction

This Gedinnian-age locality is a small sandstone quarry in the grounds of Downton Hall, near Ludlow, Shropshire (SO 525780) and is Locality 105 in Ball and Dineley (1961). The first plant fossils were recorded by Marston (1870), who interpreted what were *Parka* specimens as eurypterid egg-packets. More specimens were described by Lang (1937), who also illustrated *Cooksonia* and *Nematothallus* material. However, the most complete accounts are by Edwards and Fanning (1985) and Fanning *et al.* (1992).

## Description

### Stratigraphy

Lang (1937) stated that this exposure was in the upper Downton Group, but Ball and Dineley (1961) map it as *c.* 60 metres above the main *Psammosteus* Limestone, thus placing it in the lower Ditton Group. On the basis of the spores and fish fossils found here, the sequence is Gedinnian in age (Richardson *in* Edwards and Fanning, 1985). There is no comprehensive account of the geology of the exposure, but Edwards and Fanning state that the sediments are typical of the distal alluvial facies of the Old Red Sandstone (Allen, 1979).

### Palaeobotany

The plant fossils are preserved as coalified compressions. The following species have been found to date:

Phaeophycophyta(?):
   *Nematasketum diversiforme* Burgess and
      Edwards
   *Nematothallus pseudovasculosa* Lang

Chlorophycophyta(?):
   *Parka decipiens* Fleming
   *Pachytheca* cf. *sphaerica* Hooker

Rhyniophytoids:
   *Cooksonia hemisphaerica* Lang
   *C. pertoni* Lang
   *C. cambrensis* Edwards
   *C.* cf. *caledonica* Edwards
   *Salopella marcensis* Fanning, Edwards
      and Richardson

*Tortilicaulis transwalliensis* Edwards
*Uskiella reticulata* Fanning, Edwards and Richardson
cf. *U.reticulata*
*Tarrantia salopensis* Fanning, Edwards and Richardson

Zosterophyllopsida:
*Zosterophyllum* cf. *fertile* Leclercq

There is also a number of unnamed rhyniophytoid fructifications described from here by Edwards and Fanning (1985), and Fanning *et al.* (1992).

## Interpretation

This site yielded some of the specimens used by Lang (1937) in his description of *Nematothallus*. They show that it had a pseudoparenchymatous structure surrounded by a cuticle. Lang (1945) and Edwards and Rose (1984) suggested that there was an outer epidermal layer, but the Targrove Quarry specimens do not seem to show this. Lang (1937) also noted some cutinized spores in the Targrove Quarry specimens, although no evidence of the fruiting bodies was found.

Lang (1937) implied that *Prototaxites* occurs here, but he neither figured nor described any specimens. It is also listed by Edwards and Fanning (1985). Burgess and Edwards (1988) subsequently identified the Targrove specimens as *Nematasketum diversiforme* Burgess and Edwards, although again none from here is figured. This species and genus are briefly discussed further in Chapter 3, in the section dealing with Perton Lane.

The most significant vascular plant found here, at least from a stratigraphical standpoint, is *Zosterophyllum*. The record is based on one specimen showing the characteristic branching pattern of the form-genus, and a poorly-preserved fertile spike, neither of which has been figured (Edwards and Fanning, 1985). Its presence indicates that the flora belongs to the *Zosterophyllum* Zone of Banks (1980), and thus is probably Early Devonian in age.

Far commoner at Targrove Quarry are rhyniophytoid species. Lang (1937) initially described only *Cooksonia hemisphaerica* Lang from here, for which it is the type locality. Further collecting, however, has yielded a much wider variety of fertile rhyniophytoid specimens, which have been assigned to various other species of *Cooksonia*, as well as *Salopella, Tortilicaulis, Uskiella* and *Tarrantia* (Fanning *et al.*, 1992). There were also numerous other forms of sporangia described by Edwards and Fanning, including ellipsoidal and bifurcating types, which belong to so far undescribed form-genera. Thus it comprises by far the most diverse assemblage of fertile rhyniophytoid taxa known from anywhere in the world.

Lang (1937) was also able to demonstrate *in situ* tracheids in a slender axis from here. This was the first direct evidence of vascular tissue in these very early land plants and, until the discovery of similar specimens in the Silurian of Capel Horeb, was the oldest known evidence of *in situ* tracheids in the fossil record. Since the only fertile rhyniophytoid reported from here by Lang was *C. hemisphaerica* Lang, he concluded that *Cooksonia* must have been a vascular plant. However, the discovery by Edwards and Fanning of a much more diverse rhyniophytoid assemblage here must now cause this argument to be doubted, since it is impossible at this stage to be certain which of the species had vascular tissue and which (if any) had not.

Similar *Zosterophyllum* Zone assemblages have been reported from Caldy Island, Dyfed (Lang, 1937) and Newton Dingle, Shropshire (Edwards and Richardson, 1974), although this is not as diverse. A comparable flora comes from the Arbuthnott Group of Scotland, such as found at Turin Hill (see below), but that flora is dominated by zosterophylls and has only subsidiary rhyniophytoid elements. From outside of Britain, *Zosterophyllum* Zone floras have been described from Belgium (Leclercq, 1942), Spitsbergen (Høeg, 1942) and Czechoslovakia (Obrhel, 1968), but they differ from the Targrove Quarry assemblage by the presence of *Taeniocrada* and the more restricted rhyniophytoid composition. The assemblage from Kuznetsk in Siberia described by (Stepanov, 1975), although sharing a number of genera with the Targrove Quarry assemblage, also contains several enigmatic endemics such as *Juliphyton, Uksunaiphyton, Pseudosajania* and *Salairia*, all of unknown affinities.

The assemblage of plant fossils found at Targrove Quarry is thus unique in both a national and international sense. It is transitional between the more primitive Silurian assemblages such as are found at Perton Lane and Freshwater East (see Chapter 3) and the slightly more advanced floras of Turin Hill (see below). It is thus of key importance for understanding the earliest phases in the

# Turin Hill

diversification of land vegetation in the earliest part of the Devonian.

## Conclusion

Nowhere else in the world has yielded such a diverse assemblage of rhyniophytoid plants than Targrove Quarry. They are thought to represent the most primitive type of land plants. The rhyniophytoids first appeared about 425 million years ago in the Silurian (see previous chapter) and flourished in the earliest Devonian, such as represented by the Targrove fossils. After some 35 million years, they declined and became extinct, as more advanced plants evolved to take advantage of the land habitats. The Targrove fossils are about 410 million years old and thus represent the acme of this highly significant group of plants which, although of only very simple form, represent a key phase in the evolution of land vegetation.

## TURIN HILL

### Highlights

Turin Hill is the best site for plant fossils from the Lower Devonian Arbuthnott Group flora, the most typical *Zosterophyllum* Zone flora in Britain, and one of the best examples of its type in the world (Figure 4.10). It has yielded abundant specimens of *Parka* and fertile *Zosterophyllum* and is the type locality for *Cooksonia caledonica* Edwards. It is of international significance for the study of

**Figure 4.10** Turin Hill, Aberlemno Quarry. Strike section along flaggy deposits of the Gedinnian Dundee Formation. (Photo: C.J. Cleal.)

the *Zosterophyllum*-dominated vegetation of the earliest Devonian.

## Introduction

The famous Old Red Sandstone palaeontological sites on Turin Hill, which lie near Forfar, Tayside Region (NO 493535), were extensively worked during the eighteenth and nineteenth centuries for Arbroath Paving Stones (Mackie, 1980). Two of these quarries are particularly well known plant fossil localities: Clocksbriggs Quarry (also known as Wemyss Quarry) and Aberlemno Quarry. Early records of the plant fossils concentrated mainly on the non-vascular species (Miller, 1855, 1857; Peach, 1877; Powrie *in* Warden, 1881; Kidston, 1886, 1893, 1897; Dawson and Penhallow, 1891; Reid, 1895; Reid *et al.*, 1898; see also Niklas, 1976a). Vascular plant fossils were also recorded in some early studies (Peach, 1877; Kidston, 1886; Reid and Macnair, 1899), but were not dealt with in detail until Lang's (1927a) account of *Zosterophyllum*, and subsequently by Lele and Walton (1962a), Walton (1964a) and Edwards (1970b, 1975). In addition to the flora, a diverse freshwater fish and arthropod fauna has been found here (Hickling, 1912; Westoll, 1951).

## Description

### Stratigraphy

The geology of these quarries is described by Armstrong *et al. in* Friend and Williams (1978). Exposed here are interbedded red fluvial sandstones and grey-green lacustrine siltstones belonging to the Dundee Formation of the Arbuthnott Group (Campbell, 1913; Armstrong and Paterson, 1970; earlier called the Carmyllie Beds by Hickling, 1908). The plant fossils belong to the *Zosterophyllum* Zone of Banks (1980), indicating a Gedinnian or early Siegenian age for these strata. Palynological and fish evidence supports an early Gedinnian age (Edwards and Fanning, 1985, Table 1; Edwards, 1990).

### Palaeobotany

The plant fossils are found mainly in the laminated, lacustrine siltstones. Commonest are impressions picked out by iron staining, but some coalified compressions and petrifactions also occur. The following assemblage has been reported to date:

Phaeophycophyta(?):
 *Prototaxites forfarensis* (Kidston) Pia

Chlorophycophyta(?):
 *Parka decipiens* Fleming
 *Pachytheca* sp.

Rhyniophytoids:
 *Cooksonia caledonica* Edwards

Zosterophyllopsida:
 *Zosterophyllum myretonianum* Penhallow

## Interpretation

Although *Parka decipiens* (Figure 4.11) was first described from the Lower Devonian of Fife

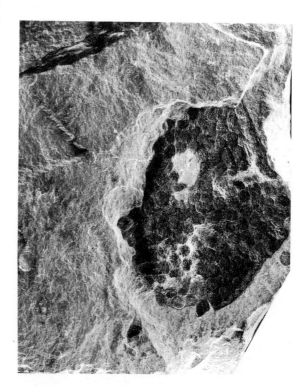

**Figure 4.11** *Parka decipiens* Fleming. Enigmatic, possible early land plant, consisting of a cutinized thallus covered with discoid sporangia; Natural History Museum, London, specimen V.24951. Dundee Formation (Gedinnian), Carmylie, Tayside. × 2. (Photo: Photographic Studio, Natural History Museum, London.)

(Fleming, 1831), Turin Hill is probably the best known locality for this species and much of the early debate about its affinities was based on specimens from here (summarized by Dawson and Penhallow, 1891 and Don and Hickling, 1917). It has been variously argued to be of animal origin (Mantell, 1852; Lyell, 1865; Woodward, 1866-1878), to be the seeds of a rush or bur-reed (Fleming, 1831), or the fruiting body of some other enigmatic plant (Dawson and Penhallow, 1891; Reid *et al.*, 1898; Edwards, 1921). However, it is now believed to have been a thalloid plant attached to the substrate by a basal holdfast, and probably spending at least part of its life out of water (Niklas, 1976a, 1976c). Possible affinities with the green algae (Niklas (1976a) and the liverworts (Neuber, 1979) have been suggested. Most recently, Hemsley (1989, 1990a) has suggested that it might have been an evolutionary intermediary between the algae and the bryophytes.

Peach (1877) identified a specimen from here as *Cyclopteris* (photographically re-figured by Høeg, 1942, pl. 46, figs 10 and 11). This form-genus is usually applied to the basal pinnules of certain Upper Carboniferous medullosan fronds, which is clearly out of the question here. More likely is that it is a specimen of *Parka* viewed from the lower surface (cf. Don and Hickling, 1917, pl. 54, fig. 9).

Niklas (1976a) described a poorly preserved specimen of *Pachytheca* from Turin Hill, and showed that it had a similar growth pattern to *Parka*. Niklas (1976b) also showed similarities in the chemical composition between the two, but pointed out, that the taxonomic significance of this similarity is uncertain.

A partially permineralized, branched axis, over one metre long, was described by Miller (1855, 1857) as part of a *Lepidodendron* stem. Most of the specimen is now believed to be lost, but Lang (1926) suggested that the holotype of *Cryptoxylon forfarense* Kidston was originally part of it. Kidston (1897) originally described the structure of the stem as being cellular, but Kidston and Lang (1924) later showed that it was pseudoparenchymatous and so transferred it to *Nematophyton* (syn. *Prototaxites*). Lang (1926) described a second species of *Prototaxites* from the Arbuthnott Group as *Nematophyton caledonianum*. To date, it has not been reported from Turin Hill but its presence might well be anticipated.

Slender branching axes occur commonly in the Arbuthnott Group (Figure 4.12), and were initially

**Figure 4.12** *Zosterophyllum myretonianum* Penhallow. Tangled mass of axes at the base of the plant that gave rise to the vertical shoots; Natural History Museum, London, specimen V.58041. Dundee Formation (Gedinnian), Balgavies Quarry, near Forfar, Tayside. × 0.5. (Photo: Photographic Studio, Natural History Museum, London.)

believed to be either part of an eel-grass type plant (Miller, 1857), or the vegetative part of a plant which bore *Parka* as its fruiting body (Dawson and Penhallow, 1891). Miller described rounded bodies attached laterally to some of these axes and argued that they might be leaves. Penhallow (1892) recognized them to be sporangia, however, and erected the name *Zosterophyllum myretonianum*. The Turin Hill specimens have not yielded such well-preserved cuticles as have been described from Myreton Quarry (Lele and Walton, 1962a), but they have included some exceptionally complete specimens which have allowed the gross morphology of the plant to be reconstructed (Lang, 1927a; Lele and Walton, 1962a; Walton, 1964a). Numerous fertile specimens have also been found here (e.g. Figure

4.13) and formed the bulk of the material used by Edwards (1975) in her study on the morphological variation of the species. Although several species from other localities have been subsequently placed in *Zosterophyllum* (reviewed by Edwards, 1975), *Z. myretonianum* remains the best understood. The form-genus is of particular importance since it is the effective type of the Zosterophyllopsida, an important class of early land plants, probably related to the lycopsids ('clubmosses'). This subdivision is generally accepted to be the evolutionary precursor that played such an important role in the later Palaeozoic floras.

Edwards (1972) described a second zosterophyll, *Z. fertile* Leclercq, from the Arbuthnott Group near Arbroath (probably Kelly Den, but locality details are not clearly recorded), but it has not so far been reported from Turin Hill.

For many years, *Zosterophyllum* was the only type of vascular plant known from the Arbuthnott Group. More recently, however, Edwards (1970b) has described specimens from Aberlemno Quarry as *Cooksonia caledonica*. These show naked, dichotomous axes with terminal, globose sporangia, typical of *Cooksonia*, but the sporangia often have a marginal rim. The latter may only be a preservational feature, but may alternatively be the remains of a dehiscence mechanism (Edwards and Edwards, 1986). If the latter eventually proves to be correct, then the generic position of this species may have to be revised.

Several localities in the vicinity of Forfar and Arbroath have yielded the Arbuthnott Group flora (Kidston and Lang, 1924; Lang, 1926, 1927a; Lele and Walton, 1962a; Edwards, 1970b, 1975). Many of these have, however, been subsequently infilled or flooded. In particular, Myreton Quarry, from where Lele and Walton (1962a) obtained cuticles of *Zosterophyllum*, has been landscaped. Of the sites still remaining, the Turin Hill quarries yield easily the most abundant and well-preserved plant fossils.

The Arbuthnott Group flora is the most typical *Zosterophyllum* Zone assemblage (*sensu* Banks, 1980) known from Britain. Other assemblages from this zone have been recorded from Caldy Island, Dyfed (Lang, 1927a, 1937), Newton Dingle, Shropshire (Edwards and Richardson, 1974) and Targrove Quarry (see above). However, these all contain a much lower proportion of zosterophylls. *Zosterophyllum* Zone assemblages also occur in Spitsbergen (Høeg, 1942), the former Czechoslovakia (Obrhel, 1968) and Kuznetsk (Stepanov, 1975), but none has been reported to yield such well-preserved specimens. Turin Hill is thus of international importance for the study of the *Zosterophyllum*-dominated vegetation, which seems to have occurred widely in the lowland terrestrial habitats during the earliest Devonian of the northern hemisphere; the radiation of this vegetation marked the first phases of the diversification of the land floras in this part of the

**Figure 4.13** *Zosterophyllum myretonianum* Penhallow. A fertile spike with sporangia arranged around the axis; Natural History Museum, London, specimen V.58047. Dundee Formation (Gedinnian), Clocksbriggs Quarry, Turin Hill. × 2. (Photo: Photographic Studio, Natural History Museum, London.)

world (see Chapter 3 for comments on the more '-advanced' floras found in Gondwana).

## Conclusion

Turin Hill yields the best example of a *Zosterophyllum* Zone flora (about 400 million years old) from anywhere in the world. Other floras of this type have been reported from South Wales, the Welsh Borders, Spitsbergen, Czechoslovakia and Siberia, but none have yielded such well-preserved specimens. Most significant here is the presence of well-preserved examples of the unusual alga *Parka*, and of the early vascular plant *Zosterophyllum*. The latter is important as it is regarded as ancestral to the club-mosses which, in the Late Carboniferous (310–300 Ma), formed extensive tropical forests that resulted in thick, economically important coal deposits (see Chapter 6). The site is also important as the type locality for *Cooksonia caledonica*, a remnant of the primitive rhyniophytoid stock that more typically occurs in the Upper Silurian, and which is thought to represent the earliest type of upright land plant.

# LLANOVER QUARRY

## Highlights

Llanover Quarry has yielded one of the most diverse examples of a *Psilophyton* Zone (Siegenian) flora from Britain. It is the type locality for *Zosterophyllum llanoveranum* Croft and Lang, the only *Zosterophyllum* for which three-dimensional internal anatomy is known. This site has also yielded important information on the form and structure of *Uskiella*, *Drepanophycus*, *Gosslingia* and *Deheubarthia*.

## Introduction

This small Old Red Sandstone exposure lies about 5 km south of Abergavenny, Gwent, Wales (SO 298079). Plant fossils were first discovered by Wickham King, who then showed the site to the palaeobotanist William Croft. Croft's collection was the basis of the classic study on the so-called Senni Beds flora (Croft and Lang, 1942). Further collecting has proved difficult, due to the small outcrop and awkwardly positioned tree roots, but some subsequent work on the assemblage has been done (Edwards, 1969a, b, 1970a, 1981; Edwards *et al.*, 1989; Shute and Edwards, 1989).

## Description

### Stratigraphy

The exposed sequence consists of about two metres of red sandstones and mudstones of the Brownstone Group (*sensu* Heard and Davies, 1924), which are correlatives of the Senni Beds of the Brecon Beacons. They were deposited on an alluvial plain crossed by fast, low sinuosity streams (Kelling *in* Owen *et al.*, 1965; Allen, 1979). No animal fossils have been reported, but the plant and microfossils suggest a Siegenian age (Croft and Lang, 1942; Mortimer, 1967; Richardson and Lister, 1969). The plant fossils occur in or near sediments of the fluvial channel (Allen, 1979), and probably originated from a variety of habitats (Edwards, 1979b).

### Palaeobotany

The plant fossils are mostly preserved as compressions and iron oxide stained impressions, with no cuticles preserved, but some limonite petrifactions also occur. The following assemblage has been described to date:

Phaeophycophyta(?):
    *Prototaxites* cf. *caledonianus* (Lang)
      Kräusel and Weyland
    *Nematothallus* sp.

Chlorophycophyta(?):
    *Pachytheca* sp.

Rhyniophytoids:
    *Sporogonites exuberans* Halle
    *Taeniocrada* sp.

Rhyniopsida:
    *Uskiella spargens* Shute and Edwards

Zosterophyllopsida:
    *Zosterophyllum llanoveranum* Croft and Lang
    *Z.* cf. *australianum* Lang and Cookson
    *Z.* cf. *fertile* Leclercq
    *Gosslingia breconensis* Heard
    *Deheubarthia splendens* Edwards, Kenrick
      and Carluccio

# *Devonian*

Lycopsida:
    *Drepanophycus spinaeformis* Göppert

Trimerophytopsida:
    *Dawsonites arcuatus* Halle

Uncertain affinities:
    *Sciadophyton* cf. *steinmannii* Kräusel and Weyland
    *Sennicaulis hippocrepiformis* Edwards
    'spherical or circular bodies *incertae sedis*'

## Interpretation

Coalified spheres of *Pachytheca* are common at Llanover. They are generally poorly preserved, but the cortical and medullary regions can sometimes be distinguished (Croft and Lang, 1942, pl. 11, fig. 78). Croft and Lang also reported some *Pachytheca*-like specimens with tuberculate surfaces, referring to them as 'spherical bodies *incertae sedis*'.

A feature of the Llanover assemblage noted by Croft and Lang is the presence of elongate, slender specimens of *Prototaxites*. The dimensions of the tubes and the presence of 'medullary spots' invite a comparison with *P. caledonianus* and with *Prototaxites* described from the Silurian of the Welsh Borderland (Lang, 1937) and the Devonian of Germany (Kräusel and Weyland, 1934).

The Rhyniaceae is represented here by *Uskiella spargens* Shute and Edwards (1989). It was first identified as *Cooksonia* sp. (Croft and Lang, 1942), but has quite a different shape and structure of the sporangia. Shute and Edwards interpreted it as a small plant of determinate growth-pattern, with naked, dichotomous axes forking at a wide angle (60–90°) and bearing ellipsoidal, terminal sporangia. Petrifactions from Llanover were particularly instructive in showing details of the sporangia, including the presence of a zone of differentiated cells around the major circumference, which may have been linked with dehiscence. This type of bivalved sporangial structure evidently had an adaptive advantage, perhaps because it maximized the area of spores exposed to the atmosphere after dehiscence, since it seems to have evolved independently in several other groups of plants, as well as the Rhyniaceae (reviewed by Shute and Edwards, 1989).

Several specimens of *Sporogonites* sporangia were described by Croft and Lang, from which they were able to prepare spores. One example was also found to have stomata preserved on the stem just below the sporangia. Their affinities are still unknown, and are currently being re-investigated by Shute and Edwards.

Croft and Lang's record of the rhyniophytoid genus *Taeniocrada* sp. is based on slender, flattened stems from a lenticular band of mudstone at Llanover. In the absence of fertile structures, it has been impossible to place them in a particular species, and even their generic assignment cannot be regarded as proven (Edwards, 1981).

This is the type locality for *Zosterophyllum llanoveranum* (Figure 4.14). It is the only species of *Zosterophyllum* for which the anatomy is known, following studies by Edwards (1969a) on petrifactions from here and Craig-y-Fro Quarry. Of particular significance was the determination of the form of the vascular strand, which is oval in cross-section and exarch, in contrast to the terete, centrarch strands of the Rhyniopsida and Trimerophytopsida. It was also possible to determine the form of the prominent sporangial dehiscence structure. A few specimens from Llanover have also been identified as *Z.* cf. *australianum* and *Z.* cf. *fertile* (Croft and Lang, 1942; Edwards, 1969b), although they may just be morphological variants of *Z. llanoveranum*.

Another member of the Zosterophyllopsida in the Llanover assemblage is *Gosslingia*. Rather better material of this form-genus has been found at Craig-y-Fro Quarry, and from these details of the anatomy have been described (Heard, 1927; Edwards, 1970a; Kenrick and Edwards, 1988a). However, the Llanover specimens have provided some important information on the gross morphology of the plant, in particular the arrangement of the sporangia.

The specimens described by Croft and Lang as cf. *Psilophyton princeps* have laterally attached sporangia similar to those of *Sawdonia ornata* (Dawson) Hueber described from the Gaspé Peninsula (Hueber and Banks, 1967; Hueber, 1968, 1971). Unlike *Sawdonia*, however, they have small bulges or curved branches just below each dichotomy, and lack the characteristic dark tip to the spines. For this reason, they have been assigned to a new genus and species, *Deheubarthia splendens* Edwards, Kenrick and Carluccio (1989).

Croft and Lang described the only fertile axes of *Drepanophycus spinaeformis* known from outside of Germany. Based on the German material, Kräusel and Weyland (1930) originally described

**Figure 4.14** *Zosterophyllum llanoveranum* Croft and Lang. A group of fertile spikes probably originating from a single plant; Natural History Museum, London, specimen V.26516a. Brownstone Group (Siegenian), Llanover Quarry. × 1.5. (Photo: Photographic Studio, Natural History Museum, London.)

the sporangia as being attached adaxially to the leaves. Later, however, Kräusel and Weyland (1935) found sporangia attached to leaf apices, which seems to be confirmed by the Llanover specimens. It is currently believed that the position of attachment of the sporangia is variable (e.g. Meyen, 1987). *Drepanophycus* is regarded by Hueber (1992) as one of the earliest and most primitive lycopsids (order Drepanophycales), being only pre-dated by the Silurian *Baragwanathia* from Gondwana (see Chapter 3).

Isolated trusses of pendant, fusiform sporangia were identified by Croft and Lang as *Dawsonites arcuatus* (Figure 4.15). They are probably the fertile parts of a trimerophyte plant, but in the absence of attached vegetative structures it is impossible to place them in one of the more natural taxa based on whole-plant morphology. They are associated with spiny axes, suggesting possible affinities with *Psilophyton princeps*. However, at least some of these axes were found by Croft and Lang to have laterally attached sporangia, which means that they belong to the zosterophyllalean genus *Deheubarthia* (see above). Until the nature of the vegetative axes which bore these sporangial trusses at Llanover

**Figure 4.15** *Dawsonites arcuatus* Halle. Terminal part of fertile truss; Natural History Museum, London, specimen V.26492. Brownstone Group (Siegenian), Llanover Quarry. × 2. (Photo: Photographic Studio, Natural History Museum, London.)

has been determined, they have to be retained within the generalized form-species *D. arcuatus* Halle. The fossils here and at Craig-y-Fro (see below) are the oldest known remains of trimerophytes from anywhere in the world.

A number of specimens were described by Croft and Lang as *Sciadophyton steinmannii*. Such structures are now believed to be gametophytes (Remy, Remy *et al.*, 1980; Remy, Schultka *et al.*, 1980), and have been compared with petrified specimens from Rhynie identified as *Lyonophyton* (Remy and Remy, 1980a, b). Schweitzer (1983a, b) has reported *Zosterophyllum*-like axes attached to German specimens of *Sciadophyton*, suggesting that the latter was a zosterophyll gametophyte. Croft and Lang noted that the Llanover specimens differ slightly from the type specimens of *S. steinmannii* Kräusel and Weyland, 1930 in the sizes of the discs and of the circular scars that they show. Contrary to Croft and Lang, therefore, they are referred to in the above species list as *S.* cf. *steinmannii*.

*Sennicaulis hippocrepiformis* Edwards, 1981 was described from specimens from both here and from Craig-y-Fro Quarry. Most of the anatomical details of this species were determined from the pyrite petrifactions from Craig-y-Fro, but the Llanover specimens showed certain details of the xylem structure particularly well. The affinities of these isolated axes remain uncertain, but probably lie either with the Rhyniopsida or Trimerophytopsida.

Llanover Quarry has yielded the most diverse *Psilophyton* Zone flora (*sensu* Banks, 1980) in Britain. Croft and Lang (1942) record comparable assemblages from the Deri Quarries near Abergavenny, along Kemeys Graig near Newport, from below the dam of Talybont Reservoir near Brecon and from a cliff near Llanthony Abbey in the Black Mountains; but none of these sites has yielded such diverse assemblages as Llanover Quarry and Craig-y-Fro.

These Welsh assemblages belong to the European phytogeographic subunit of the equatorial and low-latitude floras (as defined by Raymond *et al.*, 1985), which characterizes the southern and eastern parts of Laurussia. The closest comparison is with assemblages from Podolia in the Ukraine and the Dniester River in Moldavia (Ishchenko, 1965, 1974), particularly in the presence of *Gosslingia*, *Sciadophyton* and *Zosterophyllum*. Some comparison is also possible with floras from Belgium (Stockmans, 1940; Gerrienne, 1988, 1990a, b, 1991), although the latter have yielded fewer species.

Other *Psilophyton* Zone assemblages have been found in Britain in a borehole in Oxfordshire (Chaloner *et al.*, 1978) and in the Strathmore Group of Scotland, such as at Auchensail Quarry and Ballanucater (pp. 75–80). Raymond *et al.* (1985) assigned these to the American phytogeographic subunit and they differ markedly from the Llanover assemblage, being much less rich in species and dominated by *Pachytheca*, *Sawdonia*, *Dawsonites* and (in the Strathmore Group) *Drepanophycus*. However, this may be a function of the Llanover assemblage being marginally older than the Oxfordshire and Scottish fossils. The Llanover fossils provide a clearer reflection of the general diversity of the Siegenian vegetation of Laurussia, and are thus of considerable significance for studying the early phases of the diversification of vascular plants.

## Conclusion

Llanover Quarry has yielded one of the most diverse flora in Britain from the Siegenian Stage, and is about 400 million years old. It is particularly important for our understanding of the group of plants known as the zosterophylls, which were the ancestors of the club-mosses that dominated much of the land vegetation later in the Palaeozoic, particularly the Late Carboniferous equatorial coal swamps (see Chapter 6). Not only is this one of the most diverse assemblages of zosterophylls (*Gosslingia*, *Deheubarthia*, and three species of *Zosterophyllum*), but much important information has been discovered from here about their anatomy, which has been vital for understanding the evolutionary significance of the group. The club-mosses themselves are represented by *Drepanophycus* stems bearing reproductive organs, which are the second oldest fertile club-mosses known from anywhere in the world. Also present is one of the world's earliest examples of a trimerophyte (*Dawsonites*), which is the group thought to be ancestral to the seed plants (and thus also of flowering plants). The fossils are very similar to those found at Craig-y-Fro (see below), but are not preserved differently and thus show different aspects of the anatomy. Similar assemblages have been reported from Belgium, Moldavia and Ukraine but, except for the first of these, they have not been studied in such detail, and do not yield such well-preserved anatomical detail.

## CRAIG-Y-FRO QUARRY

## Highlights

Craig-y-Fro Quarry has yielded one of the best preserved Devonian plant fossil assemblages from Britain, second only to the Rhynie Chert assemblage. It is the type locality for *Gosslingia breconensis* Heard and *Sennicaulis hippocrepiformis* Edwards, and the only locality to yield *Tarella trowenii* Edwards and Kenrick, *Hostinella heardii* Edwards and *Krithodeophyton croftii* Edwards. The latter is of particular interest as being the only record of a member of the Barinophytales from Britain.

## Introduction

This disused quarry, cut in Old Red Sandstone clastic sediments, is in the Brecon Beacons, between the towns of Brecon and Merthyr Tydfil (SN 971208). At different times it has been called Brecon Beacons Quarry, Storey Arms Quarry and Craig-y-Fro Quarry, the latter name being adopted here because it is geographically more precise. Plant fossils from here were first described by Heard (1926, 1927, 1939), who noted that some of the specimens were petrifactions. His method of preparing these petrifactions led to their destruction, however, and the only permanent record of them was a series of photographs, most of which are now believed to be lost (Edwards, 1970a, p. 226). Plant fossils from Craig-y-Fro were also recorded by Crookall (*in* Robertson, 1932) and Cox and Heard (1937). In their classic study on the Senni Beds flora, Croft and Lang (1942) recorded material from this locality and figured a small piece of *Nematothallus*, but most of their material came from Llanover Quarry (p. 67). They made little effort to study the Craig-y-Fro petrifactions, and were mainly interested in the coalified compressions. This aspect of the assemblage has only been studied in detail in recent years, following the development of improved sectioning techniques compared with those used by Heard (Edwards, 1968, 1969a, b, 1970a, 1980a, 1981; Edwards and Kenrick 1986; Kenrick and Edwards, 1988a; see also Edwards and Banks, 1965; Shute and Edwards, 1989).

## Description

### Stratigraphy

The geology has been described by Edwards and Richardson (*in* Friend and Williams, 1978). A thickness of about 14 metres of fluviatile sandstones and siltstones belonging to the Senni Beds is exposed, and includes four horizons yielding plant fossils (Figure 4.16). The biostratigraphy has been discussed by Edwards and Kenrick (1986), who suggest that it is lower Siegenian, based on palynological evidence.

### Palaeobotany

The plant fossils are preserved mainly as impressions or heavily carbonized compressions. However, some pyritized petrifactions also occur. The following assemblage has been described to date:

Phaeophycophyta(?):
    *Prototaxites* sp.
    *Nematothallus* sp.

## *Devonian*

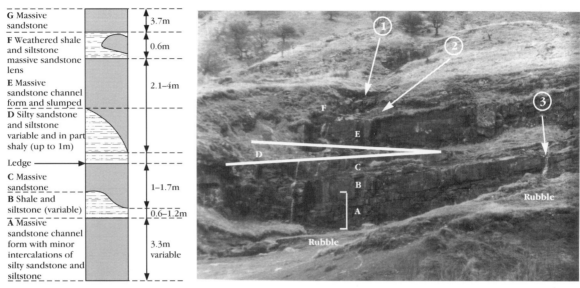

1–*Deheubarthia*
2–*Tarella*
3–*Cooksonia, Drepanophycus, Zosterophyllum*

**Figure 4.16** Craig-y-Fro Quarry. Sedimentological log through the exposed part of the Senni Beds, and a view of the quarry face showing the main beds. Also shown are three of the main plant-bearing horizons. Partly based on Friend and Williams (1978, figure 40). (Photo: D. Edwards.)

Chlorophycophyta(?):
 *Pachytheca* sp.

Rhyniopsida:
 *Uskiella spargens* Shute and Edwards

Zosterophyllopsida:
 *Zosterophyllum llanoveranum* Croft and Lang
 *Z.* cf. *fertile* Leclercq
 *Gosslingia breconensis* Heard
 *Deheubarthia splendens* Edwards, Kenrick and Carluccio

Lycopsida:
 *Drepanophycus spinaeformis* Göppert

Trimerophytopsida:
 *Dawsonites arcuatus* Halle

Barinophytales (*incertae sedis*):
 *Krithodeophyton croftii* Edwards

Uncertain affinities:
 *Tarella trowenii* Edwards and Kenrick
 *Sennicaulis hippocrepiformis* Edwards

*Hostinella heardii* Edwards
cf.*Taitia* sp.
'spherical or circular bodies *incertae sedis*'

## Interpretation

*Pachytheca* was recorded by Croft and Lang (1942). Heard (1927) had earlier described similar fossils, but with a tuberculate surface. Their internal structure is not well preserved and so they were not named by Heard, and Croft and Lang simply referred to them as 'spherical or circular bodies *incertae sedis*'.

Heard (1939) identified a pyritized specimen as cf. *Prototaxites (Nematophyton)* sp. The central part of the specimen had a pseudoparenchymatous anatomy typical of *Prototaxites*. This was surrounded by an amorphous layer referred to by Heard as a cortex, but which may be a product of the breakdown of organic tissue in the outer part of the specimen.

Heard (1939) also identified *Taitia* from here, and was supported in this by Crookall (1930), the original author of the form-genus. However, the figured specimen is poorly preserved and the record has been included in the above species list as doubtful.

The Rhyniaceae is represented by adpressions of *Uskiella spargens* Shute and Edwards (1989). Much of the key information about this species was determined from the petrifactions from Llanover. However, the Craig-y-Fro specimens have demonstrated details of the stem branching pattern and the attachment of the sporangia (Figure 4.17).

Much of the anatomical detail of *Zosterophyllum llanoveranum* described by Edwards (1969a) was based on the limonite petrifactions from Llanover Quarry, but the same features could also be seen in the Craig-y-Fro pyrite petrifactions. She noted, however, that petrified sporangia are more abundant at Craig-y-Fro, and that they yielded better preserved spores. She also found evidence of a possible tapetal structure surrounding the spores, not seen in the Llanover Quarry specimens.

Craig-y-Fro is the type locality for *Gosslingia breconensis* (Figure 4.18). The original descriptions (Heard, 1926, 1927) were brief, but covered the main features of the plant, and his conclusions have been mostly supported by subsequent studies (Croft and Lang, 1942; Edwards and Banks, 1965; Edwards, 1970a). The only significant exception is his interpretation of the small protuberances found below the branches of the axes, which he believed to be the remains of specialized 'sporangiferous branches'. The specimens described by Croft and Lang (1942) showed that the sporangia were in fact attached laterally to the axes, and were not aggregated into terminal spikes as in *Zosterophyllum*.

The most comprehensive description of *Gosslingia* is by Edwards (1970a), whose study was based mainly on specimens from Craig-y-Fro. Taxonomically significant features discussed by Edwards include the exarch xylem strand with an oval cross-section, and the laterally borne sporangia with a well developed distal dehiscence structure. Both features support its inclusion within the Zosterophyllopsida, as do details of the structure of the tracheids (Kenrick and Edwards, 1988a). Edwards discussed the nature of the axillary 'tubercle' found below each dichotomous branching point on the main axis, concluding that it was probably the remains of a third branch originally attached to the axis below the dichotomy, but which became detached either before or during preservation. The discovery by Banks and Davis (1969) of a subaxillary branch in *Crenaticaulis*, another member of the Zosterophyllopsida, lends support to this interpretation.

Another probable member of the Zosterophyllopsida has been described from Craig-y-Fro by Edwards and Kenrick (1986) as *Tarella trowenii*. *Tarella* shares many features with *Gosslingia*, particularly the organization, orientation and general distribution of the sporangia, but it differs in having sporangia distributed in two vertical rows on opposite sides of the axes, in having isotomous branching of the main axial system, and in showing no evidence of axillary tubercles. *Tarella* also has prominent protuberances on the surface of the axes, not seen in *Gosslingia*, although the exact nature of these structures is uncertain. So far, no undoubted petrified axes of *Tarella* have been reported. Consequently, the form of the xylem strand is unknown, a feature essential before the form-genus can be unequivocally placed in the Zosterophyllopsida (Edwards and Edwards, 1986).

A species unique to Craig-y-Fro is *Krithodeophyton croftii* Edwards, 1968. It has naked, dichotomous axes bearing terminal, fertile spikes. The latter comprise of two vertical rows of alternating sporangia and sterile bracts. It

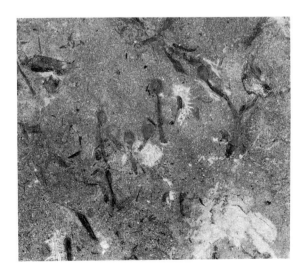

**Figure 4.17** *Uskiella spargens* Shute and D. Edwards. Branched axes bearing terminal sporangia; Natural History Museum, London, specimen V.26461a. Senni Beds (Siegenian), Craig-y-Fro Quarry. x 1. (Photo: Photographic Studio, Natural History Museum, London.)

*Devonian*

**Figure 4.18** *Gosslingia breconensis* Heard. Branched axes with some lateral sporangia; Natural History Museum, London, specimen V.26575. Senni Beds (Siegenian), Craig-y-Fro Quarry. x 0.5. (Photo: Photographic Studio, Natural History Museum, London.)

compares closely with *Protobarinophyton* described from the Lower Devonian of Siberia (Ananiev, 1957), and is generally placed in the order Barinophytales. If this assignment is correct, it is the only member of the Barinophytales to have been described from Britain, and the oldest member of the order reported from anywhere in the world.

A number of petrified axes were included in *K. croftii* by Edwards (1968), but were later transferred by Edwards (1980a) to a separate species, *Hostinella heardii*. They are of only passing botanical interest, consisting merely of naked, dichotomous axes with a terete, centrarch xylem strand. However, Edwards (1980a) used them to explain how the tracheids became petrified. This seemed to have occurred in two phases: firstly, the filling of the tracheid lumen with pyrite; and then the replacement of the cell wall by pyrite. Little or no organic material seems to have remained (see also Kenrick and Edwards, 1988a).

A second type of petrified axis has been described as *Sennicaulis hippocrepiformis* Edwards (1981). It is represented by just a few specimens of smooth surfaced axes with a terete centrarch xylem strand, consisting of tracheids with distinctive helical and annular secondary thickenings. It probably belongs to either the Rhyniopsida or Trimerophytopsida.

The 'pyritized fructification' described by Heard

(1939) is a small capsule containing 250 μm diameter spherical objects. The latter were interpreted by Heard as spores, but they are far too large for this (they are more the size of megaspores) and show no evidence of a trilete mark. It is impossible to be certain at present exactly what this structure was.

The Craig-y-Fro assemblage is part of what is known as the Senni Beds group of floras, as described by Croft and Lang (1942). It belongs to the *Psilophyton* Zone of Banks (1980), and to the European phytogeographic subunit of Raymond *et al.* (1985) (comparable assemblages are dealt with in the discussion on Llanover Quarry, earlier in this chapter). It is second in diversity only to that found at Llanover Quarry, and includes some taxa not found there (*Tarella trowenii, Krithodeophyton croftii, Hostinella heardii*). It also yields pyrite petrifactions, which show the internal anatomy of the plants in finer detail than the limonite petrifactions of Llanover. Other than the Rhynie Chert assemblage, they represent the most completely known Devonian fossil plants from Britain.

## Conclusion

Craig-y-Fro Quarry has yielded some of the best preserved Devonian plant fossils in Britain, second only in quality to those found in the Rhynie Chert (see below). They are about 400 million years old, and represent a flora typical of the southern and eastern margins of Laurussia. It is the best locality for two important types of early land plants, *Gosslingia* and *Tarella*. The first, and possibly the second of these are zosterophylls, which represent the ancestors of the club-mosses that dominated much of the land vegetation later in the Palaeozoic, including the Late Carboniferous equatorial coal swamps (see Chapter 7). It also has the earliest known example of a barinophyte (*Krithodeophyton*), which is one of the early groups of plants that adapted to the land environment during the Devonian, but which soon after disappeared, leaving no evolutionary successors. The fossils here are very similar to those from Llanover Quarry (see earlier in this chapter), but are more robustly preserved, which often makes them easier to study anatomically. Similar assemblages have also been reported from Belgium and Ukraine, but they have not been studied in such detail, and do not yield such well-preserved anatomical details.

## BALLANUCATER FARM
*R.J. Rayner*

### Highlights

Ballanucater Farm has yielded the best British assemblage of Emsian plant fossils with cuticles. They include the earliest known examples of spines with glandular tips.

### Introduction

This small stream section near Callander, Central Region (Scotland – NN 630019) exposes plant-bearing clastic sediments of the Old Red Sandstone. Although plant fossils have been reported from localities in this area since the mid-nineteenth century (history summarized by Henderson, 1932 and Lang, 1932), the earliest records of specimens from Ballanucater Farm appear to be those by Henderson (1932) and Lang (1932). Since then some of the flora has been investigated by Edwards *et al.* (1982) and Rayner (1982, 1983, 1984), particularly with reference to the finely preserved cuticles.

### Description

#### Stratigraphy

The sequence belongs to the Strathmore Group, as defined by Armstrong and Paterson (1970). It includes grey coarse flaggy sandstones and blue-green mudstones. Palynological evidence suggests that they are Emsian in age (Ford, 1974; Richardson *in* Rayner, 1983).

#### Palaeobotany

Plant fossils have been found in both the sandstones and the mudstones, the former yielding partially pyritized compressions and the latter carbonized compressions with cuticles. The following species have been described to date:

Chlorophycophyta(?):
    *Pachytheca* sp.

Zosterophyllopsida:
    *Sawdonia ornata* (Dawson) Hueber
    ?*Margophyton goldschmidtii* (Halle) Zacharova

# Devonian

Lycopsida:
*Drepanophycus spinaeformis* Göppert

Trimerophytopsida:
*Dawsonites* sp.

## Discussion

Henderson (1932), Lang (1932), Edwards *et al.* (1982) and Rayner (1984) have all described specimens of *Drepanophycus spinaeformis* from here (Figure 4.19). Although there are older known examples of this form-genus, such as the Siegenian specimens from Craig-y-Fro (p. 71), these are the oldest yet found with cuticles preserved. The plant is lycopsid-like, with broad, leafy axes up to 30 mm wide, but with a comparatively narrow vascular strand, generally no more than 3 mm wide. No sporangia have been found at Ballanucater Farm but some excellent cuticle preparations show the arrangement of epidermal cells around the stomatal pore (Lang, 1932; Edwards *et al.*, 1982; Rayner, 1984). The twin guard cells are completely enclosed by two reniform subsidiary cells. These are the oldest compression fossils of land plants from which paired guard cells and paracytic subsidiary cells have been described. Comparable cuticles of *Drepanophycus* have been described from the Eifelian of North America (Stubblefield and Banks, 1978).

The most common constituent of the assemblage is *Sawdonia ornata* (syn. *Psilophyton princeps* var. *ornatum* Dawson), a spiny zosterophyll (*sensu* Banks, 1968). It was briefly described from here as part of the form-genus *Psilophyton* by Henderson (1932) and Lang (1932), but a fuller account is provided by Rayner (1983). The axes are up to 6.5 mm wide and divide pseudomonopodially. They are covered with black-tipped, multicellular spines (or trichomes), up to 3 mm long, which are believed to have been secretory. Cuticle preparations from here (Edwards *et al.*, 1982; Rayner, 1983) indicate the papillate nature of the epidermis, details of the 'hair bases' (*sensu* Edwards, 1924), and the structure of the stomata. The 'hair bases' consist of elongate, papillate epidermal cells around a central, isodiametric cell with a thin or incomplete cuticle cover. They have been termed 'rosettes' by Rayner (1983), but their function remains unknown. The stomata are round to elongate, and consist of a pore enclosed by a guard cell area. There is no evidence of the intercellular flanges in this area, which would have suggested that a pair of guard cells was present, and compare with the stomata of *Zosterophyllum myretonianum* Penhallow (Lele and Walton, 1962a; Edwards *et al.*, 1982). Nevertheless, Rayner (1983) interpreted the *Sawdonia* stomata as originally having had a pair of guard cells, and that the intercellular flanges were just not present in life. The guard cells lay at the bottom of a shallow stomatal pit, formed by cuticular thickenings between the subsidiary and guard cells (see also Zdebska, 1972; Chaloner *et al.*, 1978).

An interesting feature of this locality is the presence of small bodies (diameter 90 µm or less) inside many of the macerated *Sawdonia* spines. Lang (1932) described them as fungal vesicles,

**Figure 4.19** *Drepanophycus spinaeformis* Göppert. Leafy shoot; Natural History Museum, London, specimen V.58185. Strathmore Group (Emsian), Ballanucater Farm. x 1. (Photo: Photographic Studio, Natural History Museum, London.)

under the name *Palaeomyces* sp., whereas Rayner (1983) interpreted them as possibly the spore resting stage of a fungus.

Within the sandstones, there are many spiny axes that appear to conform to Halle's (1916) description of *Psilophyton goldschmidtii* Halle, and were identified as such by Lang (1932). Halle erected the species to describe sterile, spiny axes with a distinctive branching pattern, found in Norway. The specimens from Ballanucater Farm are also sterile and look very similar to those illustrated by Halle. The main axes are up to 5 mm wide and divide pseudomonopodially, in contrast to the much narrower lateral branches which fork dichotomously. There is a continuous line on the axes running parallel to the edges. There are also oval regions filled with matrix at the point of divergence of the lateral branches, which may correspond to axillary tubercles. This species has since been transferred to the genus *Margophyton* by Zakharova (1981), who regarded it as a zosterophyll (see also Edwards *et al.*, 1989). However, in the absence of fertile organs, it is impossible to be certain whether or not the Ballanucater specimens also belong here.

Clusters of fusiform sporangia are found both as fossils on the rock surfaces, and in maceration residues. They are best assigned to the form-genus *Dawsonites* as, although they are sometimes connected to quite extensive dichotomously branching axes, they lack any distinguishing features which would allow identification with any species of *Psilophyton*. The fusiform sporangia are elongate, 4–6 mm long, pendulous, and borne terminally on dichotomizing axes. They occur in pairs, the individuals of which are often twisted around one another. They dehisce longitudinally, and contain trilete spores averaging 60 µm in diameter. The spores are almost completely enclosed by a finely sculptured outer exine, which shows a tendency to break away from mature individuals.

The presence of *Dawsonites* in the Ballanucater assemblage indicates that it belongs to the *Psilophyton* Zone (Banks, 1980; Edwards and Berry *in* Cleal, 1991). Similar assemblages have been reported from a series of localities in the Callander area (Jack and Etheridge, 1877; Henderson, 1932; Lang, 1932), but Ballanucater Farm has yielded easily the best preserved and most abundant specimens. A similar assemblage has also been reported from Auchensail Quarry near Cardross (p. 78), but no evidence of cuticles has been found there.

Raymond *et al.* (1985) place this assemblage in their American phytogeographic subunit of the equatorial and low latitude floras. The only other British *Psilophyton* Zone assemblage belonging to this subunit was found in Emsian beds in a borehole in Oxfordshire (Chaloner *et al.*, 1978). Those from the Siegenian Senni Beds of south-east Wales, such as from Craig-y-Fro and Llanover (pp. 67–75), belong to the European subunit of Raymond *et al.* (1985) and differ from the Strathmore Group assemblages in their more diverse composition. From outside of Britain, assemblages with a more comparable, restricted composition have been reported from Rörangen, Norway (Halle, 1916), Matringhem, France (Danzé-Corsin, 1955) and James Bay, Canada (Hueber, 1964). However, none of these places has yielded such well-preserved cuticles as found at Ballanucater. This gives Ballanucater a unique significance in Lower Devonian palaeobotany, especially for work on early epidermal structures (e.g. stomata, trichomes), whose evolution was a key factor in allowing plants to overcome the problems of desiccation inherent in living in a terrestrial environment.

## Conclusion

Ballanucater Farm has yielded an important assemblage of fossils representing plants of 390 Ma. The fossils are particularly important as they still preserve the outer, protective skin of the plant (known as the cuticle), in which details of the microscopic breathing pores (stomata) and hairs can be seen. They are some of the oldest known examples of modern-looking stomata, in which the pore is surrounded by two specialized cells (guard cells) that control the size of the opening. This was an important development for helping plants adapt to the land environment, allowing greater control over water-loss and gas-exchange with the atmosphere. A similar assemblage of fossils occurs at Auchensail Quarry (p. 78), as well as at localities in Norway, Belgium and Canada, but none has yielded such well-preserved cuticles as found at Ballanucater.

# Devonian

## AUCHENSAIL QUARRY
*R.J. Rayner*

### Highlights

Auchensail Quarry has yielded some of the best preserved plant fossils from the Emsian of Britain (Figure 4.20). These include exceptionally well-preserved examples of *Sawdonia*, and the youngest examples of *Prototaxites* known from Britain.

### Introduction

Recent quarrying near Cardross, Strathclyde Region, has exposed Lower Devonian sandstones containing abundant plant fossils (NS 345795). They were first reported by Scott *et al.* (1976) and Morton (1976). Some of the species have since been studied in detail by Rayner (1982, 1983, 1984), and revealed considerable anatomical detail.

### Description

#### Stratigraphy

The geology is described by Scott *et al.* (1976). The sequence consists of units of shallow-water, red siltstones and mudstones, up to 2 metres thick, alternating with well indurated, grey, upward-fining sandstones (Figure 4.21). They belong to the Strathmore Group (*sensu* Armstrong and Paterson, 1970), and were interpreted by Scott *et al.* as fluviatile in origin. An Emsian age has been suggested by Richardson (*in* Rayner, 1983).

#### Palaeobotany

The plant fossils occur abundantly in the sandstones. Mats of coalified compressions occur mainly within the upper part of the sandstone units, while discrete pyrite petrifactions occur throughout the beds.

**Figure 4.20** Auchensail Quarry. Emsian sandstones and shales of the Strathmore Group. Note the igneous dyke just to the left of centre of the quarry face as shown. (Photo: C.J. Cleal.)

# Auchensail Quarry

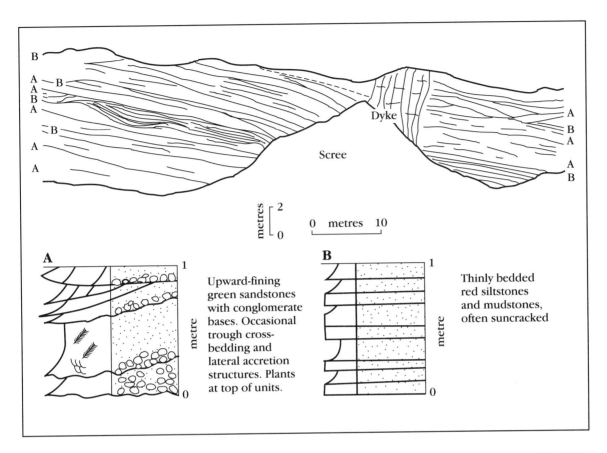

**Figure 4.21** Sedimentology of the Strathmore Group (Emsian) at Auchensail Quarry. Based on Scott *et al.* (1976).

Phaeophycophyta(?):
   *Prototaxites* sp.

Rhyniophytoid:
   ?*Sporogonites* sp.

Zosterophyllopsida:
   *Zosterophyllum* sp.
   *Sawdonia ornata* (Dawson) Hueber

Lycopsida:
   *Drepanophycus spinaeformis* Göppert

Trimerophytopsida:
   *Dawsonites* sp.

## Interpretation

*Prototaxites* is represented here by axes 5–50 mm wide. They have not been described in detail in the published literature, but are of interest as the youngest British specimens of this form-genus (younger specimens have been described from North America by Arnold, 1952).

Occurring within the mats of coalified compressions are many examples of narrow axes (<10 mm wide) that branch pseudomonopodially, have circinate vernation, a papillate epidermis and are covered with swollen-tipped spines. Despite the lack of well-preserved cuticles, these specimens have been identified as *Sawdonia ornata* (Rayner, 1983). The form of two fertile specimens, the first of such to be reported from Europe, indicate that the plant was a true zosterophyll (*sensu* Banks, 1968) with lateral sporangia apparently arranged in a terminal spike. The sporangia are sessile, oval in shape and up to 4 mm across. Several have a 'V'-shaped incision on the axial surface, which might be a dehiscence scar. Spores have been isolated from several sporangia and are small (35–40 μm), subtriangular, with an indistinct trilete mark.

The internal anatomy of the *Sawdonia* axes from here has been investigated using plastic embedding and sectioning techniques (Rayner, 1983). The most striking feature is its large vascular strand, which may be as much as three quarters of the width of the axis, and which divides unequally before branching. The strand is made up of annular or spirally thickened tracheids, and maturation is apparently exarch. The epidermis is missing on all examples so far found here, but both thick-walled cortical cells and the characteristic spines are preserved. In well-preserved coalified compressions found in the finer-grained parts of the sandstone units, the surface features of *Sawdonia* are comparable to cuticle preparations from Ballanucater Farm and elsewhere (Edwards, 1924; Zdebska, 1972; Chaloner *et al.*, 1978; Edwards *et al.*, 1982; Rayner, 1983). The papillae form prominent features on the axis surfaces, and their purpose may have been to act as a shield against excessive radiation.

*Drepanophycus spinaeformis* Göppert forms a small but important part of the Auchensail assemblage. It is mainly represented by broad, leafy axes, up to 25 mm across, and often showing a narrow zone of vascular tissue. The leaves may still be attached to the axis or, where they have been detached, be represented by oval leaf scars. A few examples show compact vascular strands made up of tracheids with annular thickenings. Petrifactions from another Strathmore Group exposure at Keltie Water (Rayner, 1984) have shown that the xylem had a deeply lobed cross-section, but this has yet to be observed in specimens from Auchensail.

Pairs of *Dawsonites*-type sporangia have been found here at the ends of dichotomizing branches. Several have yielded spores similar to those from Ballanucater Farm. The axes, which are up to 5 mm wide, are naked but have a series of longitudinal striations preserved on the upper surfaces. There is no evidence, however, of axillary structures.

Other terminal sporangia found here are similar to *Sporogonites* Halle. These sporangia have been compressed to form an oval capsule at the end of a simple striated stalk. The specimens have failed to yield spores, and they may only be compared with Halle's (1916) description on gross morphology.

This site has yielded plant fossils typical of the Strathmore Group of Scotland, and similar to those found at Ballanucater Farm (a discussion on similar plant fossil assemblages, from both Britain and abroad, is given in the section dealing with Ballanucater Farm). However, it differs from the latter site in having *Prototaxites* and *Sporogonites*, but lacking *Psilophyton*. More significant is the presence of petrifactions at Auchensail, which provide unique anatomical evidence of some of these early land plants, especially the zosterophyll *Sawdonia*.

## Conclusion

Auchensail Quarry has yielded an exceptionally well-preserved suite of fossils of Early Devonian land plants, which lived about 390 Ma on the Old Red Sandstone continent. They are similar to the fossils found at Ballanucater Farm (see earlier in this chapter), but are preserved differently, showing greater detail of the internal structure of the plant. This has been particularly important for obtaining a fuller understanding of *Sawdonia*, which belongs to the primitive group known as the zosterophylls, and which were the ancestors of the club-mosses. Also found here are some excellently preserved fructifications (sporangia) of the group of plants known as the trimerophytes, showing details of how on the living plant these sporangia split to release the spores. This group is important as the probable ancestor of the seed plants (and thus the flowering plants). The fossils of *Prototaxites* found here are the last occurrence of this large, enigmatic alga in Britain, although Late Devonian examples have been reported from North America. These algae were an important component of land vegetation in the Late Silurian and earliest Devonian, but became displaced by the better adapted vascular plants during the later part of the Early Devonian.

## RHYNIE

## Highlights

Rhynie is probably the most important palaeobotanical site in Britain, yielding the oldest known vascular plant fossils with well-preserved anatomical details (Figure 4.22). The 22 species of plant fossil (including fungi) are unique to this locality, and include the 'type-genus' of the Rhyniophytina (*Rhynia* – the best-known early land plant), the earliest well-documented lycopsid (*Asteroxylon*), and a variety of enigmatic species evidently representing early experiments in adaptation to a terrestrial environment. It is also the oldest known example of an *in situ* fossilized terrestrial ecosystem.

# *Rhynie*

**Figure 4.22** Rhynie. The field just outside of the village, under which lies the Siegenian fossiliferous chert. (Photo: C.J. Cleal.)

## Introduction

This famous locality lies beneath a field near the village of Rhynie, 14 km south of Huntly, in the Grampian Region of Scotland (NJ 495264). Probably no other single site has had such an impact on the development of palaeobotany. Walton (1959) described it as 'the most dramatic and important palaeobotanical discovery of the century'. Prior to its discovery, the structure and taxonomic affinities of Lower Devonian plant fossils (until then almost exclusively known as adpressions) were still a matter of debate, and many scientists argued that they were either algal or indeterminable fragments of 'advanced' vascular plants. The petrifactions discovered at Rhynie confirmed that vascular plants did exist during the Early Devonian and that they were truly 'primitive'. This has had a major impact on evolutionary ideas and the classification of the Plant Kingdom.

The general background to the Rhynie Chert locality and its fossilized biota is summarized by Chaloner and Macdonald (1980). There is no natural outcrop of the chert, and it was first found as loose blocks in a field and in a dry-stone wall (Mackie, 1913) (Figure 4.23). Subsequently, however, the Geological Survey excavated a series of trenches through the field, revealing the rock *in situ* and allowing a large quantity of the material to be collected (Horne, 1917; Horne and Mackie, 1917, 1920a, b; Kidston, 1922, 1923e). The specimens were the subject of a series of now classic monographs by Kidston and Lang (1917b, 1920a, b, 1921a, b), in which anatomical details of the plants were described from thin sections of the chert. Pant (1962) argued that Kidston and Lang had been able to determine such fine cellular structure that details of the Rhynie plants were 'nearly as complete as they could be even if the plants were living today'. The method used by Kidston and Lang was, however, not so good at

# Devonian

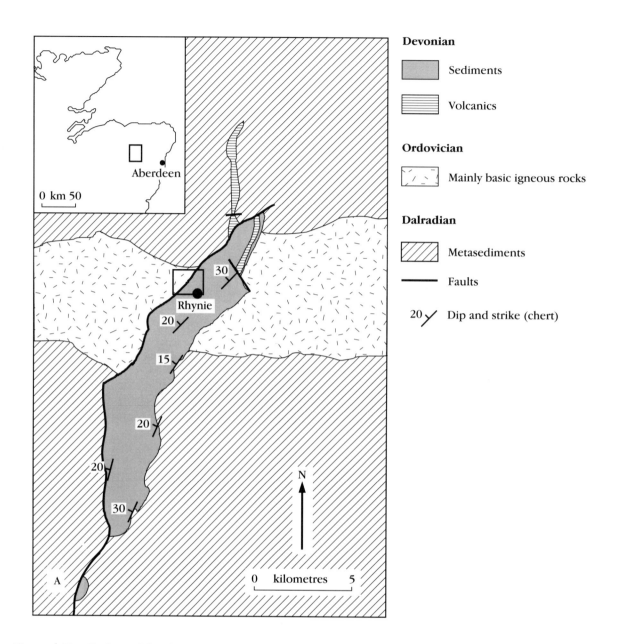

**Figure 4.23** Geology of the Rhynie area. (A) Map showing the outline of the Rhynie outlier. (B, facing page) Map showing detail of that part of the outlier containing the fossiliferous chert (marked by a rectangle on A). Based on Trewin and Rice (1992, figure 1), who provide details of the boreholes marked on (B).

revealing the gross morphology of the plants; their reconstructions (Kidston and Lang, 1921a, pls 1–2) were rather oversimplified. More refined results were only possible when the peel method (Joy *et al.*, 1956) was used, allowing more closely-spaced serial sections to be made. Examples of the use of such techniques are given by Bhutta (1972), Eggert (1974), El-Saadawy and Lacey (1979a, b), Edwards (1980, 1986), Edwards and Lyon (1983), and Lyon and Edwards (1991). Despite being known for over 70 years, new discoveries are still being made at Rhynie (e.g. the vascularized gametophyte by Remy and Remy 1980a, Remy and Hass, 1991a, b, c and Remy, 1991) and there remains considerable potential for further work here.

# Rhynie

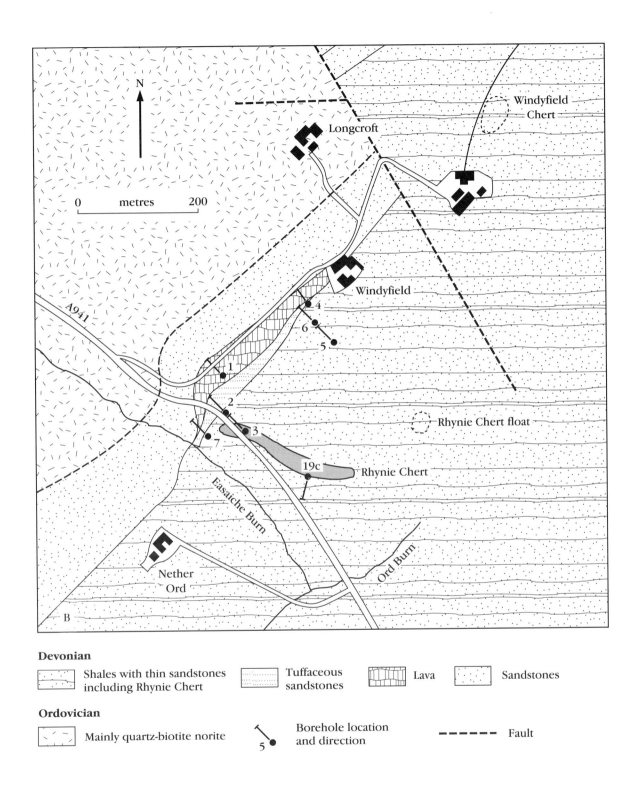

**Figure 4.23** *contd*

# Devonian

## Description

### Stratigraphy

The Rhynie Chert is usually included in the Dryden Shale Formation, the highest part of the Rhynie Old Red Sandstone outlier (Trewin and Rice, 1992). Kidston and Lang (1917b, 1921a) regarded the chert as 'not younger than the Middle Division of the Old Red Sandstone'. More recent palynological evidence has, however, suggested an Early Devonian, possibly Siegenian, age (Richardson, 1967; Richardson *in* House *et al.*, 1977). Banks (1980) included the plant fossils in his *Psilophyton* Zone, indicating a late Siegenian or Emsian age.

The most complete account of the stratigraphy of the Rhynie Chert is by Trewin and Rice (1992). They report it to consist of *c.* 3.2 metres of alternating chert, mudstone and sandstone (Figure 4.24). It is generally regarded as having been formed by hot siliceous solutions, probably from a volcanic fumerole, impregnating a peat that developed on an alluvial plain (Kidston and Lang, 1917b, 1921b; Tasch, 1957; Edwards and Lyon, 1983; Edwards, 1986; Trewin and Rice, 1992):

### Palaeobotany

The plant fossils at Rhynie are silica petrifactions, preserving very fine anatomical detail (Edwards, 1986). The following taxa have been described to date (see also Figures 4.25, 4.28 and 4.29):

Cyanochloronta:
   *Langiella scourfieldii* Croft and George
   *Kidstoniella fritschii* Croft and George
   *Rhyniella vermiformis* Croft and George
   *Rhyniococcus uniformis* Edwards and Lyon

Phaeophycophyta(?):
   *Prototaxites taitii* (Kidston and Lang) Pia
   *Nematoplexus rhyniense* Lyon

Chlorophycophyta(?):
   ?*Pachytheca* sp.
   *Palaeonitella cranii* (Kidston and Lang) Pia
   *Archaeothrix oscillatoriformis* Kidston and Lang
   *A. contexta* Kidston and Lang
   *Mackiella rotundata* Edwards and Lyon
   *Rhynchertia punctata* Edwards and Lyon

Rhyniopsida:
   *Rhynia gwynnevaughanii* Kidston and Lang

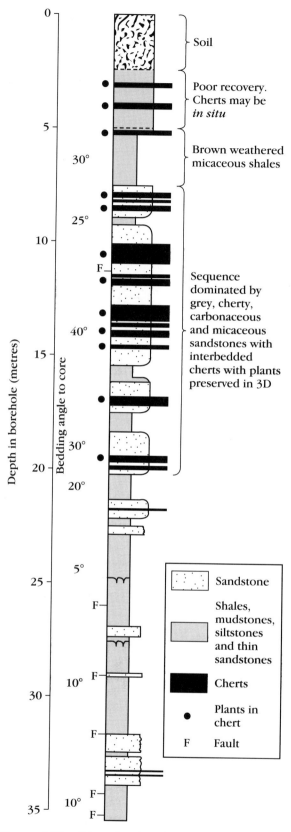

**Figure 4.24** Composite log through Rhynie Chert showing distribution of plant-bearing horizons. Based on Trewin (1989, p. 11).

*Lyonophyton rhyniensis* Remy and Remy
*Langiophyton mackei* Remy and Hess

Horneophytopsida:
  *Horneophyton lignieri* (Kidston and Lang) Barghoorn and Darrah

Zosterophyllopsida:
  *Trichopherophyton teuchansii* Lyon and Edwards

Lycopsida:
  *Asteroxylon mackei* Kidston and Lang

Fungi:
  *Palaeomyces gordonii* Kidston and Lang
  *P. agglomerata* Kidston and Lang
  *P. asteroxylii* Kidston and Lang
  *P. horneae* Kidston and Lang

Affinities uncertain:
  *Nothia aphylla* Høeg
  *Aglaophyton major* (Kidston and Lang) Edwards
  *Kidstonophyton discoides* Remy and Hess

## Interpretation

*Rhynia gwynnevaughanii* has been widely regarded as the archetypal primitive vascular plant, following the reconstruction given by Kidston and Lang (1921a, pl. 1 fig. 1). They interpreted it as having thin, naked aerial shoots, arising from a prostrate rhizome, and terminated by single sporangia. Branching of the aerial shoots was shown as predominantly dichotomous, but with some subsidiary adventitious branching. Small hemispherical bulges were distributed along the shoots.

Using the peel method, however, Edwards (1980, 1986) has shown that it was a far more complex plant (see Figure 4.27a). For instance, the aerial shoots had a much higher proportion of adventitious branching, dichotomies being relatively rare. He also found evidence of an abscission layer at the base of the sporangia, the latter being shed after the release of their spores. An adventitious branch occurs immediately below the site of many of the sporangia.

Although perceptions about the *R. gwynnevaughanii* gross morphology have changed over the years, Kidston and Lang's (1917a, b) interpretation of the anatomy of the aerial shoots is still generally accepted. The aerial shoots consist of a very slender, terete, centrarch stele, surrounded by a cylinder of phloem, a two-layered cortex and an epidermis (Figures 4.25 and 4.26). The relative slenderness of the stele has attracted some comment, Filzer (1948) suggesting that it reflects a primitive and inefficient water metabolism control system. Looking at it from another point of view, however, Kevan *et al.* (1975) argued that the relative thickness of the cortex was a strategy to protect the stele from arthropod predation. Speck and Vogellehner (1988a, b) have shown that the stele contributed little to supporting the axes in

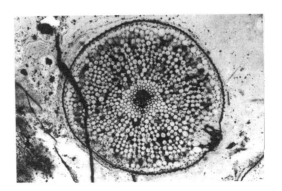

**Figure 4.25** *Rhynia gwynnevaughanii* Kidston and Lang. Transverse section through stem showing central vascular strand; Natural History Museum, London, specimen SC.3132. Rhynie Chert (Siegenian), Rhynie. × 25. (Photo: Photographic Studio, Natural History Museum, London.)

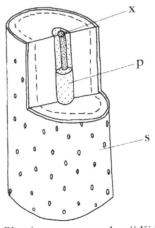

**Figure 4.26** *Rhynia gwynnevaughanii* Kidston and Lang. Cut-away reconstruction of stem showing cylindrical protostele of xylem (x), phloem (p), and stomata on the stem surface (s). From Thomas and Spicer (1987, figure 3.2D; after Chaloner and Macdonald).

such early land plants, which was mainly controlled by turgor in the cortex.

Satterthwaite and Schopf (1972) described what they regarded as sieve-like structures in the cylinder of phloem (Satterthwaite and Schopf identified their specimens as '*Rhynia*' *major* Kidston and Lang, but Lemoigne and Zdebska (1980) have argued that they almost certainly belong to *R. gwynnevaughanii*). If correctly interpreted, this remains the oldest unequivocal evidence of phloem tissue in the fossil record.

*R. gwynnevaughanii* sporangia have traditionally been interpreted as lacking a dehiscence structure, a view which was supported by Edwards (1980). However, Remy (1978) has described a band of thin-walled cells, probably lying longitudinally along the sporangium, which may have functioned as a line of weakness to facilitate rupturing (a stomium). No evidence of an annulus was reported, but opposite the stomium was a flexible, so-called joint region, which again may have helped in the dehiscence process. Remy's interpretation has not received universal acceptance (Edwards and Edwards, 1986) but, if proved correct, it will require a significant modification to our view of *Rhynia* as a simple land plant.

Another contentious issue surrounding *R. gwynnevaughanii* is its possible status as a vascularized gametophyte. Considering the fine preservation in the chert, the apparent absence of readily identifiable gametophytes of the Rhynie plants had puzzled palaeobotanists (e.g. Chaloner, 1960; Pant, 1962). Merker (1958, 1959, 1961) argued that perhaps some of the prostrate axes of *R. gwynnevaughanii* were gametophytes, from which the sporophytic aerial shoots arose. Pant (1962) went further, suggesting that the *R. gwynnevaughanii* plant as a whole was a vascularized gametophyte, and Lemoigne (1968a) has argued that it was the gametophyte of the plant then known as *Rhynia major* Kidston and Lang (now included in the form-genus *Aglaophyton* – see p. 89). The argument was based largely on structures on *R. gwynnevaughanii* axes, which were interpreted by Pant and Lemoigne as archegonia and antheridia (see also Lemoigne, 1968b, c, 1969a, b, 1970, 1975, 1981). However, the photographic record of these structures is not entirely convincing, and Bhutta (1969) has argued that the 'archegonia' are merely stomata damaged by fungal attack. Edwards (1980) made an extensive search for similar structures in his specimens of *R. gwynnevaughanii*, without success, and furthermore found unequivocal evidence that at least the aerial shoots of the plant were sporophytic. Edwards (1979b) argued that the gametophytes of the primitive vascular plants were fast growing and quick to mature, and thus had a very low preservational potential. Alternatively, Edwards (1986) suggested that gametophytes of these plants only developed under relatively rare and favourable conditions, and that they normally spread by the development of extensive growth of the sporophytic generation. Whatever the true explanation, the nature of the *R. gwynnevaughanii* gametophytes is unknown.

Based mainly on Kidston and Lang's observations, *Rhynia* has been used as the effective type of the most primitive class of vascular plants – the Rhyniopsida (Banks, 1975b). In this role, it has tended to be central to any discussion on the early evolution of vascular plants. However, the new evidence provided by Edwards (1980) has required the concept of the Rhyniopsida to be modified and enlarged (Edwards and Edwards, 1986). It still probably represents the most primitive class of vascular plants, but now includes forms such as *Renalia* showing a wider diversity of morphologies, particularly in the branching of the axes.

*R. gwynnevaughanii* has also played a key role in understanding the early evolution of the stele and leaves. It had been argued that the stele first evolved in leaves, and only later developed in stems by the coalescence of the leaf traces (for example see comments by Arber, 1921). The evidence from *Rhynia* clearly indicates that the primitive condition is a simple protostele in the stem (Scott, 1924a). Zimmermann's (1926) discovery that the stomata are associated mainly with the hemispherical bulges on the aerial shoots suggests that these bulges may represent the early phases in the development of non-vascularized, leaf-like emergences, similar to those of *Asteroxylon* (see below). Further aspects of the epidermal structure of *Rhynia* are discussed by Edwards *et al.* (1982).

Another primitive vascular plant found at Rhynie was described as *Hornea lignieri* Kidston and Lang (later re-named *Horneophyton lignieri* (Kidston and Lang) Barghoorn and Darrah, 1938). It has been reconstructed, with thin, dichotomous axes arising from a corm-like rhizome (Kidston and Lang, 1921a; Eggert, 1974) (Figures 4.27B, C and 4.28). The axes have a very irregular surface, which Eggert (1974) argued to be a taphonomic effect, but which El-Saadawy and Lacey (1979b)

# Rhynie

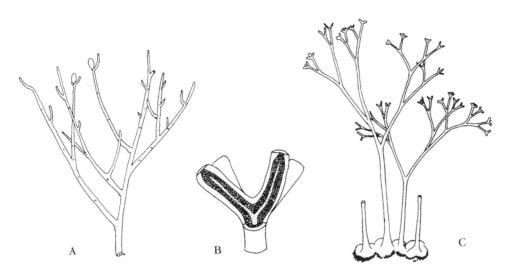

**Figure 4.27** (A) *Rhynia gwynnevaughanii* Kidston and Lang; reconstruction of aerial parts of plant including terminal sporangia. (B-C) *Horneophyton lignieri* (Kidston and Lang) Barghoorn and Darrah; (B) cut-away reconstruction of dichotomizing sporangia showing the central columella; (C) reconstruction of whole plant. From Thomas and Spicer (1987, figure 3.2C, E, and F; after Eggert and D.S. Edwards).

 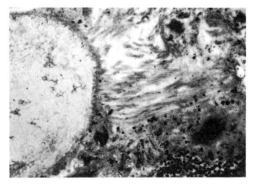

**Figure 4.28** *Horneophyton lignieri* (Kidston and Lang) Barghoorn and Darrah. Longitudinal sections through corm-like rhizome; Natural History Museum, London, specimen V.15648. Rhynie Chert (Siegenian), Rhynie. (left) × 8, (right) × 5. (Photos: Photographic Studio, Natural History Museum, London.)

interpreted as being due to small emergences, as in *Nothia* (p. 89). The anatomy of the aerial shoots is essentially similar to that of *Rhynia*, except that the terete, centrarch stele is rather thicker. The nature of the transverse breaks in the protostele observed by Kidston and Lang (1920a) has never been properly established, but might be a preservational feature. The limited available evidence of the cuticles of *Horneophyton* is discussed by Edwards *et al.* (1982).

In addition to its corm-like rhizome, *Horneophyton* is distinctive because of the structure of its apical sporangia. These are little more than cavities in the apical parts of the aerial shoots, into which extends the vascular tissue, forming a columella-like structure (Eggert, 1974, fig. 28). These fertile shoot-terminations are often branched, resulting in lobed sporangia. Kidston and Lang (1920a) failed to recognize any dehiscence structures, but an apical slit has since been reported in the apex of each sporangial lobe (Bhutta, 1972; Eggert, 1974; El-Saadawy and Lacey, 1979b). Considerable variation in the size and ornamentation of the spores was found by

Bhutta (1973a), which he suggested might represent incipient heterospory. Alternatively, however, the smaller spores may merely be immature.

The taxonomic position of *Horneophyton* has never been firmly established. Similarities with the bryophytes have been noted (Bower, 1920; Scott, 1924a, 1928; Church, 1926), and Smith (1955) has argued that it provides an evolutionary link between the bryophytes and the more 'advanced' vascular plants. However, it is normally regarded as a true vascular plant and is often included in the Rhyniaceae, because of its elongate sporangia and centrarch stele (Banks, 1968). As pointed out by Edwards and Edwards (1986), however, its lobed sporangia with apical dehiscence structures are not really compatible with it belonging to the Rhyniopsida. Eames (1936) placed it in its own family, the Horneaceae, whereas Meyen (1978, 1987) included it in a monotypic class, the Horneophytopsida. The latter view has been followed here.

Lemoigne (1966) has described a possible spore-producing capsule of a bryophyte from Rhynie. It contains spores of only about half the size, but is otherwise indistinguishable from a *Horneophyton* sporangium. Its bryophytic affinities must remain in doubt until more complete specimens are found.

For many years, the zosterophyllopsids appeared to be notable by their absence from Rhynie. Recently, however, some fragments described by Lyon and Edwards (1991) as *Trichopherophyton teuchansii*, clearly belong to that class (they had earlier been mentioned by Lyon *in* Lindley (1968) and Gensel *et al.* (1975)). Particularly distinctive features are their spiny axes with a terete, exarch stele, circinately-tipped shoots, and marginally dehiscent sporangia. The main difficulty with interpreting the fossils was that many features of the gross-morphology could not be determined, in particular the configuration of the sporangia on the axes; most other zosterophyllopsids are identified on such features of gross morphology. However, that the spines on the axes are unicellular structures seems to be unique in the class; the emergences found in the other genera such as *Sawdonia* and *Crenaticaulis* are multicellular. The more or less terete stele also contrasts with that present in most other members of the class, in which it tends to be oval in cross-section. These factors alone are enough to justify placing this Rhynie species in a separate form-genus. However, much more information on its gross morphology will be needed before its detailed taxonomic position within the Zosterophyllopsida can be established.

The most 'advanced'-looking of the Rhynie plants is *Asteroxylon mackei*, which in many ways resembles a lycopsid (Figure 4.4a). Like *Rhynia*, it has a prostrate, creeping rhizome, but the aerial shoots are clothed in microphyll-like leaves. The xylem in the aerial shoots is fluted, giving a stellate cross-section, and is mesarch with annular tracheids (Kidston and Lang, 1920b; Lang, 1952; Figure 4.29). Such an actinostele was probably a more efficient water-conducting structure than the simple protostele of the rhyniopsids, but apparently gave little extra support to the plant (Speck and Vogellehner, 1988a, b). The actinostele has spirally arranged leaf traces, but they only enter the very base of the leaves, which are thus not the same as true lycopsid microphylls. Two types of leaves were recognized by Edwards *et al.* (1982), those with a smooth surface, and those with an undulate surface caused by epidermal cells with a strongly convex outer wall.

Kidston and Lang (1921a) reconstructed *Asteroxylon* with naked, dichotomous axes attached to the 'leafy' aerial shoots, on which were borne terminal sporangia. This was based on the frequent association of such naked shoots with the more typical *Asteroxylon* axes. Subsequently, however, Lyon (1964) and Bhutta (1969) found examples of *Asteroxylon* with

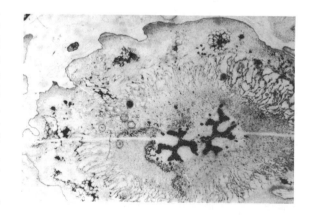

**Figure 4.29** *Asteroxylon mackei* Kidston and Lang. Transverse section through dichotomizing stem, showing two vascular traces with the characteristic stellar cross-section; Natural History Museum, London, specimen V.15643. Rhynie Chert (Siegenian), Rhynie. × 6.66. (Photo: Photographic Studio, Natural History Museum, London.)

zosterophyll-like reniform sporangia attached laterally to the leafy shoots (the naked axes with sporangia have since been assigned to a separate species *Nothia aphylla* Høeg – see below).

Kidston and Lang (1920b) regarded *Asteroxylon* as being a possible link between the simple rhyniopsids and the lycopsids, a view which is still broadly held today (Hueber, 1992; Gensel, 1992). The discoveries by Lyon and Bhutta of attached sporangia have further refined this view. Kidston and Lang argued that it should be regarded as an advanced member of what they called the psilophytes (essentially equivalent to the class Rhyniopsida of Banks, 1968, 1975b), placing it in a family Asteroxylaceae. Other authors have, however, regarded it as a primitive lycopsid, including it in the Protolepidodendrales (e.g. Taylor, 1981). A third solution has been to assign it to a separate taxon, intermediate between the zosterophylls and lycopsids, such as the class Drepanophycopsida of Rayner (1984). Whatever solution is found to the problem of classifying this plant, it holds a key position in helping us to understand the origins and early evolution of the lycopsids.

As stated above, the fertile axes originally assigned to *Asteroxylon* by Kidston and Lang, are now recognized to belong to a quite separate plant. Lyon (1964) introduced the name *Nothia aphylla* for it, but this remained a *nomen nudum* until Høeg (*in* Boureau *et al.*, 1967) provided the first validly published diagnosis. Only the terminal parts of the plant are known; these consist of slender, dichotomous axes, covered with small, tuberous emergences. Stomata are usually situated on these emergences but, unlike those of *Rhynia* and *Horneophyton*, have distinctive broad guard cells (Edwards *et al.*, 1982). Lyon (1964) originally regarded it as rhyniophytoid. Subsequent work by El-Saadawy and Lacey (1979a) has shown, however, that it has what appear to be a combination of rhyniopsid characters (thin dichotomous axes and a centrarch 'stele') and zosterophyll characters (rhizomes with 'H'-type branching, aerial shoots with circinately curved tips, oval cross-section to the 'stele', and reniform sporangia with a dehiscence slit). There are also certain characters which place it outside either of these taxa, such as the mixture of both terminally and laterally borne sporangia, and the absence of thickening of the sporangial wall near the dehiscence slit. Perhaps most significantly, however, El-Saadawy and Lacey failed to find any evidence of thickenings or pitting on the cells in the so-called xylem, calling into question the status of *Nothia* as a true vascular plant (Edwards and Edwards, 1986). The taxonomic position of this curious plant must thus remain in doubt.

Some similarity has been noted between *Nothia* and the adpression species *Sartilmania jabachensis* (Kräusel and Weyland), both in the structure and the position of attachment of the sporangia (Fairon-Demaret, 1986b). The *Sartilmania* sporangia are more elongate and attached to longer lateral branches, and have never been found terminally attached, as sometimes occurs in *Nothia*. These differences, as well as those of the preservation, make it unwise to assign them to the same form-genus. As pointed out by Fairon-Demaret, however, they may both represent a group of Early Devonian plants in the process of diversifying to form the major taxonomic groups more clearly recognizable in later floras.

Another Rhynie species whose status as a vascular plant has recently been questioned is *Aglaophyton major*. Kidston and Lang (1920a) regarded it as a second species of *Rhynia*, and they reconstructed it as looking similar to *R. gwynnevaughanii* except that it was larger (more than half a metre high), and had aerial shoots with no adventitious branching or hemispherical bulges. They noted the absence of any clear thickenings on the cells in the central conducting tissue of the aerial shoots, but put this down largely to taphonomy. However, Edwards (1986) regarded this as unlikely, considering the fine preservation of other tissues in this plant, and concluded that the central conducting tissue was not composed of tracheids, but of tissue similar to the hydroids of certain bryophytes. As pointed out by Speck and Vogellehner (1988a, b), the central conducting strand in the early land plants provided little strength to the axes, and so the development of thickenings on the 'tracheids' would give them little immediate advantage. Edwards also reconstructed the plant rather differently, giving it an essentially decumbent habit, with vertical shoots attaining a height of no more than 0.18 metres (less than one third of the height suggested by Kidston and Lang) and with a much wider angle of branching (Figure 4.30). Germinating spores of *Aglaophyton* showing immature gametophytes have been described by Lyon (1957) and Bhutta (1973b).

It is difficult at present to classify non-vascular rhyniophytoids such as *Aglaophyton* and *Nothia*, there being no established high-ranked taxa to receive them. Edwards suggested that

## Devonian

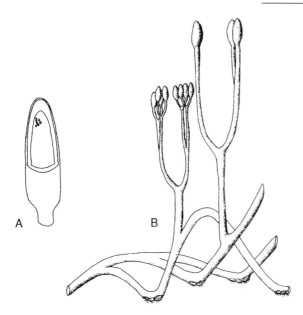

**Figure 4.30** *Aglaophyton major* (Kidston and Lang) D.S. Edwards. (A) cut-away reconstruction of sporangium. (B) reconstruction of whole plant. From Thomas and Spicer (1987, figure 3.2A-B; after D.S. Edwards).

shoots (similar to *Horneophyton*) but with a non-vascular conducting strand (similar to *Aglaophyton*), and terminated by discoidal gametangiophores with archegonia and antheridia. The fact that in its vegetative form it is probably isomorphous with the sporophytic generation gives this oldest unequivocal gametophyte particular significance. It provides a possible stem-condition from which both the bryophytic and more typical pteridophytic heteromorphic generations could be derived (Remy, 1980a). It has also given added impetus to the search for the missing *Rhynia* gametophyte, discussed above.

Other non-vascular plants represented at Rhynie include the so-called nematophytes. The best documented to date is *Prototaxites taitii* (Kidston and Lang, 1921b). The species was originally based on two fragments, one of which shows the typical pseudoparenchymatous tissue with smooth-walled thick tubes normally associated

*Aglaophyton* might represent a group of primitive land plants, ancestral to both bryophytes and vascular plants, but he was reluctant to propose a formal taxon (which would have to be at the rank of division at least) in the absence of any other definite members of the group (he did not regard *Nothia* as part of it). Whatever formal classification is eventually adopted, however, these taxa will play an important role in helping unravel the early evolutionary history of land plants, and perhaps the relationship between the vascular plants and the bryophytes (the earliest unequivocal bryophyte fossil is from the Middle Devonian of Podolia – Ishchenko and Shlyakov, 1979).

Some of the more remarkable discoveries in recent years at Rhynie have been vascularized gametophytes, which were named *Lyonophyton rhyniensis*, *Langiophyton mackei* and *Kidstonophyton discoides* (Remy and Remy, 1980a, b; Remy, 1991; Remy and Hass, 1991a, b, c). The former was thought to be probably the gametophyte of either *Horneophyton* or *Aglaophyton*, with which it was found closely associated (Figure 4.31A). Germinating spores of *Horneophyton* showing immature gametophytes had been earlier described by Bhutta (1973a), but the specimens described by Remy and Remy are clearly in a mature condition. They have dichotomous aerial

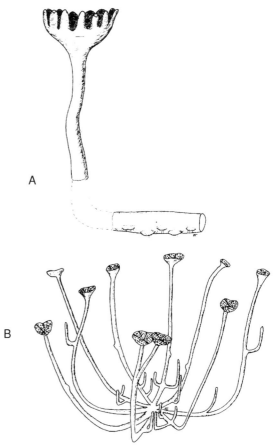

**Figure 4.31** Putative Devonian gametophytes. (A) *Lyonophyton rhyniensis* Remy and Remy, based on Rhynie fossils. (B) *Sciadophyton steinmannii* Kräusel and Weyland, based on Lower Devonian fossils from Germany. From Thomas and Spicer (1987, figure 3.9; after W. Remy and R. Remy).

with the form-genus, but the second has thick tubes with spiral thickenings (further material has been described by Lyon, 1962). Burgess and Edwards (1988) have suggested that if Kidston and Lang's two specimens represent the same species, they might not be true *Prototaxites*, but belong to their new form-genus *Nematasketum*. Another type of nematophyte from Rhynie was described as *Nematoplexus* by Lyon (1962). This also has tubes with spiral or occasionally annular thickenings, but which are characterized by localized areas of branching known as 'branch knots'. An unpublished thesis by D.S. Edwards (1973) provides the most extensive account of this group from Rhynie, and this makes it evident that other species occur here. He also demonstrated possible evidence of the anatomy of the outer parts of these prototaxitoid axes (also mentioned by Kidston and Lang, 1921b) and of appendages (?rhizomorphs) to a holdfast. It is clear that Rhynie may hold the key to establishing the true nature of these enigmatic plants, although considerably more work is required.

D.S. Edwards (1973) described from Rhynie a poorly preserved spherical object, with a structure reminiscent of *Pachytheca*. Lang (1945) also recorded *Pachytheca* from the Rhynie outlier, but did not state if it came from the chert.

Examples of fine, septate filaments with distinctive nodal structures have been described by Kidston and Lang (1921b) and Edwards and Lyon (1983) as *Palaeonitella cranii* (Kidston and Lang). They have been interpreted as probably the vegetative parts of a charophyte, but no reproductive organs have yet been found.

A variety of other green and blue-green algae have been described from here (Kidston and Lang, 1921b; Croft and George, 1959; Edwards and Lyon, 1983), including filamentous, palmelloid and unicellular forms. It is by far the most diverse, non-marine algal assemblage to be described to date from the Devonian, but there remains little information on their reproductive structures and thus their taxonomic positions.

Also abundant in the Rhynie Chert are mycelia, vesicles and resting spores of fungi (Kidston and Lang, 1921b; Harvey et al., 1969; Boullard and Lemoigne, 1971; Taylor et al., 1992a, b). In the absence of detailed information on their reproductive organs, they have been assigned to the generalized form-genus *Palaeomyces*. *P. gordonii* and *P. agglomerata* were found throughout the chert and were undoubtedly saprophytic. Others, however, occur only in the rhizomes of particular vascular plants, such as *P. asteroxylii* in *Asteroxylon* and *P. horneae* in *Horneophyton*. Boullard and Lemoigne argued that these were mycorrhizal and, if this is correct, they represent the oldest known examples of such a symbiosis. It has been suggested that symbiosis between vascular plants and mycorrhizal fungi was essential to the early colonization of the land, allowing the vascular plants to obtain sufficient nutrients from a relatively impoverished environment (Pirozynski and Malloch, 1975; Pirozynski, 1981). Most recently, Taylor et al. (1992a, b) have argued that, at least in some cases (e.g. the putative charaphyte *Palaeonitella*), the fungi were parasitic, resulting in a reaction by the host plant.

None of the vascular plant fossil species from Rhynie has been found anywhere else in the world. Even at the rank of form-genus, the plants are unique. Two other species were included for a time in *Asteroxylon*: *A. elberfeldense* Kräusel and Weyland, 1926, from the Middle Devonian of Germany (see also Scott, 1926); and *A. setchellii* Read and Campbell, 1939, from the Mississippian of the USA. However, the former has been transferred to *Thursophyton* because of its medullated stele and scalariform tracheids (Lyon, 1964; Høeg in Boureau et al., 1967); and the latter is now included in the ?pteridosperm form-genus *Stenokoleos* (Hoskins and Cross, 1951, 1952). A possible correlation between *Lyonophyton* and the adpression form-genus *Sciadophyton* (Figure 4.31B) has been suggested (Remy et al., 1980a), but the limited anatomical detail preserved in the latter makes it difficult to confirm this. A possible correlation between *Rhynia* and *Salopella* has also been suggested (e.g. Edwards and Richardson, 1974; Taylor, 1981) but again the preservation of the latter is a problem.

In addition to the significance of its individual plant fossil taxa, the Rhynie Chert is important as a preserved *in situ* ecosystem (Edwards, 1986), the oldest of its kind discovered to date (Figure 4.32). The palaeoecology of the chert has been discussed by Kidston and Lang (1917b, 1921b), Tasch (1957), Kevan et al. (1975) and Trewin and Rice (1992). Of particular interest is the evidence that it provides of the triangular interaction between the 'vascular' plants, the fungi and an arthropod fauna. The relationship between the 'vascular' plants and the fungi has already been mentioned, but it is also worth noting that the arthropods may have been a significant dispersal vector for the fungi (Kevan et al., 1975). There is considerable evidence, particularly in the aerial

# Devonian

**Figure 4.32** Reconstruction of the Early Devonian vegetation at Rhynie. Based on Trewin (1989, p. 12).

shoots of *Rhynia*, of lesions possibly produced by sap-sucking animals. There is also some evidence that the smaller arachnids occupied empty sporangia and cavities in the aerial shoots of *Rhynia*, which may have represented a more equable micro-environment than conditions outside. There is considerable potential for further elucidating the structure of this early terrestrial ecosystem, particularly if observations on the *in situ* chert become possible.

## Conclusion

Rhynie is the most important single Devonian palaeobotanical site, and one of the most important of any age in the world. It yields the oldest examples of land plants, *c.* 390 million years old, in which anatomical details are still largely preserved intact. It provided the first indisputable evidence that Devonian plants were truly primitive, and not just badly preserved remains of more advanced groups seen in younger rocks. Its discovery just before the First World War catalysed the search for other Devonian (and subsequently Silurian) plant fossils throughout the world, and has resulted in the model for the early evolution of land plants that is accepted today (summarized in Chapter 3). Practically all that is known about the early evolution of land plants is a direct consequence of the discoveries at Rhynie, and this in turn provided a focus for the rest of palaeobotanical research. The 22 species of plant fossil (including fungi) are unique to this locality, and include what is widely regarded as the archetypal early land plant (*Rhynia*), the earliest club-moss for which a detailed anatomy has been described (*Asteroxylon*), and a variety of enigmatic species evidently representing early experiments in adapting to a terrestrial environment. Rhynie is also important as the oldest known example of an entire terrestrial ecosystem preserved in place, showing how plants and animals interacted with each other and with the abiotic environment at this early stage in the development of life on land.

# Bay of Skaill

## BAY OF SKAILL

### Highlights

The Bay of Skaill has yielded the most diverse *Hyenia* Zone plant fossil assemblage in Britain. It is the type locality for *Protopteridium thomsonii* (Dawson), the oldest and most primitive known progymnosperm (Figure 4.33).

### Introduction

This exposure of the Sandwick Fish Bed, on the west coast of Mainland Island, Orkney (HY 233196) has yielded some of the best examples of Middle Devonian plant fossils from Britain. Although heavily out-numbered by the fish fossils, plant remains have been recorded from this horizon in Orkney for nearly 150 years (Clouston, 1845; Miller, 1849; Carruthers, 1873; Dawson, 1870; 1871; 1878; Lang, 1925, 1926, 1927a, b; Lang *in* Wilson *et al.*, 1935).

### Description

#### Stratigraphy

The geology has been briefly described by Wilson *et al.* (1935). The fish bed is *c.* 3 metres thick here and consists of thinly laminated, 'varved' sandstones and siltstones. It is believed that it is part of an extensive lacustrine deposit originally extending over large areas of north-eastern Scotland, including Cromarty (the Cromarty Fish Bed), Caithness (the Achanarras Fish Bed) and Shetland (Melby Fish Bed). The 'varved' structure of the bed probably reflects an annual cyclicity, of either algal blooms (Rayner, 1963) or climatically induced variations in sediment input (Trewin, 1985). The chronostratigraphical position of the

**Figure 4.33** Bay of Skaill. Beds associated with the Sandwick Fish Bed, at the Eifelian–Givetian boundary. (Photo: C.J. Cleal.)

bed appears to be near the Eifelian–Givetian boundary based on fish and spores (Westoll, 1951; Westoll in House *et al.*, 1977; Richardson, 1964).

### Palaeobotany

The plant fossils are preserved mainly as compressions. To date, the following species have been reported:

Lycopsida:
*Thursophyton milleri* (Salter) Nathorst

Progymnospermopsida:
*Protopteridium thomsonii* (Dawson) Kräusel and Weyland

Uncertain affinities:
*Barrandeina pectinata* Høeg
'Fern' *sensu* Miller (1849)

## Interpretation

The most significant element in the assemblage is *Protopteridium thomsonii*, for which Bay of Skaill is the type locality (Dawson, 1878). The nomenclature of this species has undergone a number of changes but, according to Matten and Schweitzer (1982), *P. thomsonii* is the correct combination. It has been most extensively investigated by Leclercq and Bonamo (1971), who have shown that it has a combination of trimerophyte-like sporangial trusses and gymnosperm-like secondary wood (Figure 4.34). Details of the vegetative parts of the plant are not well shown in the specimens found to date from Bay of Skaill, but material from elsewhere suggests that they consist of helically arranged, dichotomous branches. The ultimate sterile appendages may be bi- or trifurcate and often show an incipient, narrow lamina. The combination of trimerophyte- and gymnosperm-like features has resulted in *Protopteridium* being assigned to the progymnosperms, usually to the order Aneurophytales. It is the oldest and most primitive known progymnosperm, and provides valuable evidence as to how they evolved from the trimerophytes, probably in the Middle Devonian.

According to Leclercq and Bonamo (1971), *Milleria pinnata* (Lang) from the Cromarty Fish Bed (Lang, 1925, 1926) should be included in *Protopteridium thomsonii*. It has also been suggested by Kidston (1903a and *in* Hinxman and Grant Wilson, 1902) that *Caulopteris? peachii* Salter, described by Salter (*in* Murchison, 1859) from the Achanarras Fish Bed, was the trunk of this plant. Consequently, *P. thomsonii* has often been reconstructed as a semi-arborescent plant (e.g. Seward, 1931, fig. 45). As pointed out by Leclercq and Bonamo (1971), however, the connection has never been proved.

*Thursophyton milleri* refers to branching axes with microphyllous leaves or spines, but without leaf cushions (Salter, 1858; Lang, 1925). Specimens figured by Penhallow (1892) and Reid and Macnair (1896, 1899) show what appear to be sporangia borne in the axils of the leaves, but Nathorst (1915) and Lang (1925) regarded the evidence as doubtful. Little is known of the anatomy of the axes, other than that they contained annular tracheids (Lang, 1925). The general aspect of the plant suggests affinities with the lycopsids, but further evidence of its anatomy and fertile structures is needed to confirm its taxonomic position.

Lang (*in* Wilson *et al.*, 1935) reported *Barrandeina pectinata* from Bay of Skaill, but the Lang specimens have never been figured. This species belongs to an enigmatic group of Devonian plants with apparently fan-shaped leaves (also including *Enigmophyton* Høeg, 1942 from the Middle Devonian of Spitsbergen), which Høeg (*in* Boureau *et al.*, 1967) has referred to the order Palaeophyllales. The taxonomic position of these Devonian megaphyllous plants is not known.

The 'fern?' figured by Miller (1849) and refigured by Lang (1925, pl. 4, fig. 66) is an extremely faint impression of what appears to be a small, pinna-like structure. Kidston (*in* Lang, 1925) reported markings on its surface suggesting the presence of spores, and thus it may be some sort of fructification. However, nothing more is known about it.

In addition to the above taxa, *Hostinella racemosa* Lang, *H. globosa* Lang, *Protolepidodendron karlsteinii* Potonié and Bernard, and *Pseudosporochnus krejcii* Potonié and Bernard have been described from other localities in north-east Scotland at this horizon (Lang, 1925, 1926, 1927a). Although not yet reported from Bay of Skaill, further work there may well reveal them.

Although of rather restricted composition, the Bay of Skaill assemblage appears to belong to the *Hyenia* Zone of Banks (1980). Similar assemblages have been reported from other exposures of this fish bed and its correlatives in north-east Scotland,

# Bay of Skaill

**Figure 4.34** *Protopteridium thomsonii* (Dawson) Kräusel and Weyland. Fertile spike of the oldest known progymnosperm; Natural History Museum, London, specimen V.9425. Sandwick Fish Bed (Eifelian–Givetian boundary), Bay of Skaill. × 2. (Photo: Photographic Studio, Natural History Museum, London.)

including Lyking Quarry in Orkney, Achanarras Quarry in Caithness, and further south at Coal Heugh and Navity in the Cromarty Black Isle (Peach, 1877; Kidston and Lang, 1923a; Lang, 1925, 1926, 1927a). However, many of these localities have been filled-in and, of those remaining, Bay of Skaill yields the most diverse plant fossils of this age.

Coeval assemblages are also known from Germany (Kräusel and Weyland, 1929, 1932, 1938), the former Czechoslovakia (Obrhel, 1968) and Spitsbergen (Høeg, 1942). These are mostly more diverse than the Scottish assemblages, and include a number of taxa not yet reported from Scotland, such as *Aneurophyton*, *Hyenia*, *Pectinophyton* and *Duisbergia*. Bay of Skaill is nevertheless of considerable international significance as the type and one of the most important localities for *Protopteridium*, the earliest progymnosperm, and thus probably the remote ancestor of all seed plants, including the angiosperms.

## Conclusion

Bay of Skaill has yielded an important assemblage of Middle Devonian plant fossils, about 380 million years old. It cannot compare in diversity with similar aged floras abroad, especially Germany, the former Czechoslovakia and Spitsbergen, but it is the best that is known in Britain. It is particularly important as the best locality for the oldest and most primitive known progymnosperm (*Protopteridium*), which is one of a group of plants thought to be the immediate ancestors of the seed plants (and thus the flowering plants). It is thus of great significance for charting the development of the seed as a reproductive organ, which was probably the single most important evolutionary event that allowed plants to spread from the lowland, coastal areas into drier, inland habitats.

## SLOAGAR

### Highlights

Sloagar provides the best example of a *Svalbardia* Zone flora in Britain. It is also the type locality for *Svalbardia scotica* Chaloner, the only known species of this form-genus known from Britain (Figure 4.35).

### Introduction

Plant fossils are abundant throughout the Old Red Sandstone of Shetland but have been studied little by palaeobotanists. Stems known as 'corduroy plant' occur extensively and have been noted by

**Figure 4.35** Sloagar. Steeply-dipping Givetian lacustrine beds of the North Gavel Formation. Plant fossils occur in the shales in the middle of the picture. (Photo: C.J. Cleal.)

Hooker (*in* Tufnell, 1853), Miller (1857), Murchison (1859), Gibson (1877) and Peach (1877), and *Thursophyton* and *Hostinella* axes have been recorded by Mykura (1972) and Mykura and Pheimister (1976). In none of these cases, however, were the specimens described. During the early twentieth century, W.H. Lang made an extensive collection of Shetland plant fossils (now stored in the Natural History Museum) but the results of his work were never published. The only sites to have been subject to any detailed work are those yielding *Svalbardia* (Chaloner, 1972; Allen and Marshall, 1986), the most prolific of which is this one high in the Middle Devonian on the coast of Fair Isle, Shetland Islands (HZ 228726; see Figure 4.36).

## Description

### Stratigraphy

The sequence at Sloagar has been described by Mykura (1972), Marshall and Allen (1982) and Allen and Marshall (1986). It consists of alternating lacustrine shales and sandstones of the North Gavel Formation (Figure 4.37). Palynological evidence indicates a late Givetian age (Marshall and Allen, 1982), which is also supported by the plant fossils, which belong to the *Svalbardia* Zone of Banks (1980).

### Palaeobotany

The shales here have yielded heavily carbonized compressions of just one species, *Svalbardia scotica* Chaloner. The sandstones contain casts and adpressions of wide axes known as 'corduroy plant', which may have been the stems of the *Svalbardia* plant.

## Interpretation

Other than this locality, *Svalbardia scotica* has only otherwise been recorded from Voe of Cullingsburgh and Leebotten on Shetland (Allen and Marshall, 1986). It consists of 20 mm wide, ribbed axes with ?spirally attached, subsidiary axes. Mostly wedge-shaped, digitate leaves are attached probably spirally to both orders of axes. Associated with the vegetative shoots are fertile axes bearing sporophylls, again probably in a spiral arrangement. The sporophylls bear 8–12 fusiform sporangia on their adaxial surface.

These fructifications associated with *Svalbardia* are very similar to those of the Upper Devonian *Archaeopteris*, and the two form-genera were traditionally distinguished only by the more deeply digitate leaves of the former. Scheckler (1978) argued that *Svalbardia* was merely a developmental stage of *Archaeopteris*, but Matten (1981) has maintained that it is a morphological and evolutionary precursor of the latter. Marshall and Allen (1986) concluded that the evidence for separating the form-genera was doubtful, although they provisionally retained the name *Svalbardia* for the Shetland material until a more thorough analysis had been made of all the species. Whatever the taxonomic results of such an analysis, the *Svalbardia*-type species will hold an important evolutionary position between the more primitive progymnosperms which did not have dorsiventral, laminate leaves (e.g. *Protopteridium* from Bay of Skaill) and the more typical *Archaeopteris* species with flattened, fully laminate leaves.

As to the specimens recorded from Sloagar by Mykura (1972), ?aff. *Zosterophyllum* and *Hostinella* sp. are probably conspecific with *S. scotica* Chaloner.

The early records of *Calamites* from Shetland have subsequently been referred to as 'corduroy plant' (Finlay, 1926). They are axes up to 150 mm wide, which occasionally show monopodial branching (Allen and Marshall, 1986). They are characterized by uninterrupted longitudinal ribs (hence, their original assignment to *Calamites*), similar to those found on the *S. scotica* axes, with which they are often associated. Allen and Marshall (1986) concluded that the 'corduroy plant' may well represent the main stem that bore *Svalbardia*, although evidence of attachment has yet to be found. If correct, this would call into question Scheckler's (1978) view that *Svalbardia* was a developmental form of *Archaeopteris*.

Mykura (1972) recorded cf. *Prototaxites* sp. from Sloagar. This would be of considerable interest as the youngest record of this group of enigmatic, non-vascular plants from Britain. However, it is possible that they refer simply to axes of 'corduroy plant'.

Although yielding only one species, Sloagar is of considerable palaeobotanical importance. It has yielded the best specimens for showing the overall morphology of the *Svalbardia* plant, as well as some well-preserved fructifications and foliage. It

## Devonian

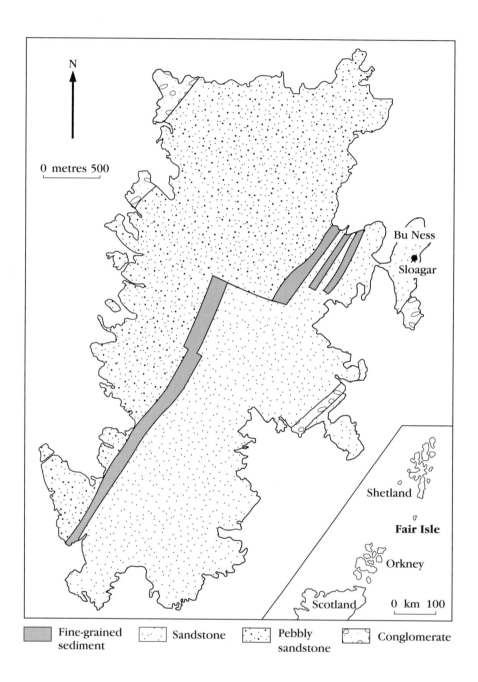

**Figure 4.36** Geological map of Fair Isle, showing position of Givetian palaeobotanical site at Sloagar. Based on Marshall and Allen (1982, text-figure 1).

# Sloagar

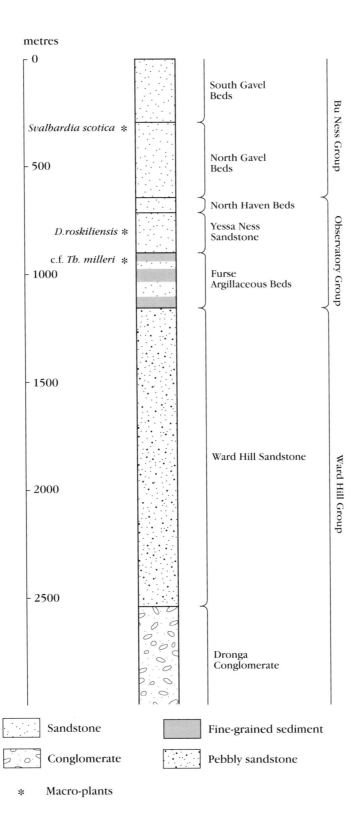

**Figure 4.37** Stratigraphical section of the Middle Devonian of Fair Isle, showing the main plant fossil-bearing horizons. Based on Marshall and Allen (1982, text-figure 2).

is second only to Spitsbergen and New York State for understanding the detailed structures of this phylogenetically important progymnosperm.

## Conclusion

Sloagar has yielded the only species of the progymnosperm *Svalbardia* known from Britain. This plant, which lived about 375–380 Ma, had typically progymnosperm-like sporangia, and flattened, deeply-digitate leaves. This foliage suggests that *Svalbardia* holds an intermediate evolutionary position between the primitive progymnosperm *Protopteridium* which did not have flattened leaves (see Bay of Skaill, above) and the more advanced and 'typical' progymnosperm *Archaeopteris*, which had flattened, more fully laminate leaves. The Sloagar specimens are of great importance for charting the evolutionary history of the progymnosperms, which eventually gave rise to the seed plants and flowering plants.

## PLAISTOW QUARRY

### Highlights

Plaistow Quarry yields the best *Rhacophyton* Zone plant fossils in Britain, and includes the oldest evidence of seed plants in this country.

### Introduction

Plant fossils have been known from the Upper Devonian Baggy Formation of north Devon since the mid-nineteenth century (e.g. Williams, 1838; Hall, 1867; see Arber and Goode, 1915 for a more complete historical account). They have been recorded from a number of localities (see Hall, 1867 for the most complete list) but the most abundant are found at Plaistow Quarry (also known as Sloley Quarry), 4 km north of Barnstaple, Devon (SS 568373). The only detailed account of the assemblage is that given by Arber and Goode (1915), although Fairon-Demaret and Scheckler (1987) have recently commented on one of the species.

### Description

#### Stratigraphy

The Baggy Formation consists of a mixture of shallow marine and deltaic non-marine strata (Goldring, 1971), the plant fossils usually occurring in the latter (Figures 4.38 and 4.39). There has been some disagreement as to the chronostratigraphical position of the formation (Goldring, 1970, 1971; House *et al.*, 1977). In terms of the currently accepted Devonian–Carboniferous boundary (Paproth, 1980), however, it is probably upper Famennian (Fairon-Demaret, 1986a).

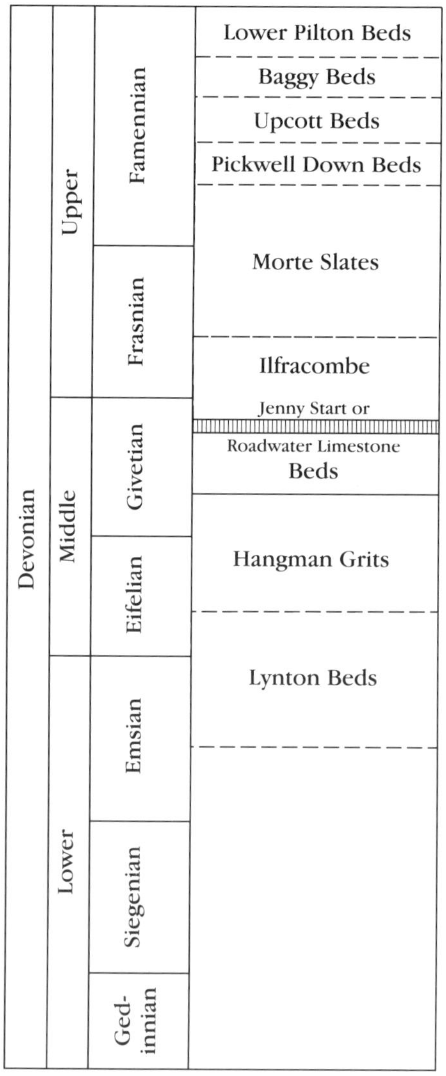

**Figure 4.38** Devonian stratigraphy of North Devon, showing position of the Baggy Beds.

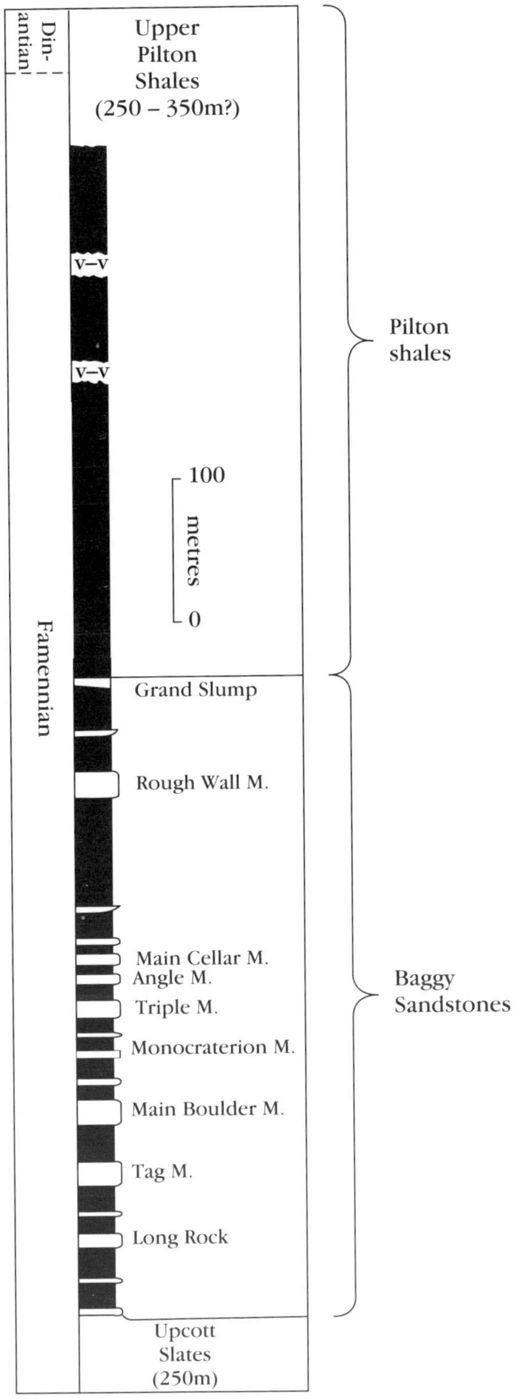

# Plaistow Quarry

### Palaeobotany

The plant fossils are impressions, sometimes iron stained, found in thin and impersistent mudstone lenses. The following species are known:

Lycopsida:
  *Knorria* sp.

Equisetopsida:
  *?Archaeocalamites radiatus* (Brongniart) Stur

Lagenostomopsida(?):
  *Sphenopteridium rigidum* (Ludwig) Potonié

## Interpretation

The most interesting aspect of the Baggy Beds assemblage is the presence of probable gymnosperms, the oldest evidence of such plants known from Britain. At Plaistow Quarry, they are represented by *Sphenopteridium* frond fragments (Figure 4.40). At Croyde Hoe Quarry at Baggy Point (some 15 km to the west), similar foliage was found in association with cupulate structures identified by Arber and Goode as *Xenotheca* (see also Rogers, 1926). These have recently been reinvestigated and found to contain ovules with an open, lobate integument, thus confirming that they were gymnospermous fructifications (Fairon-Demaret and Scheckler, 1987; Rothwell and Scheckler, 1988). They are similar in general morphology to other primitive cupulate structures (e.g. *Moresnetia* of Stockmans, 1948; *Elkinsia* Rothwell *et al.*, 1989) but Fairon-Demaret and Scheckler argued that they are generically distinct. Also found at Croyde Hoe are fragmentary sporangial clusters identified by Arber and Goode as *Telangium* (syn. *Telangiopsis* Eggert and Taylor). These are extremely small specimens, but appear to compare with the lyginopterid sporangial clusters described by Eggert and Taylor (1971).

Also found at Croyde Hoe is pinnate foliage with more swollen lobes, described by Arber and Goode (1915) as *Sphenopteris* sp. There is a comparison with certain species of *Triphyllopteris* (Read and Mamay, 1964) and *Eusphenopteris foliolata* (Stur) (van Amerom, 1975), although neither of these usually range below the Tournaisian. Since Arber and Goode's material is very fragmentary, their original identification as *Sphenopteris* sp. should probably be retained.

Hall (1867) and Etheridge (1867) record *Adiantites hibernicus* Forbes (syn. *Archaeopteris*

**Figure 4.39** Details of the stratigraphy of the Baggy and Pilton formations in North Devon. Based on Scrutton (1978, figure 6).

# Devonian

**Figure 4.40** *Sphenopteridium rigidum* (Ludwig) Potonié. Fragments of early pteridosperm fronds; Natural History Museum, London, specimen V.3562. Baggy Formation (upper Famennian), Plaistow Quarry. x 2. (Photo: Photographic Studio, Natural History Museum, London.)

*hibernica*) from Plaistow Quarry. Arber and Goode (1915) attempted to verify the records, but could find neither the original specimens, nor any new ones from the locality. It may be significant that small fragments of Arber and Goode's *Sphenopteris* sp. bear a passing resemblance to *Archaeopteris*.

Specimens from Plaistow Quarry, identified by Arber and Goode (1915) as *Knorria*, were regarded by them as decorticated axes of arborescent lycopsids. Alternatively, they may be examples of more primitive lycopsids without leaf-scars. The specimens clearly need reappraisal.

Arber and Goode (1915) mention ribbed axes from here. These possibly represent internodes of *Archaeocalamites radiatus*. No evidence of nodes was found to confirm the identification, however, and Arber and Goode suggested that they might be poorly preserved lycopsid axes.

The presence of *Sphenopteridium* and *Xenotheca* in the Baggy Formation indicates the *Rhacophyton* Zone of Banks (1980), and supports a late Famennian age. Although of limited composition, this formation nevertheless has the most diverse *Rhacophyton* Zone assemblage known from the British Isles. The nearest comparison is with fossils found in the Lower Limestone Shale along the Taff Gorge near Cardiff (Evans and Cox, 1956; Gayer *et al.*, 1973), which includes cf. *Xenotheca devonica* Arber and Goode and another type of cupulate fructification, *Telangiopsis* sp., but which has no recorded specimens of megaphyllous foliage. The best documented assemblages of this age are from Kiltorcan in Ireland (Chaloner, 1968) and the lost site of Prestonhaugh in southern Scotland (Miller, 1857; Crookall, 1939; Long, 1973), but these are restricted mainly to *Archaeopteris* and *Cyclostigma*. From outside of the British Isles, the best known assemblages are from Belgium (Stockmans, 1948) and Bear Island near Spitsbergen (Schweitzer, 1967, 1969), which are significantly more diverse. Typically, such assemblages include (in addition to the form-genera mentioned above) *Protolepidodendropsis*, *Sublepidodendron*, *Pseudobornia*, *Sphenophyllum* and *Aneurophyton*.

Amongst the Baggy Formation sites that have been reported to yield plant fossils, Plaistow Quarry now yields the best material. The Croyde Hoe Quarry, from where Arber and Goode (1915) obtained much of their material, including the fertile pteridospermous organs, is no longer available. This site may thus be regarded as of national significance for understanding the vegetation in this country towards the end of the Devonian Period, as it evolves into the more advanced floras found in the Carboniferous.

## Conclusion

Plaistow Quarry is the best site in Britain for Upper Devonian plant fossils, some 360 million years old, which are otherwise very poorly represented in this country. The site is particularly important in providing at least some reflection of the vegetation growing in the tropical belt towards the end of the Devonian. It is transitional between the more primitive Devonian vegetation, dominated by zosterophylls, trimerophytes and progymnosperms, and the more modern-looking Carboniferous vegetation, which is dominated by ferns, club-mosses, horsetails and seed plants. This particular site has yielded the earliest horsetail and seed plant remains known from this country.

# Chapter 5

# *Lower Carboniferous*

# Stratigraphical background

The Early Carboniferous saw few really significant phylogenetic changes in terrestrial plants; rather it was a time of consolidation of the developments that had occurred in the Devonian (Figure 5.1). In particular, there was the diversification of early ferns and fern-like plants (Galtier and Scott, 1985), lycopsids (Thomas, 1978a), equisetopsids (Thomas and Spicer, 1987) and gymnosperms (Rothwell, 1986; Rothwell and Scheckler, 1988). The development of the latter was a major advance in the evolution of terrestrial vegetation, as they were the first plants that were not constrained by their reproductive biology to moist habitats, and could thus take advantage of the extra-basinal or 'upland' areas. The Early Carboniferous also saw the development of the earliest extensive forests, first in equatorial latitudes (Long, 1979a), and then in northern high latitudes (Meyen, 1982); forests did not develop in the southern high latitudes until the end of the Late Carboniferous, probably due to the influence of a polar ice-cap (Retallack, 1980).

There are marked differences in the plant fossils from the Lower and the Upper Carboniferous. This not just at the rank of species or even form-genus: taxa such as the Calamopityales and Archaeocalamitaceae have not been identified with certainty from above the Lower Carboniferous (although see Mamay and Bateman, 1991 for a possible exception); whereas the Trigonocarpales, the Callistophytales and the Cordaitanthales are more or less restricted to the Upper Carboniferous (Cleal, 1993). For this reason, the palaeobotany of the two subsystems of the Carboniferous are treated separately in this volume.

## PALAEOGEOGRAPHICAL SETTING

As in the Devonian, Britain in the Early Carboniferous was on the southern margins of the Laurussian continent and was very close to the equator (Figure 5.2). The Gondwanan continent was progressively drifting north and eventually, in the Late Carboniferous, collided with Laurussia to form the Pangaea 'super-continent', which extended from the south pole to high northern latitudes; but, in the Early Carboniferous, the two landmasses were still separate.

It was once thought that Lower Carboniferous plant fossils indicated a globally uniform *Lepidodendropsis-Rhacopteris-Triphyllopteris* flora for that time (Jongmans, 1952). Subsequent work by Raymond (1985) and Raymond and Parrish (1985) has revealed some evidence of provincialism. The extent of floristic differentiation was exaggerated in these studies, because most of the southern latitude (Gondwanan) assemblages used by Raymond and her co-workers are in fact from the Upper Carboniferous (Wagner *et al.*, 1985). It is nevertheless possible to recognize two discrete palaeokingdoms: the Angaran Palaeokingdom, representing northern high latitude vegetation, and the Euramerian Palaeokingdom for equatorial and southern high latitude vegetation (Figure 5.2). The British assemblages belong to the latter.

Much of Britain at this time was covered by shallow seas, which resulted in extensive carbonate deposits that contain few well-preserved terrestrial plant fossils. There were also some areas of land, however, such as the St George's island extending over much of Wales and the English Midlands, and the Caledonian landmass in northern Scotland. Deltaic, lagoonal and shallow marine deposits on the margins of these land areas have yielded a variety of plant fossil assemblages. The Scottish deposits are particularly significant, as they were influenced by contemporaneous volcanic activity, resulting in the extensive suite of petrifaction sites in this region.

## STRATIGRAPHICAL BACKGROUND

Kidston (1894a) and Gothan (1913, 1952) argued that there is a sharp Floral Break or 'Florensprung' between the Lower and Upper Carboniferous. Subsequently, there have been disagreements as to exactly how sharp the change really is and at what stratigraphical level it occurs (see, for instance, papers by Havlena, Wagner, and Pfefferkorn and Gillespie *in* Ramsbottom *et al.*, 1982). That there is at least a broad distinction between Lower and Upper Carboniferous plant fossil assemblages is, however, undeniable and provides a convenient division for the discussion of the palaeobotany of this period. In this volume the boundary has been drawn at the Arnsbergian-Chokierian ($E_2$-$H_1$) stage boundary. It thus corresponds approximately with the mid-Carboniferous boundary currently being investigated by the IUGS Subcommission on Carboniferous Stratigraphy (Lane *et al.*, 1985), rather than with the European Dinantian–Silesian subsystem boundary, which cannot be correlated with any significant palaeobotanical change.

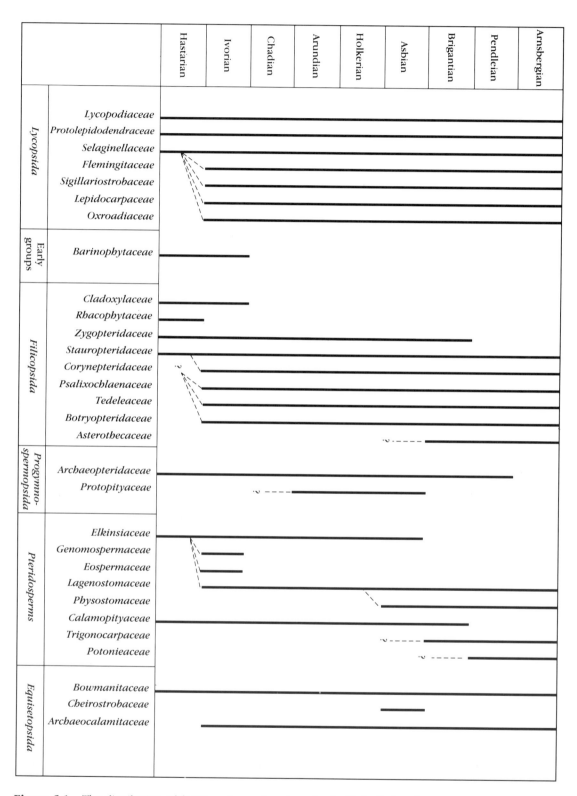

**Figure 5.1** The distribution of families of vascular plants in the Early Carboniferous. Based on data from Cleal (1993).

# Early Carboniferous vegetation

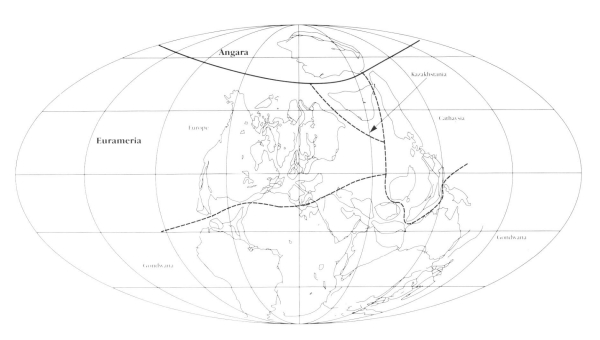

**Figure 5.2** The palaeogeography of the Early Carboniferous, showing the distribution of the major floristic zones (phytochoria). Based on Scotese and McKerrow (1990) and Cleal and Thomas *in* Cleal (1991).

| Chronostratigraphy | | Plant Fossil Biostratigraphy | GCR Adpression sites | Other significant Adpression sites (Palaeoequatorial Belt) |
|---|---|---|---|---|
| **Series** | **Stages** | | | |
| Namurian | Arnsbergian | *Lyginopteris larischii* | | South Wales[6] Belgium Upper Silesia[5] Zonguldak[7] |
| | Pendleian | *Lyginopteris bermudensiformis/ Lyginopteris stangeri* | | Upper Limestone Group[6] Belgium Upper Silesia[5] Zonguldak[7] |
| Visean | Brigantian | *Lyginopteris bermudensiformis/ Neuropteris antecedens* | Puddlebrook Quarry Moel Hirradug Teilia Quarry Wardie Shore Glencartholm Loch Humphrey Burn | Horton Group[4] Upper Silesia[5] |
| | Arundian | | | |
| | Chadian | | | |
| Tournaisian | Ivorian | *Triphyllopteris* | | Pocono Formation[1] Geigen near Hof[2] Doberlug-Kirchhain[2] Valdeinfierno[3] Horton Group[4] |
| | Hastarian | "*Adiantites*" | | Pocono Formation[1] Geigen near Hof[2] |

[1] USA
[2] Germany
[3] Spain
[4] Nova Scotia, Canada
[5] Poland and Moravia
[6] UK
[7] Turkey

**Figure 5.3** Chronostratigraphical and biostratigraphical classification of the Lower Carboniferous, and the positions of the GCR and other major palaeobotanical sites in this subsystem (adpressions).

| Chronostratigraphy | | GCR Petrifaction Sites | Other Petrifaction Sites |
|---|---|---|---|
| Series | Stages | | |
| Namurian | Arnsbergian — Pendleian | Victoria Park | |
| Visean | Brigantian — Asbian | Weak Law, Pettycur, Kingswood End, Laggan | Glätzisch-Falkenberg[2], Roannais[1], Esnost[1] |
| Visean | Holkerian — Chadian | Kingwater, Loch Humphrey Burn, Glenarbuck | |
| Tournaisian | Ivorian | Lennel Braes, Whiteadder, Oxroad Bay, Loch Humphrey Burn | |
| Tournaisian | Hastarian | | New Albany Shale[3], Saalfeld[2], Montagne Noire[1] |

[1] France
[2] Germany
[3] USA

**Figure 5.4** Chronostratigraphical classification of the Lower Carboniferous, and the positions of the GCR and other major palaeobotanical sites in this subsystem (petrifactions).

The series and stages referred to in this work (see Figures 5.3 and 5.4) are based on the European chronostratigraphy, and are discussed in detail by George *et al.* (1976) and Ramsbottom *et al.* (1978). These references also provide details of the lithostratigraphy of the British Lower Carboniferous.

Figure 5.3 shows the relationship between this chronostratigraphy and the plant fossil biostratigraphy established by Wagner (1984) for use throughout the equatorial belt (see also Cleal, 1991). Unlike the biostratigraphy used in the Devonian (see Chapter 4), Wagner's scheme is based exclusively on assemblage biozones, established on the ranges of the plant taxa. It nevertheless provides a useful framework for the discussion of changes in plant fossil assemblages through time.

# EARLY CARBONIFEROUS VEGETATION

There is little evidence of an abrupt change in terrestrial vegetation between the Late Devonian and Early Carboniferous; the fossil record of most of the orders of terrestrial plants that lived in the Early Carboniferous extend back at least to the Upper Devonian (Figures 4.1 and 5.1). There appears, however, to have been a steady increase in abundance of the vegetation, particularly in the equatorial latitudes. The reason for this is not clear. At the beginning of the Carboniferous, there was a rise in sea-level, which flooded many of the extensive deltas that had developed during the Devonian. This would have reduced the area of swampy lowlands, which were probably still the habitat occupied by most terrestrial vegetation. However, the increased area of sea could also have helped increase precipitation, which would favour more lush vegetation. Another factor may simply

**Figure 5.5** Reconstruction of an Early Carboniferous herbaceous lycopsid, *Oxroadia gracilis*. Based on Bateman *et al.* (1992, figure 2D).

have been that vascular plants had become better adapted to a wider range of terrestrial habitats.

Lycopsids were probably the most abundant land plants in the Early Carboniferous (Thomas, 1978a). They are represented in the fossil record by a variety of morphologies, including arborescent (Flemingitaceae, Sigillariostrobaceae) and herbaceous (Lycopodiaceae, *Oxroadia* – Figure 5.5) forms. The Lepidocarpaceae also show the development of a 'seed-like' reproductive structure, in which each female sporangium contained just a single functional megaspore (Thomas, 1981b). The arborescent forms, in particular, were restricted to swampy habitats, due to limitations in the efficiency of their rooting structures. Within these habitats, however, they appear to have formed extensive forests, although they rarely developed thick peat deposits, as in the Upper Carboniferous coal-forming environments. For taphonomic reasons, lycopsid remains are not particularly abundant in the British Lower Carboniferous record, although there are exceptions such as at Glenarbuck, near Glasgow in Scotland, where there is also, at Victoria Park, the best preserved example of an *in situ* lycopsid fossil forest floor.

The progymnosperms also remained important in the Early Carboniferous, and had a fairly uniform global distribution (Beck, 1976; Stewart, 1981). In contrast to the lycopsids, they seem to have favoured somewhat drier habitats, which they shared with the fern-like plants and pteridospermous groups (see below). Arborescent forms were common, such as the Archaeopteridaceae and Protopityaceae, the latter being best known from Britain. However, one of the commonest types of fossil foliage found in the British Lower Carboniferous (*Rhacopteris*) may well represent small progymnosperms.

The abundance of *Archaeocalamites* stems and pith casts in Lower Carboniferous rocks suggests that the equisetopsids were common at that time (Crookall, 1969). They were not a particularly diverse group, however, consisting almost exclusively of the Archaeocalamitaceae; the Bowmanitaceae (e.g. *Sphenophyllum*) and Calamostachyaceae (e.g. *Calamites*) are found in the Lower Carboniferous, but only rarely. The archaeocalamitids appear to have been shrubby plants, probably forming thickets around areas of standing water (this may partly explain their abundance in the fossil record). There have been few studies on their fructifications, and as a consequence the position of the family within the Equisetopsida is still a matter of debate, but most of what is known has been determined from British fossil material, both adpression and petrifaction.

The so-called 'pre-ferns' reached their acme in the Early Carboniferous. Their fossil remains are best known from the equatorial latitudes, particularly of Britain and France (Galtier and Scott, 1985; Scott and Galtier, 1985). However, they can only be reliably distinguished from true ferns and some pteridosperms if evidence of the stem and fructification anatomy is available. Since virtually no Lower Carboniferous petrifactions have been reported from Gondwana or Angara, their presence there cannot be entirely ruled out. They seem to have been exclusively herbaceous or possibly shrubby plants. Their foliage approached the form of discrete fronds, but did not develop a fully laminate form as in true ferns.

The earliest occurring order of true ferns, the Botryopteridales, make their first rare appearances in the Early Carboniferous, although did not become important elements of terrestrial vegetation until the Late Carboniferous (Galtier and Scott, 1985; Scott and Galtier, 1985). Most of these early ferns were small, rambling plants. Although their fronds were small, they had started to develop a clearly laminate form. Towards the top of the Lower Carboniferous, the first evidence of arborescent marattialean ferns is found (Pfefferkorn, 1976, although see also Meyer-Berthaud and Galtier, 1986b). These are the oldest known ferns to have developed the arborescent habit, which they achieved by adopting a polystelic strategy to build thick trunks, rather than by the development of thick zones of secondary wood, as in the progymnosperms and gymnosperms. The marattialeans are also the oldest known order of ferns still living today (Cleal, 1993).

# Lower Carboniferous

Although seed plants first appear in the Late Devonian (Chapter 4), it was not until the Early Carboniferous that they underwent a significant phylogenetic radiation and became important components of terrestrial vegetation (Niklas *et al.*, 1980; Rothwell and Scheckler, 1988). The plant fossils from the Scottish Cementstone Group have been particularly important in demonstrating this radiation, and have shown that a number of families had already appeared by the late Tournaisian. The Early Carboniferous gymnosperms appear to have been mostly of the type traditionally known as pteridosperms (Figures 5.6, 5.7 and 5.8). The concept of pteridosperms is now thought to be polyphyletic (e.g. Crane, 1985; Meyen, 1987), but it nevertheless provides a convenient descriptive term for the early gymnosperms with large, fern-like fronds. In the Lower Carboniferous, two orders have been identified, the Lagenostomales and Calamopityales, both of which included arborescent and herbaceous plants (see reconstructions by Retallack and Dilcher, 1988, based mainly on British material).

## LOWER CARBONIFEROUS PLANT FOSSILS IN BRITAIN

Plant fragments occur sporadically throughout the Lower Carboniferous of Britain, but identifiable

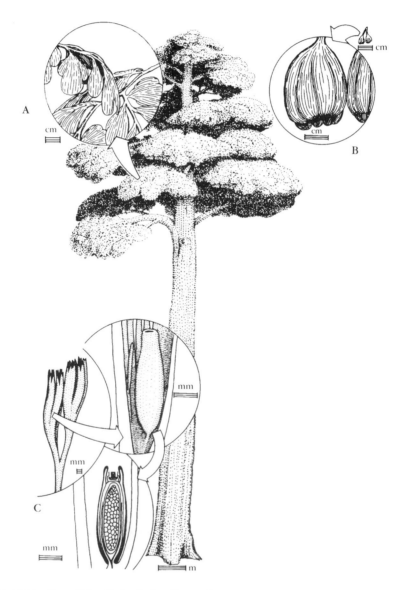

**Figure 5.6** Reconstruction of the Early Carboniferous lagenostomalean pteridosperm tree *Stamnostoma*, with insets showing details of foliage (A), pollen organs (B) and seeds (C). Based on Retallack and Dilcher (1988).

**Figure 5.7** Reconstruction of the Early Carboniferous lagenostomalean pteridosperm *Diplopteridium*. Based on Rowe (1988b, figure 35).

assemblages are mainly restricted to sites in Scotland, Wales and the Welsh Borders. These sites fall into two broad categories: those that yield petrifactions sometimes in conjunction with adpressions, and those that yield adpressions alone.

Petrifaction sites can themselves be divided into two main types: those in fluviolacustrine and those in volcanogenic facies (Scott *et al.*, 1984; Scott and Rex, 1987). The former are best represented in Britain by the Cementstone Group of Scotland, which has yielded some of the best known Tournaisian fructifications from anywhere in the world. The volcanogenic sites are also mainly restricted to Scotland (Figure 5.9). By the nature of their genesis, such deposits tend to be of restricted extent. Nevertheless, they include some of the most important Lower Carboniferous palaeobotanical sites in the world, such as the Pettycur Limestone in Fife, the Oxroad Bay Tuff in the Lothians, and the Clyde Plateau Volcanic Formation in Strathclyde.

Very few adpressions have been described from the Tournaisian of Britain. Only one lower Tournaisian site has been recorded, in the Avon Gorge near Bristol (Utting and Neves, 1970), but the material from here has yet to be described in detail. In the Upper Tournaisian, adpressions are known sporadically from the Scottish Cementstone Group, but again there are few published descriptions, other than from Foulden (Scott and Meyer-Berthaud, 1985).

Visean adpressions occur more widely in Britain. In Scotland, some of the best material originated from the Burdiehouse Limestone in the middle Oil Shale Group, but there are no longer any exposures of this horizon that yield plant fossils. However, adpressions can still be found in other parts of the Oil Shale Group, such as the Wardie Shales (Scott, 1985). The Visean strata in the Borders Region of Scotland have also yielded adpressions. On the whole, assemblages from the Borders Region tend to be of restricted composition, although there are exceptions, such as in the Glencartholm Volcanic Group.

Elsewhere in Britain, Visean plant fossils are best known from North Wales, in particular from the Lower Brown Limestone and Upper Black Limestone groups. There is also the well-known Drybrook Sandstone 'flora' from the Forest of Dean.

Basal Namurian (i.e. Pendleian and Arnsbergian) adpressions are known from southern Scotland (e.g. Walton *et al.*, 1938), Avon (Moore, 1941) and South Wales (Dix, 1934). These assemblages are on the whole restricted in composition, and full descriptions of the material have yet to be published.

## LENNEL BRAES

### Highlights

Lennel Braes is of considerable historical interest as being the origin of some of the earliest described thin sections of fossil wood. The techniques developed here were of fundamental significance to the development of palaeobotany and petrology.

### Introduction

This locality, in the bed of the River Tweed, near Coldstream, Borders Region, Scotland (NT 855410), is one of many Cementstone Group sites in the Scottish borders to yield plant petrifactions.

## Lower Carboniferous

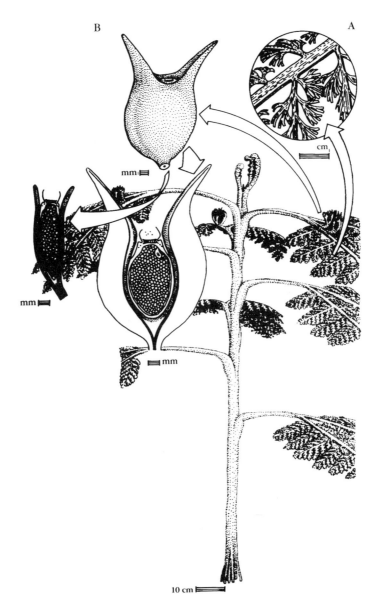

**Figure 5.8** Reconstruction of the Early Carboniferous calamopityalean pteridosperm *Lyrasperma*, with insets showing details of foliage (A) and seeds (B). Based on Retallack and Dilcher (1988).

It is of particular historical interest, however, as the origin of some of the specimens used in Witham's pioneering work on fossil wood thin sections (Witham, 1831, 1833; see also Scott, 1899). Subsequently, adpressions were recorded from here by Kidston (1901a, 1923c, 1924), and further petrifactions by Long (1963, 1964a, 1975).

## Description

### Stratigraphy

The only account of the stratigraphy is that given by Witham (1831). It is unclear where precisely in the Cementstone Group the concretion-bearing shales lie. Not all the 14 metres of rock described by Witham can still be seen, but the shales at the base of the section are still exposed and are

# Lennel Braes

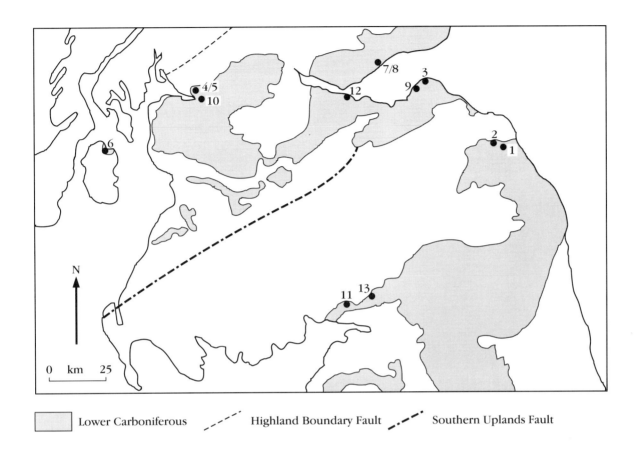

**Figure 5.9** Distribution of Lower Carboniferous rocks in Scotland, showing location of GCR palaeobotany sites. 1 - Lennel Braes; 2 - Whiteadder; 3 - Weak Law; 4 - Loch Humphrey Burn; 5 - Glenarbuck; 6 - Laggan; 7 - Pettycur; 8 - Kingswood End; 9 - Oxroad Bay; 10 - Victoria Park; 11 - Glencartholm; 12 - Wardie Shore; 13 - Kingwater.

where Long (1963) reported plant fossils. Stratigraphically diagnostic faunas have not been reported from here and so its exact age is uncertain. However, evidence from other localities suggest that the Cementstone Group is mainly late Tournaisian (Scott *et al.*, 1984). The environment represented was probably lacustrine.

### Palaeobotany

Only a limited assemblage has been reported from here to date, including the following petrifactions, all of which are pteridosperm (probably Lagenostomopsida and Calamopityales) remains.

*Pitus antiqua* Witham
*Lyginopteris papilio* Kidston
*Stenomyelon tuediana* Kidston and
    Gwynne-Vaughan
*Dolichosperma pentagonum* Long
*Eurystoma burnense* Long

In addition, the following adpressions have been found:

Lycopsida:
*Lepidodendron* sp.
*Stigmaria ficoides* (Sternberg) Brongniart

Lagenostomopsida:
*Diplothmema patentissima* (Ettingshausen) Stur

Calamopityales (*incertae sedis*):
*Alcicornopteris convoluta* Kidston

## Interpretation

The preparation of thin sections of fossil wood was first developed in the 1820s by George Sanderson and an early example of its application was by Sprengel (1828). However, it was not until the method was refined by William Nicol, to enable the sections to be mounted on glass slides (Nicol, 1834), that its full potential could be realized. One of the most widely read studies on sections of fossil wood was by Nicol's friend Henry Witham (1831, 1833), who included specimens from Lennel Braes. Although he did not himself invent the technique, Witham was the first person to popularize it and bring it to the attention of the wider palaeontological community (Scott, 1911; Gordon, 1935b; Long, 1959a and Andrews, 1980 for an historical account). It had important consequences for the future of palaeobotany since, without the use of thin sections, the discoveries in the Rhynie Chert (see Chapter 4) and the Pennsylvanian coal-balls would have been impossible. It was also of major importance in petrology, since the experience obtained from sectioning such fossils was later applied to making rock sections (Sorby, 1858). As pointed out by Gordon (1935b), however, a more regrettable side-effect was that it tended to cause palaeobotany to be divided into biological and geological camps, a problem which still hinders the development of the subject today.

Witham described sections from a number of other Carboniferous and Jurassic localities, as well as Lennel Braes. The latter is the only one for which he gives a detailed account of the geology, however, and the only one from which material can still be collected.

Witham concluded that the fossil wood from Lennel Braes was coniferous, which appears to explain his use of the generic name *Pitus* (Witham, 1833). However, this has since been recognized to be incorrect, and at various times it has been assigned to the cordaites (Scott, 1902) and progymnosperms (Beck, 1960); but the currently available evidence now points to it being from lagenostomalean pteridosperm trunks (Long, 1963, 1979a).

## Conclusion

Lennel Braes is of considerable historical interest as being the origin of some of the earliest described thin sections of fossil wood. The techniques of making thin sections developed here were of fundamental significance to the development of palaeobotany and petrology. Stratigraphically the locality is the same age as the Whiteadder limestones (see below) with an age of some 350 Ma.

## WHITEADDER

### Highlights

The Whiteadder has yielded the most diverse Lower Carboniferous plant petrifaction assemblage in Britain (Figure 5.10). Particularly important is the wide range of early gymnosperm ovules showing details of cell structure, including the morphologically most simple ovule so far found in the fossil record (*Genomosperma*).

### Introduction

The Whiteadder River is a tributary of the Tweed, which it joins about 3 km south-west of Berwick-upon-Tweed, Borders Region (Figure 5.11). Finely preserved plant petrifactions have been reported from the limestones of the Tournaisian Cementstone Group exposed between Hutton Bridge (NT 921546) and Cumledge (NT 792564), and particularly from Edrom (NT 821560). Fossils were occasionally recorded from here during the nineteenth century (e.g. Witham, 1830), but it was not until the site was investigated during the early years of the twentieth century by A. Macconochie for the Geological Survey that its real palaeobotanical potential started to be realized. Initially, only adpressions were found, although petrifactions were known from homotaxial strata in the Tweed valley and at Langton Burn (Kidston, 1901a), but petrifactions were soon also found at the Whiteadder (Kidston, 1902a). Kidston clearly recognized the importance of the Berwickshire

# *Whiteadder*

**Figure 5.10** Edrom, Whiteadder River. Shales and cemented mudstones of the upper Tournaisian Cementstone Group. The main plant bed is in the bed of the river. (Photo: C.J. Cleal.)

petrifactions and had numerous thin sections prepared. One paper, intended to be the first of a series, was written in 1912 with Gwynne-Vaughan. However, the project was prematurely terminated, partly because of the death of the latter in 1915, and partly because of the discovery of plant fossils at Rhynie (see Chapter 4), which diverted Kidston's attention. Some of Kidston's Whiteadder slides were described by Solms-Laubach (1910), Gordon (1912) and Calder (1934, 1938), and there was also some interest in the adpressions (Chaloner, 1953). It was not until Long started to re-investigate the fossils from here in 1957, that the significance of the assemblage became evident. In a series of papers published since 1959, Long has described 40 form-species, many of which were new to science. Long's work, which for the first ten years he effectively did as an amateur (Long, 1976c; Andrews, 1980), ranks as one of the most significant contributions to British palaeobotany this century.

## Description

### Stratigraphy

The strata exposed along this stretch of the Whiteadder belong to the Cementstone Group, and comprise mainly alternating shales and harder bands of calcareous rocks, known as cementstones. The majority of the cementstones are calcareous mudstones, probably of diagenetic origin. However, they also include lacustrine, shelly limestones, and it is these that contain the plant petrifactions (Scott *et al.*, 1984; Scott and Rex, 1987). A measured stratigraphical log for the section has never been published, although general descriptions are given by Scott *et al.* (1984). Miospores obtained from both the shales and the lacustrine limestones at Edrom belong to the CM Biozone, thus indicating a late Tournaisian age. The plant adpressions (see next section) appear to belong to the upper *Triphyllopteris* Biozone, which according to Wagner (1984) indicates an

## Lower Carboniferous

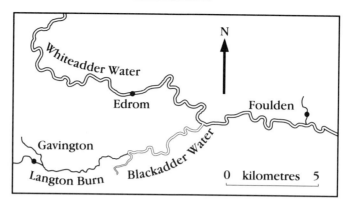

**Figure 5.11** Location map for the Whiteadder GCR palaeobotany site. Based on Scott *et al.* (1984, figure 14).

early Visean age. However, few of the adpressions have been studied in recent years and a revision is necessary to verify their biostratigraphical positions.

### Palaeobotany

The following taxa have been described from here as petrifactions:

Lycopsida:
  *Lepidodendron calamopsoides* Long
  *Paralycopodites brevifolius* (Williamson) DiMichele
  *Stigmaria* sp.
  *Oxroadia gracilis* Alvin
  *Mazocarpon pettycurense* Benson

Filicopsida:
  *Cladoxylon kidstonii* Solms-Laubach
  *C. waltonii* Long
  *Hierogramma mysticum* Unger
  *Clepsydropsis antiqua* Unger
  *Protoclepsydropsis kidstonii* (Bertrand) Hirmer
  *Psalixochlaena berwickense* Long
  *Stauropteris berwickensis* Long

Lagenostomopsida:
  *Genomosperma kidstonii* (Calder) Long
  *G. latens* Long
  *Hydrasperma tenuis* Long
  *Stamnostoma huttonense* Long
  *S. bifrons* Long
  *Salpingostoma dasu* Gordon
  *Calathospermum fimbriatum* Barnard
  *Deltasperma fouldenense* Long
  *Eosperma edromense* Long
  *Eccroustosperma langtonense* Long
  *Camptosperma berniciense* Long
  *Telangium* sp.
  *Tristichia ovensii* Long
  *Rhetinangium arberi* Gordon
  *Lyginorachis kidstonii* Long
  *L. arberi* Long
  *L. papilio* Long
  *Pitus primaeva* Witham

Calamopityales (*incertae sedis*):
  *Eurystoma angulare* Long
  *E. burnense* Long
  *Lyrasperma scotica* (Calder) Long
  *Dolichosperma sexangulatum* Long
  *D. pentagonum* Long
  *Alcicornopteris convoluta* Kidston
  *Stenomyelon tuedianum* Kidston
  *S. heterangioides* Long
  *S. primaevum* Long
  *Kalymma tuediana* Calder

Uncertain affinities:
  *Mitrospermum bulbosum* Long
  *Cystosporites devonicus* Chaloner and Pettit
  *Triradioxylon primaevum* Barnard and Long
  *Lyginorachis whitadderensis* Barnard and Long

In addition, a number of taxa have been described from adpressions found in the shales:

Lycopsida:
  *Lepidodendron veltheimii* Sternberg
  *L. rhodeanum* Sternberg
  *L. spetsbergense* Nathorst
  *L. nathorstii* Kidston
  *L. berwickense* Crookall
  *Flemingites allantonensis* (Chaloner) Brack-Hanes and Thomas
  *Stigmaria ficoides* (Sternberg) Brongniart

Lagenostomposida:
  ?*Sphenopteridium pachyrrachis* (Göppert) Schimper

Calamopityales (*incertae sedis*):
   *Samaropsis bicaudata* Kidston
   *Triphyllopteris collumbiana* Schimper

## Interpretation

The diversity of Lower Carboniferous petrified plant remains found along the Whiteadder is unrivalled anywhere in the world. However, the significance of the site lies not just in the number of taxa that it yields, but also in the importance of many of the individual species for understanding the evolution of Early Carboniferous plants. In order to assess this, the following discussion will deal separately with each of the classes of plant represented.

### *Lycopsida*

Arborescent lycopsids are amongst the commonest plant remains found along the Whiteadder, but have been relatively ignored. Long (1964c, 1971) briefly described petrified aerial stems of *Lepidodendron calamopsoides* (believed to be equivalent to the adpression species *L. spetsbergense*, *L. nathorstii* and *L. berwickense* - Long 1971) and *Paralycopodites brevifolius*, and stigmarian rooting structures. Long (1971) also described some petrifactions of the herbaceous lycopsid *Oxroadia gracilis* from here, providing the first evidence of its rooting structures.

Few lycopsid fructifications have been reported here. Chaloner (1953) described a compression now known as *Flemingites allantonensis* (see also Crookall, 1966), which is very similar in general structure and dimensions to the petrified *Flemingites scottii* (Jongmans) Brack-Hanes and Thomas, the presumed strobilus of *Paralycopodites brevifolius* from Pettycur (see below). Long (1968a) also described two incomplete petrified sporophylls as *Mazocarpon pettycurense*. Little can be determined from these latter specimens, but they are of interest as the oldest known examples of the Sigillariaceae, one of the lycopsid families that came to dominate the equatorial swamp forests later in the Carboniferous.

*Cystosporites devonicus* was described from here by Long (1968a), who argued that they were megaspores of either lycopsids (e.g. *Lepidocarpon* or *Achlamydocarpon*) or of progymnosperms. Pettitt and Beck (1968) have also linked them with the early gymnosperm *Archaeosperma* (see also comments by Hemsley, 1990b).

### *Filicopsida*

This group of plants is represented here by several taxa. Particularly common are distal parts of vegetative organs, assignable to the artificial form-genera *Hierogramma* and *Clepsydropsis* (Long, 1967, 1968b).

More diagnostic are the main stems of these plants. The Whiteadder has yielded two species of cladoxylids, both of which are unique. The best known is *Cladoxylon waltonii*, 35 specimens of which were described by Long (1968b), demonstrating a range of forms. They fall into two groups: radially symmetrical stems, 8-12 mm in diameter, with a slightly dorsiventral actinostele; and dorsiventral stems, 2-7 mm in diameter, with a 'U'-shaped stele. It is far from certain that they all in fact belonged to the same species, but Long united them in the absence of evidence to the contrary. If they are, then they provide a valuable insight into the anatomical variability of this unusual group of plants, particularly with reference to the anastomosis and dissection of the stele.

*Cladoxylon kidstonii* is only known from a single imperfect specimen, and is inadequately understood (Solms-Laubach, 1910; Seward, 1917; Scott, 1920-1923).

A single species of zygopterid (Coenopteridopsida) stem has been described - *Protoclepsydropsis kidstonii*. The fragmentary holotype originated from Langton Burn near Gavinton (Bertrand, 1911a, b; Hirmer, 1927), but all other specimens have come from the Whiteadder (Long, 1967). It has been interpreted as a creeping, herbaceous plant with a deeply dissected stele that produced *Clepsydropsis* petioles in a 2/5 phyllotaxy. The anatomy of the stele and the petioles is somewhat similar to *Cladoxylon*, which Long (1967) used as evidence for the close affinity between the Cladoxylopsida and the Coenopteridopsida.

A second type of coenopterid found at the Whiteadder is *Stauropteris berwickense* Long (1966). It has similar vegetative structures to *Stauropteris burntislandica* Bertrand from Pettycur (p.147), except for having a more symmetrical vascular cross-section, but it does not have the significant swelling of parenchymatous tissue at the base of the sporangium. Long (1966) argued that these two features point to *S. ber*-

*wickense* being less specialized and more primitive than both the Pettycur species and *Stauropteris oldhamia* Binney from the Westphalian.

The only possible true fern described to date from the Whiteadder is *Psalixochlaena berwickense* Long (1976b). It is based on 21 slender stem fragments (although Long suggested that some might be parts of the same stem) and some isolated petioles. In the absence of fructifications or of evidence as to the orientation of the petioles in relation to the stems, its generic position is provisional (contrast the situation with the Westphalian species *P. cylindrica* (Williamson) Holden – Holmes, 1981). However, a point of interest is the possible presence of trifurcate branching of the petioles, produced by a foreshortened double dichotomy, which Taylor (1981) has compared with similar branching in the trimerophytes.

## Lagenostomopsida

About two thirds of the form-species described from the Whiteadder belong to the group of gymnosperms with large, divided leaves, known informally as the pteridosperms. The 28 form-species probably originated from 15 whole-plant species, belonging to two classes: the Lagenostomopsida, and a so-far unnamed class that includes the Calamopityles. They are, scientifically, probably the most significant components of the Whiteadder flora. This is not so much because they were dominant in the life assemblage (lycopsids were probably dominant – Long, 1964c), but instead probably reflects a combination of greater species diversity in a group occupying less stable habitats and the greater attention that they have received from palaeobotanists. Although the earliest known ovules originated from the Upper Devonian of the USA (Gillespie *et al.*, 1981; Rothwell and Scheckler, 1988; Rothwell *et al.*, 1989), the Whiteadder has yielded the greatest variety of different primitive types showing details of their anatomy. The site has also yielded key information in relating them to whole-plant reconstructions (e.g. Retallack and Dilcher, 1988). It has thus been central to the development of ideas about early gymnosperm taxonomy and evolution.

The Lagenostomopsida includes pteridosperms with protostelic or siphonostelic stems (e.g. *Heterangium*, *Rhetinangium*, *Lyginopteris*). Long (1975) characterized the ovules by a number of features, but this has to be modified as he included within the Lagenostomales ovules now thought to belong to the Calamopityales and the enigmatic Eospermaceae. The ovules of this order appear now to be characterized as follows: integument free from nucellus at least above the plinth; a single set of vascular bundles in the integument; no vascular tissue in the nucellus; and radial symmetry. The male reproductive organs are of the *Telangium/Telangiopsis* type.

Until recently, the Lower Carboniferous lagenostomalean species were only known from isolated organs, which made it difficult to classify the group. However, studies by Long, based mainly on Whiteadder material, have gone a long way towards relating the various organs to one another and it is now possible to think more in terms of whole plants. It must be emphasized that the proposed reconstructions are based mainly on histological similarities of the various organs, and have not been confirmed by organic connection between them (cf. comments under Oxroad Bay – see p. 134). They nevertheless provide a starting point from which to consider these early lagenostomaleans as whole plants.

The most completely known is that which bore the *Stamnostoma* ovules (Long, 1979a; Retallack and Dilcher, 1988). Retallack and Dilcher state that the Foulden site is the main locality for developing this reconstruction, but nearly half the specimens on which Long (1960a, 1962, 1963, 1979a) based his primary studies in fact originated from the Whiteadder. The plant is interpreted as arborescent, over 25 metres tall, with a trunk about one metre in diameter, and of a habit similar to that of the extant Scots pine (*Pinus sylvestris* L.) (see Figure 5.6). The trunk is believed to be of the type commonly described as *Pitus primaeva* Witham (Long, 1963, 1979a). The fronds were bipartite, with at least two pairs of sub-opposite pinnae attached below the dichotomy, and *Lyginorachis papilio* Kidston petioles. There has been disagreement as to the form of these leaves and what they would be called if preserved as adpressions. Long (1963) argued that they would be called *Sphenopteris affinis* Lindley and Hutton, but such fronds are in fact quadripinnate. Based on evidence of association at Foulden (Scott and Meyer-Berthaud, 1985), Retallack and Dilcher (1988) argue that they were *Aneimites acadica* Dawson; but fronds with such flabellate pinnules were probably from calamopityalean plants. Since the details of the *Aneimites* frond architecture are unknown, it is unwise to link it with the *Stamnostoma*-bearing plant. A more convincing link is with the bipartite adpression fronds known as *Sphenopteridium pachyrrachis* (Göppert) Schimper, which are of a similar size and have pinnae attached below the dichotomy (Kidston, 1923b).

The fertile organs of this plant are borne on an axillary branch attached to the dichotomy of some fronds. Such fertile branches differ from the vegetative part of the frond in having a radial symmetry, and were identified with axes described by Long (1962, 1963) as *Tristichia ovensii*. Scott and Meyer-Berthaud (1985) have argued that *Tristichia* was part of a quite different plant with non-cupulate ovules (see also Galtier, 1977), but Long's description of the fertile axillary branch of the *Stamnostoma*-bearing plant nevertheless remains valid. In the female fronds, the fertile branches are terminated by clusters of *Calathiops*-like campanulate cupules (microcupules *sensu* Long, 1977a), which are formed from a sheaf of dichotomously branched, terete telomes. Each cupule contains four *Stamnostoma* ovules. No direct evidence of the male fronds has been reported, but Long (1979a) noted the frequent association of *Telangium* synangia bearing pre-pollen.

The lagenostomalean plant that bore *Genomosperma* ovules was probably rather different. Although the sort of detailed reconstruction outlined above has not been achieved for this plant, there is enough circumstantial evidence to suggest that it was significantly smaller, perhaps only one metre or so high (Long, 1959b, 1964b). The main stem is probably of a type known as *Rhetinangium*, which is only about 20 mm wide, and the longest known example is 260 mm long (Gordon, 1912). Based on association, Long (1964b) argued that the stem bore *Lyginorachis arberi* (syn. *L.* cf. *trinervis* Calder of Long, 1959b) petioles and *Genomosperma latens* ovules. The overall form of the fronds is unknown, but they were probably bipartite, as in many other lagenostomaleans. Again based partly on association, Long (1964b) suggested that a second species of ovule (*Genomosperma kidstonii*) was borne on fronds with *Lyginorachis kidstonii* (syn. *Lyginorachis* sp. Crookall, 1931a) petioles. In all likelihood, these were also probably borne on *Rhetinangium* stems, although supporting evidence has yet to be forthcoming.

Another well-known Lower Carboniferous lagenostomalean plant was that which bore *Salpingostoma dasu* ovules in *Calathospermum fimbriatum* cupules (Retallack and Dilcher, 1988). However, the reconstruction described by Retallack and Dilcher is based mainly on material from Oxroad Bay (p. 134), and the Whiteadder has provided little additional information.

Isolated ovules from the Whiteadder have been described by Long (1961b) as *Hydrasperma tenuis*. However, such ovules have been described from other localities in two quite distinct types of cupules, which have been named *Kerryia* and *Pullaritheca* (Rothwell and Wight, 1989) and it is impossible to be certain which (if either) contained the Whiteadder ovules. Long (1975) included *Hydrasperma* in his new family the Eurystomaceae, most of whose member species probably belong to the Calamopityales (see below). However, *Hydrasperma* does not have the characteristic broad, barrel-shaped lagenostome (or salpinx) of most of the other Eurystomaceae ovules, and is more comparable to the lagenostomalean *Stamnostoma*. Furthermore, *Pullaritheca longii* from Oxroad Bay (see below) was borne on *Sphenopteris bifida* fronds, which is probably lagenostomalean (Long, 1979b).

One of the most significant aspects of the lagenostomaleans from the Whiteadder is the primitive form of the ovules, particularly in the form of the integument (Figure 5.12). The following levels of development of the integumentary sheaf can be recognized.

1. *Genomosperma kidstonii*. The integument comprises a tubiform sheaf of eight more or less terete lobes, which are only attached to the nucellus at the chalaza (Long, 1959b). Niklas (1981) has argued that this in fact is better termed a 'pre-integumentary truss', although most palaeontologists continue to refer to it as a true integument.
2. *Genomosperma latens*. The integumentary telomes are fused to one another in the proximal part of the ovule, and converge at the distal end of the ovule to form a canopy over the lagenostome.
3. *Hydrasperma* and *Salpingostoma*. The integumentary telomes are even more fused, forming a sheath around the entire ovule below the plinth.
4. *Stamnostoma*. The integument is a completely fused sheath forming an open collar around the lagenostome. Long (1975) interpreted the latter as a micropyle but, as it does not cover the lagenostome fully, it cannot have had the same pollen-capturing function as the micro-

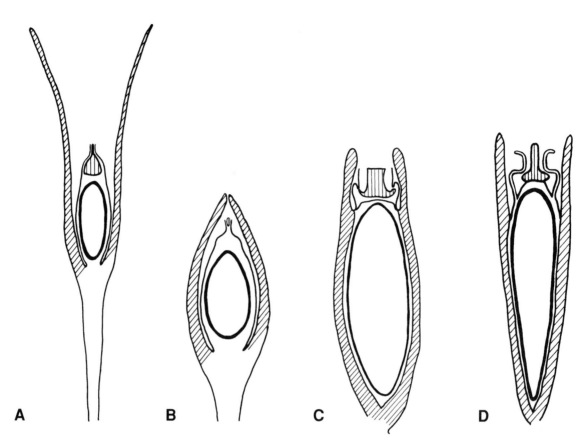

**Figure 5.12** Sections through Early Carboniferous seeds from the Whiteadder. (A) *Genomosperma kidstonii* × 5. (B) *G. latens* × 5. (C) *Stamnostoma huttonense* × 20. (D) *Hydrasperma tenuis* x 20. Based on Long (1959b, 1960a, 1961b).

pyles of later lagenostomalean ovules. It may nevertheless represent an early stage in the evolution of the micropyle.

Most of the lagenostomalean ovules from the Whiteadder belong either to the families Genomospermaceae or Elkinsiacaea (Cleal, 1993). However, there is a third group consisting of *Eosperma*, *Deltasperma*, *Eccroustosperma* and *Camptosperma* (Long, 1961a, b, 1966, 1975). There is no evidence as to which plant-type bore them, or even if they were contained in cupules. However, they are all platyspermic and show the same type of short, distally-tapered lagenostome, containing a conical central plug (Figure 5.13) The most distinctive aspect of the group is the variation in curvature of the ovules: *Eosperma* is flat, *Deltasperma* slightly curved, *Eccroustosperma* 'S'-shaped and *Camptosperma* fully campylotropous (i.e. the chalaza and lagenostome are on the same side of the ovule). The latter is particularly unexpected, as it is a condition most usually associated with angiosperms. Long (1961a) argued that it might indicate that the *Camptosperma* ovules were attached directly to a dorsiventral frond, and was to assist in the better exposure of the lagenostome (and presumed pollen drop) for pollination. Long (1975) assigned these ovules to a separate family within the Lagenostomales, the Eospermaceae (see also comments by Barnard, 1959). This is still widely accepted (e.g. Cleal, 1993), although no other lagenostomalean ovule is known to be platyspermic, and there are also differences in the structure of the lagenostome.

The Whiteadder lagenostomalean ovules thus provide strong support to the telomic hypothesis for early ovule evolution, whereby a sporangium containing a single viable megaspore becomes surrounded by a sheath of fused lobes, that eventually fuse to become the integument (Walton, 1940, 1954, 1964b; Smith, 1959; Long,

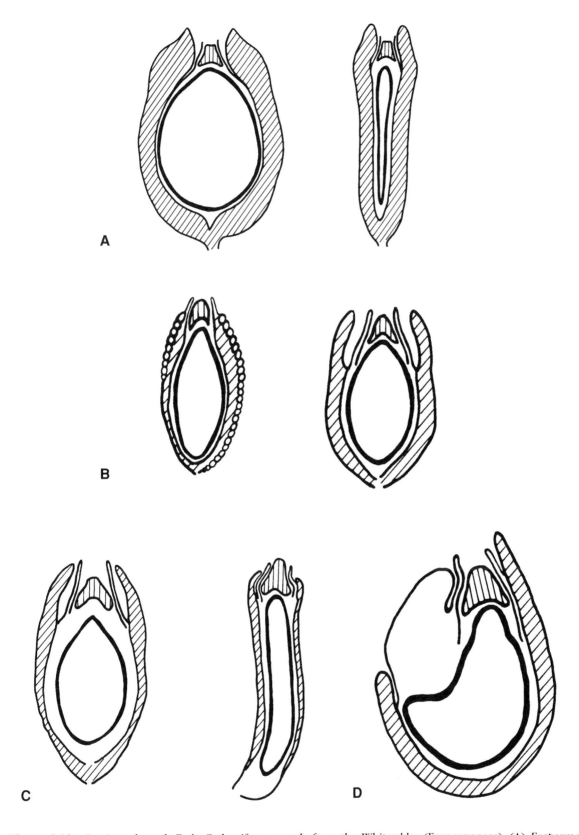

**Figure 5.13** Sections through Early Carboniferous seeds from the Whiteadder (Eospermaceae). (A) *Eosperma edromense*. (B) *Deltasperma fouldenense*. (C) *Eccroustosperma langtonense*. (D) *Camptosperma berniciense*. All × 18. Based on Long (1961a, b, 1966).

1960a; Andrews, 1961). However, it is less certain from the available evidence whether the surrounding telomes were originally all fertile (Benson, 1904) or were sterile.

This is the most diverse known assemblage of lagenostomalean ovules showing the primitive hydrasperman reproductive mechanism, in which pollen-capture prior to fertilization is achieved by the nucellus, rather than by the integument and micropyle as in most other gymnosperms (Rothwell, 1986). Pollen was probably wind transported (Niklas, 1981), and pollen capture achieved by a pollen-drop secreted by the lagenostome. There is little evidence that arthropods played a part in this process.

### Calamopityales

This major order of Lower Carboniferous pteridosperms is also well represented in the Whiteadder assemblage. The best known reconstructed plant from here is that known as the *Lyrasperma scotica* plant of Retallack and Dilcher (1988; see also Long, 1960b, 1964a) (Figure 5.8). It is thought to have been up to about 1.5 metres high, with a succulent trunk. Isolated stems are identified as *Stenomyelon tuedianum* and have a distinctive protostele, which is divided into three strands by radiating bands of parenchyma. Long (1964a) described petioles known as *Kalymma tuediana* attached to such stems. The overall form of the fronds borne by this plant has caused some difficulties. Retallack and Dilcher state that they probably correspond to foliage known as *Sphenopteridium pachyrrachis* (Göppert) Schimper when preserved as adpressions, but, as correctly pointed out by Long (1964a), there are serious discrepancies in the frond architecture, especially in the absence of pinnae attached to the primary rachis below the main dichotomy. The report by Long (1964a) of *S. pachyrrachis* pinnules associated with *Stenomyelon/Kalymma* at nearby Foulden was probably coincidental, the former being more likely to have lagenostomalean affinities (see above). Other types of calamopityalean have been shown to have fronds with rounded, non-digitate pinnules (Sebby and Matten, 1969; Skog and Gensel, 1980). In the Scottish Cementstone Group, foliage with similar pinnules belong to *Aneimites acadica* Dawson and it is thus more likely that this is the adpression form of the foliage in the *Lyrasperma scotica*-bearing plant.

Retallack and Dilcher state that ovules of this plant have been found with pre-pollen of the *Colatisporites*-type in the pollen chamber. However, there is no evidence of the pre-pollen-bearing organs themselves. The *Lyrasperma scotica* ovules are platyspermic and slightly curved, with prominent apical horns on either side of the exposed salpinx (Calder, 1938; Long, 1960b). The integument is fused to most of the nucellus, being free only above the plinth. They probably correspond to the ovules known as *Samaropsis bicaudata* Kidston when preserved as adpressions (Kidston, 1901a, 1902a, 1902b), in which the megaspore has been found to show a prominent trilete mark, a feature generally regarded as primitive.

Although direct evidence is not available, Long (1969) argued that the *Lyrasperma* ovules were probably borne in multiovular cupules, of the type known as *Alcicornopteris*. Such cupules contain a group of ovules surrounded by a sheaf of terete telomes, which in turn is enclosed by a second sheaf, made up of more laminar elements. These are quite different from the single-layered lagenostomalean cupules, such as bore *Stamnostoma* (see above). This lends support to the view of Smith (1964a) that cupules evolved independently in more than one group of gymnosperms. It remains uncertain as to where the cupules were borne on the plant. However, if the outer sheaf of the cupule is the laminate part of a fertile frond (i.e. the structure is a megacupule *sensu* Long, 1977a), then it would probably have been attached directly to the main stem, perhaps near the top of the plant. There is no direct evidence that they were attached to the fronds near the dichotomy of the main rachis, as suggested by Retallack and Dilcher (1988).

A distinctive aspect of *Lyrasperma* ovules is the squat, barrel-shaped lagenostome, with a thin, biconcave central plug at the base, quite different from the more elongate lagenostome, and thicker central plug of the lagenostomalean ovules. Almost identical lagenostomes have also been described in *Eurystoma angulare*, *E. burnense* (syn. *Anasperma burnense* Long), *Dolichosperma pentagonum* and *D. sexangulatum* (Long, 1960b, 1961b, 1965, 1966, 1969, 1975) (Figure 5.14). These are also presumably calamopityalean ovules, although there is no evidence of their attachment to identifiable stems/petioles. They are of particular interest in the great variation that they show in symmetry, varying from bilateral (in *Lyrasperma*) to six-fold (in *D. sexangulatum*). Ovule symmetry, which is controlled largely by the number of vascular bundles entering the integument at the chalaza, has been given considerable weight in certain phylogenetic analyses (e.g. Meyen, 1984; Crane, 1985). The results from the Whiteadder calamopityalean ovules, however, clearly indicate that the feature has to be used in this context with caution (Rothwell, 1986).

In addition to *S. tuedianum*, two other types of calamopityalean stem have been described from the Whiteadder by Long (1964a): *Stenomyelon heterangioides* and *S. primaevum*.

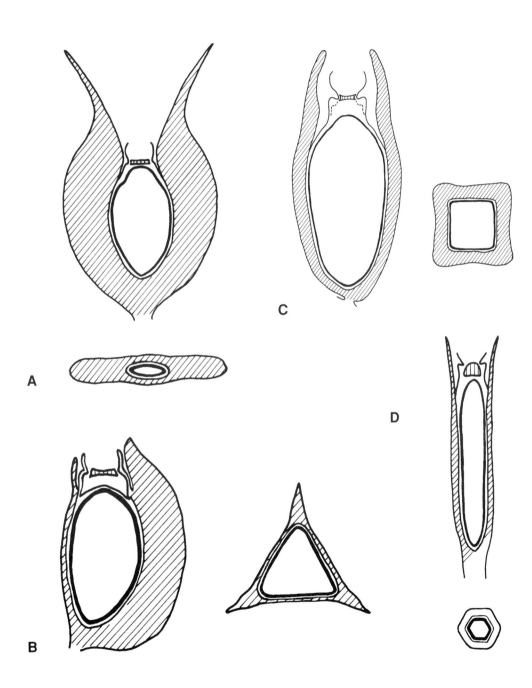

**Figure 5.14** Sections through Early Carboniferous seeds from the Whiteadder (Calamopityales). (A) *Lyrasperma scotica* × 4.5. (B) *Eurystoma trigona* × 13.5. (C) *E. angulare* × 9. (D) *Dolichosperma sexangualtum* × 4.5. Based on Long (1960b, 1961b, 1969).

They are distinguished mainly by the distribution of parenchyma in the stele, although it is possible that it merely represents variation in structure at different positions in the plant. A fourth type of stem reported from here, *Stenomyelon tripartum*, Kidston is now regarded simply as a form of *S. tuedianum* (Calder, 1938).

### Problematic gymnosperms

*Triradioxylon primaevum* has a progymnosperm-like, trilobed protostele and secondary wood; but the sclerotic nests and sparganum structure in the cortex, and the dorsiventral symmetry of the 'fronds', suggest possible affinities with the early pteridosperms. Barnard and Long (1975) placed this and the stem *Buteoxylon* in the *incertae sedis* family Buteoxylonaceae, and Taylor (1981) suggested that it might be intermediate between the progymnosperms and pteridosperms. Recent research (e.g. Rothwell and Erwin, 1987), however, suggests that the Buteoxylonaceae may not be a 'natural' group, and *Triradioxylon* may have affinities with the lagenostomaleans.

An ovule from here which stands out from the rest was described by Long (1977c) as *Mitrospermum bulbosum* (Figure 5.15). Like *Lyrasperma* it is platyspermic with prominent apical horns, but it is not curved, has no basal plug in the lagenostome, and has a free nucellus. Long (1977c) argued that it has many features normally associated with the Cordaitales. Other evidence of cordaites in the Lower Carboniferous is equivocal. Crookall (1970, pl. 155 figs 6-7) figures some possible leaves from the Lower Carboniferous of Scotland, but their identity is far from certain. Lacey (1953) and Barnard (1962) have described some petrified stems and roots from elsewhere in Scotland (Oxroad Bay, Glenarbuck, Loch Humphrey Burn - see p. 135) as having possible cordaite affinities, but they have since been interpreted as probably pteridospermous (Long, 1987; Bateman and Rothwell, 1990). The Whiteadder ovules appear to be the best available evidence of Early Carboniferous cordaites, but on their own cannot be used to dismiss Rothwell's (1986) argument that the group probably did not appear before the Westphalian.

### Angiosperm origins

Long (1966, 1975, 1977b, 1985) has used the evidence from the Whiteadder to argue that

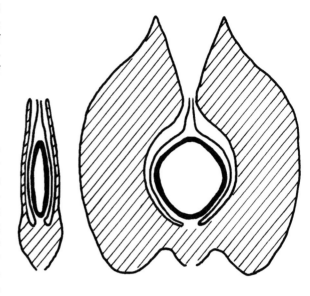

**Figure 5.15** Section through the Early Carboniferous seeds from the Whiteadder, *Mitrospermum* × 4.5. Based on Long (1977c).

angiosperms may have evolved from Carboniferous pteridosperms, particularly by the development of carpels from cupules (see also Delevoryas, 1962; Andrews, 1963). This view has not met with widespread acceptance (e.g. Stewart, 1983; Crane, 1985), although one of the often mentioned objections to it (the long stratigraphical gap between the Lower Carboniferous and where angiosperms appear in the fossil record) may have to be re-assessed in the light of recent work on DNA sequencing in extant angiosperms (Martin *et al.*, 1989; Cleal, 1989).

### General remarks

This is by far the best locality for plant fossils in the Lower Carboniferous Cementstone Group. It has yielded 44 petrifaction form-species, including lycopsids, ferns, progymnosperms and pteridosperms, which is far more than any other site. Of the other Cementstone Group localities, the most comparable are Langton Burn and Cove, which have yielded 20 and 11 form-species respectively (Scott *et al.*, 1984), but neither has yielded taxa not found in the Whiteadder. The Oxroad Bay (see below) assemblage has some taxa in common with the Whiteadder, but there are also significant differences. The other major Lower Carboniferous petrifaction sites in Britain, such as Pettycur (p. 144), Loch Humphrey Burn (p. 155) and Glenarbuck (p. 164), yield quite distinct assemblages from Whiteadder, a function of differences of both environment and age.

From outside Britain, the best known Tournaisian petrifaction assemblages are from the New Albany Shales of the USA (Cross and Hoskins, 1951) and the Lydienne Formation in the Montagne Noire, France (Galtier, 1970). Both have yielded a wide variety of taxa, but mainly of stems and rachides; neither has such a variety of pteridosperm ovules as the Whiteadder. Also similar to the Montagne Noire assemblage is that from Saalfeld (Bertrand, 1935), but as no modern study has been undertaken its flora is relatively poorly understood and a full comparison with the Whiteadder is difficult. The basal Tournaisian (topmost Devonian) assemblage from Kerry Head, Irish Republic (Matten *et al.*, 1975, 1980; Bridge *et al.*, 1980) has some taxa in common with the Whiteadder, but is of far more restricted composition. The Whiteadder assemblage clearly stands out as the most significant Tournaisian plant petrifaction site in the world.

## Conclusion

The Whiteadder section is one of a series of sites in basal Carboniferous rocks in southern Scotland yielding anatomically-preserved plant fossils, about 350 million years old. Nowhere else in the world has so many sites of this age yielding such well preserved fossils. Of the Scottish sites, the Whiteadder has yielded the most diverse assemblage, with 44 species having been described in the literature. Particularly important is the wide range of 'seeds' of early higher (seed-bearing) plants, whose anatomy has been studied in great detail, revealing features such as their unusual and distinctive pollination structures (later replaced by the micropyle in modern 'seeds'). They include the most morphologically primitive 'seed' so far found in the fossil record (*Genomosperma*). They are marginally pre-dated by the earliest known 'seeds' from West Virginia, USA, but the latter are not preserved in such fine detail. Also, the American site yields just one type of 'seed', whereas the Whiteadder contains 16 distinct types belonging to at least two major plant groups (lagenostomaleans and calamopityaleans). This site is thus of outstanding significance for understanding the early evolution of the seed plants, and thus the origin of most modern groups of plants (including flowering plants).

## OXROAD BAY
*R.M. Bateman, G.W. Rothwell and C.J. Cleal*

## Highlights

Oxroad Bay has yielded a wide variety of plant petrifactions that give a remarkable insight into Early Carboniferous plants, particularly among the lycopsids, sphenopsids and pteridosperms. They are yielding the first rigorous whole-plant reconstructions for the Early Carboniferous, as well as establishing details of the plant communities and their environments. This is clearly a palaeobotanical site of international importance (Figure 5.16).

## Introduction

This remarkable palaeobotanical locality in Tournaisian volcaniclastic strata on the coast near North Berwick (Figure 5.17), Lothian Region (NT 599848), was discovered in 1930 by W.T. Gordon, who provided the earliest accounts of its plant fossils (Gordon, 1938, 1941). P.D.W. Barnard then applied the rapid acetate peel technique to the petrifactions collected by Gordon. His doctoral thesis (Barnard, 1960a) resulted in several publications (Barnard, 1959, 1960b, 1962). Barnard's work prompted A.G. Long to collect a considerable amount of new material from the site (276 new blocks by 1976 – Long, 1977b), described in a series of important studies (Barnard and Long, 1973, 1975; Long, 1962, 1975, 1976a, 1977a, b, 1979b, 1984, 1986, 1987). Alvin (1965) described a new lycopsid from here, and Alvin (1966) and Pettitt (1969) undertook maceration studies on lycopsids and pteridosperms respectively. Most recently, the site has been investigated by R.M. Bateman, G.W. Rothwell and A.C. Scott, who have integrated palaeobotanical (including maceration and SEM analyses) and sedimentological work. Their main aims have been to reassess the diversity of dispersed organs, reconstruct them into biologically meaningful plants, and establish details of plant communities and their environments of growth and deposition. The project has resulted in the collection of a further 590 blocks

## Lower Carboniferous

**Figure 5.16** Oxroad Bay. View south across the foreshore, with Dinantian lacustrine sediments in the foreground overlain by reworked volcanigenics exposed in the cliffs. Two of the four horizons yielding anatomically-preserved plants are shown: (A) was discovered by W.T. Gordon in 1930, (C) by R.M. Bateman in 1984. Hammer is 0.3 metres long. (Photo: R.M. Bateman.)

# Oxroad Bay

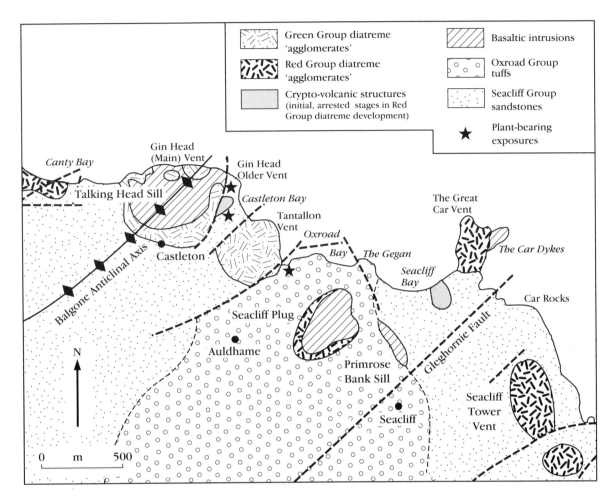

**Figure 5.17** Geological map of the Tantallon area of East Lothian, showing location of palaeobotany sites including Oxroad Bay. Based on Bateman and Scott (1990, figure 2).

containing petrifactions, together with numerous adpressions. Some of the results are summarized by Bateman (1988, 1991, 1992), Bateman *in* Cleal (1991), Bateman and Rothwell (1990) and Bateman and Scott (1990).

## Description

### Stratigraphy

The Oxroad Bay sequence is about 38 metres thick and belongs to the North Berwick Member of the Garleton Hills Volcanic Formation (Figure 5.18; McAdam and Tulloch, 1985; Davies *et al.*, 1986). Its complex sedimentology is discussed by Scott and Rex (1987), Bateman (1988; *in* Cleal, 1991) and Bateman and Scott (1990). The sediments are reworked volcaniclastics deposited along the margins of an active volcano. The plant fossils occur mainly on several discrete bedding-planes, each representing a distinct terrain. Three of these were dominantly lacustrine, the fourth volcanically-induced mass-flow. Palynological evidence suggests that the strata are upper Tournaisian (Tn3, Upper Courceyan – Scott *et al.*, 1984), although some of the plant megafossils are more characteristic of the lower Visean (Bateman, 1988, Appendix 4).

## Lower Carboniferous

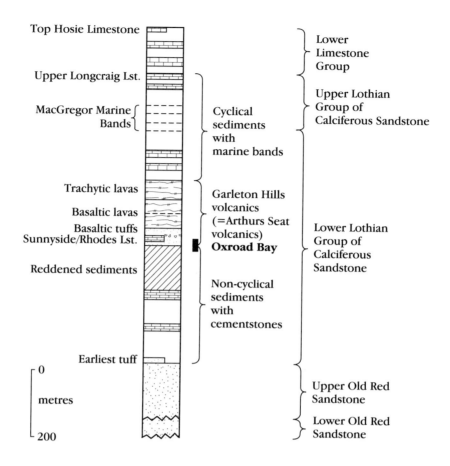

**Figure 5.18** Devonian and Lower Carboniferous stratigraphy in the Lothian Region, showing position of the Oxroad Bay exposures. Based on Bateman and Scott (1990, figure 1).

### Palaeobotany

All but the most recent species list published for this locality (Barnard and Long, 1973; Long, 1984; Scott *et al.*, 1984) included taxa identified from loose blocks, which were transported into Oxroad Bay from the nearby Castleton Bay and Gin Head plant beds (described by Scott and Galtier, 1988; see also Figure 5.17). The following list of petrifactions is based on material recognized as *in situ* by Bateman (1988) and Bateman and Rothwell (1990), although it uses a different supra-generic nomenclature.

Lycopsida:
 *Oxroadia gracilis* Alvin
 *O. conferta* Bateman

Equisetopsida:
 *Protocalamites longii* Bateman
 *Protocalamostachys farringtonii* Bateman

Filicopsida:
 *Cladoxylon* cf. *waltonii* Long
 *Stauropteris* cf. *berwickensis* Long

Gymnosperms:
 *Amyelon bovius* Barnard
 *Tetrastichia bupatides* Gordon
 *Bilignea* cf. *solida* D.H. Scott
 *Eristophyton beinertianum* Zalessky
 *Buteoxylon gordonianum* Barnard and Long
 *Triradioxylon primaevum* Barnard and Long
 *Oxroadopteris parvus* Long
 *Calathopteris heterophylla* Long
 aff. *Tristichia ovensii* Long
 *Lyginorachis waltonii* Calder
 *Lyginorachis* sp. nov.
 *Calathospermum fimbriatum* Barnard
 *Pullaritheca longii* Rothwell and Wight
 *Salpingostoma dasu* Gordon
 *Dolichosperma* cf. *sexangulatum* Long

*Tantallosperma setigera* Barnard and Long
*Hydrasperma tenuis* Long
cf. *Sphaerostoma* sp. nov.
*Stamnostoma oliveri* Rothwell and Scott
cf. *Eurystoma burnense* Long
*Eosperma oxroadense* Barnard
cf. *Eccroustosperma langtonense* Long
*Deltasperma fouldenense* Long
cf. *Telangium* sp.
cf. *Melissiotheca* sp.

Two species of lycopsid megaspore have also been described (Alvin, 1966). Bateman (1988) recognized at least eight other (probably new) species of petrifactions from Oxroad Bay that have yet to be named. He also described, but did not name, several adpression taxa, including a lycopsid (stem), a probable filicopsid (stem and foliar organ) and pteridosperms (one stem, two petioles, four foliar organs, four ovules and one pollen-organ).

## Interpretation

According to Bateman and Scott (1990) the vegetation was growing in a hostile, volcanic environment in which the substrate was unstable. It resulted in a patchwork of subcommunities that consequently yielded an unusually wide range of plants.

### Lycopsida

*Oxroadia* axes are distinguished from most pene-contemporaneous lycopsids by their consistently small size and prominent protoxylem strands (Figures 5.5 and 5.19b). *Paurodendron* is similar but has unbranched vasculature in its rootstock and much smaller tracheids in the central metaxylem of the axes. Bateman (1992) and Bateman *et al.* (1992) argued that *Oxroadia* and *Paurodendron* are small-bodied lepidodendrids derived by paedomorphosis from relatively primitive arboreous species. Bateman recognized two form-species each of stems, strobili and megaspores (*Setispora*), and reconstructed them to form two whole-plant species of *Oxroadia*. These two lycopsids are mutually exclusive at Oxroad Bay, both stratigraphically and (probably) ecologically. Together with *Protocalamites longii* (Bateman, 1991), they are of particular interest as the first fully reconstructed Lower Carboniferous plants.

Alvin's (1965) original description of *Oxroadia gracilis* Alvin was based on a loose block containing a strobilus, seven megaspores and about 12 axes. Additional loose-block material from both here and Berwickshire was described by Long (1971, 1986) to provide details of the leaf bases and rhizomorphs. A further several hundred axes and 43 strobilus fragments from Oxroad Bay were used by Bateman (1992) to reconstruct the bizarre 'pseudoherbaceous' gross morphology of the plant, which possessed a very short stem, strongly and basally concentrated secondary thickening and wide-angle axial dichotomies that suggest a scrambling mode of growth. The leaves were ligulate but lacked sophisticated abscission zones (details of their anatomy are shown in Figure 5.20a,c). The rhizomorph resembled a small, compact *Stigmaria* (Long, 1986; Bateman, 1992). The strobili were *c.* 120 mm long, of the bisporangiate *Flemingites*-type (cf. Brack-Hanes and Thomas, 1983), and were probably borne upright. The megasporangia typically contained four megaspores of the *Setispora subpalaeocristata*-type with prominent, anastomosing laesural crests and *c.* 60 μm-long spines (Alvin, 1965, 1966; Spinner, 1984; Bateman, 1988, 1992).

*Oxroadia conferta* has a similar rhizomorph, but the vegetative axes exhibit fewer protoxylem strands and more closely-spaced branching (Bateman, 1988, 1992). The single known strobilus was shorter (*c.* 30 mm long) and its sporangia contained more megaspores (?16) of smaller diameter (Figure 5.20d). Dispersed megaspores of this type were described as '*Triletes' pannosus* by Alvin (1966), and later reclassified as *Setispora pannosa* by Spinner (1984). They have coarser, more profoundly fused laesural fimbriae, and longer (*c.* 200 μm) spines than the megaspores of *O. gracilis*. All of the aerial organs of these two *Oxroadia* species can be distinguished, and they are mutually exclusive at individual stratigraphical levels within the Oxroad Bay sequence (Bateman, 1992). This almost certainly reflects an early example of ecological control on the spatial distribution of two closely-related species.

Stems of a third lycopsid were described by Barnard (1960a) as cf. *Paurodendron*; small fragments occur sparsely throughout the sequence here (Bateman and Rothwell, 1990). The xylem lacks wood and the stele is deeply incised into 7–10 rounded lobes. It thus resembles *Asteroxylon* (from the Lower Devonian Rhynie Chert – see Chapter 4) and *Leclercqia* (Banks,

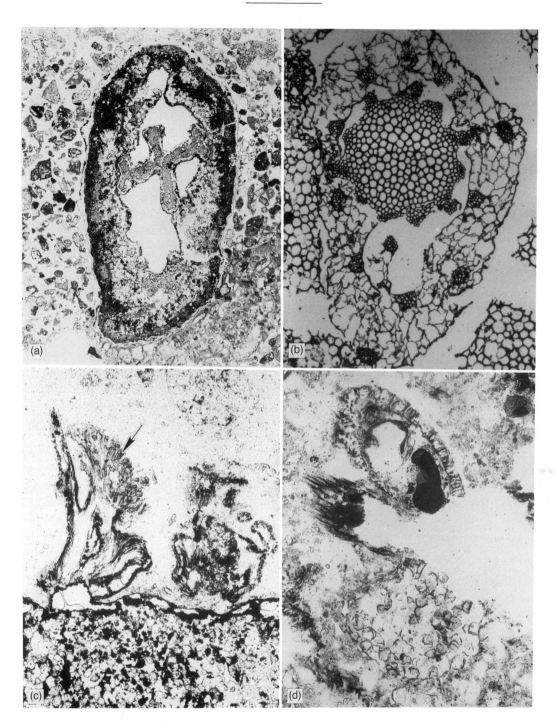

**Figure 5.19** Petrified plants from North Berwick Member of the Garleton Hills Volcanic Formation (Courceyan), Oxroad Bay. (A) *Tetrastichia bupatides* Gordon. Transverse section of stem of shrubby pteridosperm, showing the characteristic four-lobed stele emitting paired leaf traces; Natural History Museum, London, Gordon Collection (holotype). × 8.4. (B) *Oxroadia gracilis* Alvin. Transverse section of a branch of the pseudoherbaceous rhizomorphic lycopsid, showing the coronate stele emitting leaf traces through the inner cortex; Bateman Collection, specimen OBD(?2.15)038bT/2. × 118. (C) *Pullaritheca longii* Rothwell and Wight cupule with *Hydrasperma tenuis* Long ovules attached to the placental margin. Rare example of a developmental anomaly in a fossil plant, where the abortive ovule to the right is normal, whereas the ovule to the left has a deformed, proliferated and non-functional pollen chamber (arrowed); Long Collection, Hancock Museum, Newcastle upon Tyne, specimen HM 11718. × 38. (D) *Protocalamostachys farringtonii* Bateman. Sporangiophore axis (left centre) emitting two of four sporangia, the upper being a megasporangium, the lower a microsporangium; Bateman Collection, specimen OBC084gB/5. × 71. (Photo A: G.W. Rothwell; photos B-D: R.M. Bateman.)

# Oxroad Bay

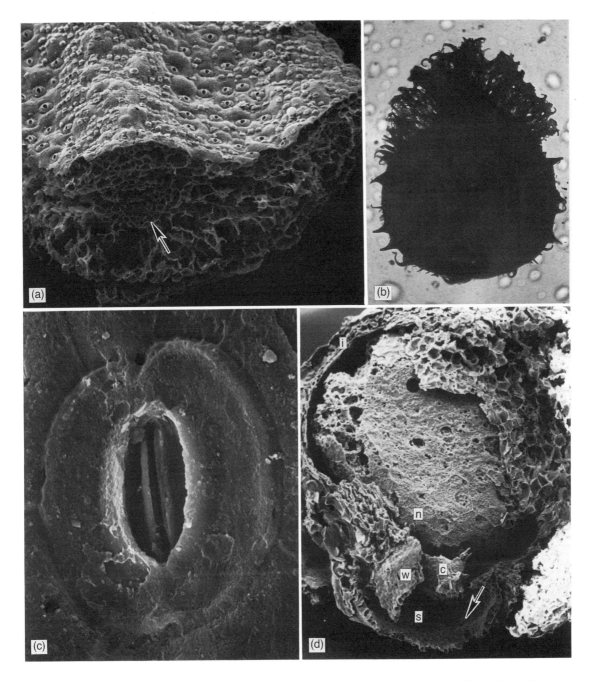

**Figure 5.20** Fusainized plants from North Berwick Member of the Garleton Hills Volcanic Formation (Courceyan), Oxroad Bay. (A) *Oxroadia gracilis* Alvin. Transverse section through microphyll leaf of this rhizomorphic lycopsid, showing terete vascular strand (arrowed) surrounded by mesophyll; highlighted abaxial surface shows stomata restricted to two shallow furrows (left, centre right); Bateman Collection, specimen OBD(2.17)176:CP21. × 108. (B) *Setispora pannosa* (Alvin) Spinner. Elaborately ornamented megaspore of the pseudoherbaceous lycopsid *Oxroadia conferta* Bateman; Natural History Museum, London, specimen V.52016e. × 37. (C) *Oxroadia gracilis* Alvin. Detail of stomata of leaf, showing paired sunken guard cells surrounded by putative subsidiary cells; Bateman Collection, specimen OBD(2.17)176:CP21. × 1634. (D) *Deltasperma fouldenense* Long. Platyspermic pteridosperm ovule with one half of the integument (i) removed to reveal the nucellus (n); at the bottom of the photograph is a typical apical pollen chamber; a cylindrical wall (w) surrounds the central column (c) and salpinx (s), containing a single prepollen grain (arrowed); Bateman Collection, specimen OBD(2.26)190:CP17. × 52. (Photos A, C and D: R.M. Bateman. Photo B: C.H. Shute, Natural History Museum, London.)

1960; Grierson and Bonamo, 1979), rather than supporting Barnard's comparison with *Paurodendron*.

## *Equisetopsida*

Petrified equisetopsid remains were first reported here by Bateman (1988). Vegetative axes were named *Protocalamites longii* and reproductive structures *Protocalamostachys farringtonii* (Bateman, 1991). They represent a single whole-plant species with short, slender stems. The tuberoid stem bases generated dense root-balls and were interconnected by rhizomes. The woody stems subtended at least one order of branching and bore repeatedly dichotomizing leaves. Strobili were probably small, with three vertical rows of paired sporangiophores. The sporangia were small and the plant heterosporous, this being the earliest evidence of heterospory amongst the sphenophytes (Figure 5.19d). The thick-walled megaspores vary greatly in diameter from 100–320 μm. They resemble *Protocalamostachys pettycurensis* Chaphekar (1963) more closely than *P. arranensis* Walton (1949b).

## *Filicopsida*

Two filicopsid species have been reported from here (Bateman, 1988; Bateman and Rothwell, 1990). *Cladoxylon* cf. *waltonii* Long (1968b) is represented by a few polystelic stem bases that are greatly expanded and exhibit weakly-developed secondary xylem. Other stem fragments are smaller than the largest petiole-like primary branches, which are adaxially concave and appear to undergo initial equal dichotomies. The more distal branches exhibit typical *Hierogramma*-type anatomy and bear pinnae with clepsydropsid vascular traces. Possible stem and foliar correlatives occur in the adpression assemblages, and resemble those of *Cladoxylon scoparium* (Leclercq, 1970) and *Pseudosporochnus nodosus* (Leclercq and Banks, 1962). These cladoxylalean fragments re-assemble to form a short, stocky, infrequently branched, upright plant.

Several rachises with a characteristic four-lobed stele have been identified as *Stauropteris berwickensis* Long (1966). Also tentatively placed in this species was a single fragment of a larger axis with a five-lobed stele. No filicopsid reproductive organs have been found at Oxroad Bay.

## *Pteridosperms*

There has been much speculation concerning possible whole-plant restorations of pteridosperms from the assemblages of isolated, petrified organs at Oxroad Bay. A hypothetical *Calathopteris–Calathospermum–Salpingostoma* plant postulated by Long (1976a) is in danger of becoming enshrined in the literature as a genuine reconstruction (cf. Retallack and Dilcher, 1988). Other postulated vegetative–reproductive (stem–ovule) correlations were *Buteoxylon* with *Tantallosperma* (Barnard and Long, 1973), and *Tetrastichia* with *Eosperma* (Barnard, 1959, 1960a). Unfortunately, these hypotheses are based only on histological similarity and comparisons with preconceived bauplans. The abundance of pteridosperm organs of all types at Oxroad Bay permits an enormous number of theoretical combinations.

More recent studies (Bateman, 1988; Bateman and Rothwell, 1990) have yielded much information on how to interpret associations of organs from different exposures and different horizons within exposures, thus reducing the number of likely combinations of organs. Individual plants are being reconstructed by organic connection, though the most difficult correlation (megasporangiate and microsporangiate reproductive organs to stem via petiole) has yet to be achieved unequivocally for any Lower Carboniferous pteridosperm. However, sufficient evidence has accumulated to suggest that five or six species of whole plant were present. This is more consistent with the numbers of petioles (4), pollen-organs (4) and ovulate cupules (4) currently recognized at Oxroad Bay. The reported number of taxa of stem (9) has probably been inflated by the use of different names for different ontogenetic stages of the same species (Long, 1984; Wight, 1987; Bateman and Rothwell, 1990). Despite many previous assertions to the contrary, all of these reconstructed plants will probably prove to be lagenostomalean. The large number of recorded isolated ovules (10) probably reflects greater dispersal potential of disseminules (Bateman and Rothwell, 1990; Bateman and Scott, 1990; Bateman *in* Cleal, 1991). In the absence of an unequivocal correlation between them, the following discussion will treat each of the pteridosperm organ-types separately.

Such evidence has stimulated ideas concerning early cupule organization and structure (Matten and Lacey, 1981), the reproductive biology of early pteridosperm ovules (Andrews, 1940;

Walton, 1954; Smith, 1964), and the ontogeny of these ovules (Bateman, 1988; Rothwell and Wight, 1989). An important result of this work has been the development of the 'hydrasperman reproduction' concept, in which pollination was facilitated by an elongate extension of the pollen chamber wall (the lagenostome or salpinx) rather than by a micropyle formed by the integument, and the presence of a specialized structure (the central column) that sealed the pollen chamber after fertilization had occurred (Rothwell, 1986). This work also prompted the controversial ideas on angiosperm origins, put forward by Long (1966, 1975, 1977b, 1985; see discussion on the Whiteadder, above).

*Rooting structures*

Barnard (1962) attributed certain gymnosperm roots to the form-species *Amyelon bovius* Barnard. The diversity of pteridosperms at the locality obscures their affinities.

*Stems*

Bateman and Rothwell (1990) list nine pteridosperm stem species, which probably represent five or six whole-plant species. The first four are clearly distinct:

*Tetrastichia bupatides* Gordon (1938) has a distinctive, cruciform protostele (Figure 5.19a). They are mostly small stems, probably from shrubby plants, and only the largest exhibit secondary xylem (Gordon, 1938; Barnard, 1960a). Dichotomously branching, *Lyginorachis*-like petioles with few abaxial corrugations were borne in four vertical columns of sub-opposite pairs (Gordon, 1938; Wight, 1987). The fronds have been compared with *Sphenopteris affinis* Lindley and Hutton (Gordon, 1938; Andrews, 1948) and *Adiantites*, but the evidence is equivocal. Elsewhere, the form-genus *Tetrastichia* has only been reported from the lower Tournaisian at Ballyheigue, Irish Republic (Matten *et al.*, 1984b).

*Bilignea* cf. *solida* Scott, consisting of stems up to 70 mm in diameter with a distinctive pith of short tracheids, was first reported by Long (*in* Scott *et al.*, 1984). Attached lyginopterid petioles reported by Bateman (1988), confirm the lagenostomalean affinities of this pteridosperm tree.

The assignment of *Eristophyton beinertianum* Zalessky to the lagenostomalean pteridosperms similarly resulted from studies of Oxroad Bay material. Originally described as calamopityalean by Gordon (1941), it was first recognized as *Eristophyton* by Barnard (1960a). Long (1987) demonstrated that the stems reach 100 mm in diameter and possess a parenchymatous pith that extends into the lateral branches. Associated *Lyginorachis waltonii* petioles are large, with several stelar corrugations that separate near the base; Long (1987) suggested correlation with the adpression frond genus *Diplothmema*.

Stem I of Bateman and Rothwell (1990) resembles a similarly unnamed pteridosperm axis with three sympodia recovered from Visean strata at Kingswood End by Scott *et al.* (1986). A single short length of narrow fusainized axis exhibits parenchymatous pith in intimate association with five mesarch sympodia. They are surrounded by a secondary xylem cylinder of narrow tracheids interspersed with abundant small, uniseriate rays. In the absence of attached petioles, its familial affinities are unclear.

It is uncertain whether *Buteoxylon gordonianum* Barnard and Long represents an additional whole-plant species or should be synonymized with the taxa listed below. The single specimen described from here by Barnard and Long (1973, 1975) was the apical portion of a small stem with a mixed pith and numerous attached petioles. The petioles possessed single protoxylem strands in the centre of both arms of the characteristic Y-shaped stele, and apparently did not dichotomize. A much larger axis of this species with well developed wood was reported by Bateman and Rothwell (1990). Barnard and Long (1975) used *Buteoxylon* and *Triradioxylon* to erect a new family, the Buteoxylonaceae, which they suggested might belong to the progymnosperms. The discovery of protostelic, triradiate stems associated with pteridosperm ovules (Rothwell and Erwin, 1987) renders this less probable.

Although originally described as separate genera, the four remaining pteridosperm stem taxa probably belong to only one whole-plant species (Bateman and Rothwell, 1990; Rothwell and Scott, 1992b). Although not yet fully reconstructed, it has been tentatively interpreted as scrambling or lianascent.

1. *Oxroadopteris parvus* Long (1984) is a small axis with a pith of elongate tracheids. It bears *Lyginorachis*-type petioles that bifurcate close to the stem and possess few stelar corrugations.
2. *Calathopteris heterophylla* Long (1976a) is also small, has a mixed pith and bears two

types of petiole. The more numerous, *Lyginorachis*-type petioles have few stelar corrugations and were probably vegetative. The remainder have three-ribbed xylem strands that trichotomize within the cortex of the stem, and were probably fertile (Long, 1976a).

3. Stem H of Bateman and Rothwell (1990) was ascribed to *Tristichia ovensii* by Long (1962). However, *T. ovensii* as defined by Long includes several different species of protostelic axes, and is clearly in need of revision. The Oxroad Bay stems of this type emit in 1/3 phyllotaxis *Lyginorachis*-type petioles that dichotomize close to the stem.

4. *Triradioxylon primaevum* Barnard and Long (1975) (cf. '*Aneurophyton*' of Barnard, 1960a) resembles Stem H in stelar anatomy. Attached petioles are widely spaced along the stem; each has a Y-shaped stele and a single, central protoxylem strand. Barnard and Long (1975) asserted that these petioles did not dichotomize, and were thus similar to *Rhacopteris* fronds with possible progymnosperm affinities. However, Bateman and Rothwell (1990) have shown that a dichotomy is present and that *Triradioxylon* is thus a pteridosperm.

*Other foliar organs*

Only four isolated, pteridospermous foliar organs have been named from Oxroad Bay (Bateman and Rothwell, 1990):

1. Barnard's (1960a) *Lyginorachis* sp. A probably represents secondary racheis and therefore is not strictly comparable with the petioles that were used to characterize most other *Lyginorachis* species.
2. Barnard's (1960a) *Lyginorachis* sp. B has since been identified with *L. waltonii* Calder and shown to be attached to *Eristophyton* stems (Long, 1987). It possesses dorsiventrally shallow vascular traces with several abaxial corrugations that divide into discrete bundles near the base.
3. *Lyginorachis* sp. nov. of Bateman (1988) occurs in organic connection with *Bilignea* cf. *solida* Scott. It lacks sclerotic nests, and possesses vascular traces with four corrugations that remain connected at the base.
4. Foliar organ A of Bateman and Rothwell (1990) encompasses a wide range of triradiate axes formerly attributed to *Tristichia ovensii* Long.

They have yet to be partitioned into biologically meaningful species.

Compressed petioles and frond segments have not been successfully correlated with the more informative petrified material.

*Ovulate cupules*

Two types of ovulate cupule found at Oxroad Bay have generated more discussion than any other species present. *Calathospermum fimbriatum* Barnard (1960b) is a large, apparently solitary and bilaterally symmetrical cupule that is dissected into many lobes and probably contained numerous ovules. It was interpreted by Barnard (1960b) and Smith (1964a) as a modified frond borne on an apparently unbranched but otherwise *Lyginorachis*-like petiole. Long (1975) interpreted *Calathospermum* as a 'megacupule', resulting from the fusion of a pair of radially symmetrical cupules. He also argued that it is a possible precursor of the angiosperm carpel (Long, 1966, 1975, 1977b, 1985). Matten and Lacey (1981) interpreted branching of the cupule lobes distal to the basal-most dichotomy as pseudomonopodial. *In situ* ovules reported by Barnard (1960a, b) and Long (1975) resemble *Salpingostoma dasu* Gordon in pattern of organization but are much smaller; they may have been abortive. However, Bateman and Rothwell (1990) report *in situ* ovules more closely resembling *Tantallosperma setigera* Barnard and Long.

A.G. Long (pers. comm. 1988) argued that specimens with unusually long axes that appear to lack pinnae may constitute a second species of *Calathospermum* here. Study of similar compressed cupules from Oxroad Bay may help resolve this question.

*Pullaritheca longii* Rothwell and Wight (1989) represents cupule aggregates containing *Hydrasperma tenuis* Long ovules (Figure 5.19c). They compare with adpression cupules usually identified as *Sphenopteris bifida* Lindley and Hutton (Long, 1977b; Bateman and Rothwell, 1990). Identical ovules have also been reported in *Kerryia* cupules from the Irish Republic (Matten *et al.*, 1980), which differ from *Pullaritheca* in symmetry, branching pattern and number of ovules enclosed (Rothwell and Scheckler, 1988; Rothwell and Wight, 1989). Nomenclatural instability in the past (cf. Long, 1977a; Matten *et al.*, 1975; Matten *et al.*, 1980; Matten *et al.*, 1984a) reflects reluctance to take full account of ontogenetic variation in the ovules and to treat separately the taxonomy of the ovules and of the cupules. Many of the *Pullaritheca* cupules collected by Bateman and Rothwell (1990) contained several abortive ovules, and a few contained more mature individuals (Bateman, 1988; Rothwell and Wight, 1989).

Long (1977a, b, 1979b) recognized two modes of arrangement of the cupules (paired 'hemicupules' and unpaired 'megacupules') that were both considered to represent an entire megasporophyll, and necessitated revision of his 'cupule-carpel' theory. However, Rothwell and Wight (1989) and Bateman and Rothwell (1990) reported a gradation between these two extremes. Comparison of *Pullaritheca* with adpressions assigned to *Sphenopteris bifida* Lindley and Hutton led Long (1979b) to suggest that both bore cupules in similar fashion, as aggregates attached to the median axes of trichotomous petioles. Indeed, clusters of up to eight *Pullaritheca*-like cupules occur in adpression assemblages at Oxroad Bay (Bateman, 1988). One petrified *Pullaritheca* cupule found by Long (1977a) contained both ovules and deformed sporangia. This suggests a developmental anomaly reflecting the homologous architecture of pollen-organ aggregates that were probably borne on other trifurcating petioles of the same plant (Bateman, 1988; Bateman and Rothwell, 1990; Bateman and DiMichele, in press).

An open, frequently and irregularly dichotomous, hirsute cupule contains *Dolichosperma* cf. *sexangulatum* Long ovules (Bateman, 1988). Its branching pattern resembles that of the cupules bearing the ovule *Eurystoma burnense* (Long, 1960b, 1965).

A similarly open, but pseudomonopodially branched, cupule bears *Stamnostoma oliveri* ovules in aggregates of up to four (Bateman and Rothwell, 1990; Rothwell and Scott, 1992a).

*Isolated ovules*

Bateman (1988) and Bateman and Rothwell (1990) list ten species of isolated petrified ovule. All appear to be distinct, though an ontogenetic relationship between *Tantallosperma* and *Dolichosperma* is credible. Some potential correlatives occur in the adpression assemblages, notably the ovule megaspore studied by Pettitt (1969), which probably represents *Eosperma oxroadense* Barnard (Bateman, 1988).

*Salpingostoma dasu* Gordon (1941) is the largest known Lower Carboniferous ovule, up to 50 mm long and 6 mm wide (Bateman and Rothwell, 1990). It usually had six integumentary vascular bundles extending into lobes that project well beyond the tall, narrow lagenostome. The integument also bore large, antapically concentrated trichomes. Gordon (1941) erroneously endowed *Salpingostoma* with an atypical pollen-receiving apparatus (Rothwell, 1986).

*Dolichosperma* cf. *sexangulatum* was first reported from Oxroad Bay by Bateman (1988). It typically had six integumentary bundles and lobes that extend well beyond the lagenostome and was covered with evenly distributed, coarse trichomes. Very similar but rather smaller is *Tantallosperma setigera*, which possessed four (rarely five or six) vascular bundles, integumentary lobes that extend well beyond the lagenostome, and an even covering of coarse trichomes. Since the two are essentially distinguished only on size and number of integumentary lobes, they may conceivably be ontogenetic stages of the same species.

*Hydrasperma tenuis* Long was reported by Long (1977a, b, 1979b). It had 8-12 integumentary lobes that extended only slightly beyond the lagenostome, which is the widest part of the ovule. Dispersed *H. tenuis* ovules exhibit a wide range of ontogenetic variation at Oxroad Bay, and were erroneously segregated as *H. longii* by Matten *et al.* (1980).

Cf. *Sphaerostoma* sp. nov. (Bateman, 1988; Bateman and Rothwell, 1990) is characterized by an integument with eight vascular bundles, but no lobes. It differs from *S. ovale* Benson (1914; see also Long, 1961b) in its thicker, denser integument and sparser, but coarser, trichomes.

The 'cf. *Sphaerostoma* sp.' of Barnard (1960a) differs from bona fide *Sphaerostoma* in having a morphologically distinct pollen chamber, and an integument that forms an open collar surrounding the lagenostome. It was therefore assigned to *Stamnostoma huttonense* Long by Barnard and Long (1973; see also Rothwell, 1986). However, it differs from *S. huttonense* Long in having a smaller length:breadth ratio and papillae on the integument. Consequently, it has been referred to a new species, *Stamnostoma oliveri* Rothwell and Scott (1992a).

Cf. *Eurystoma burnense* Long has been reported by Bateman (1988). Elsewhere, Long (1966, 1969) has recognized both pre-prothallial and prothallial ontogenetic phases. Its distinctive, triangular transverse section reflects its three vascular bundles and associated wide lateral keels, which terminate as short lobes projecting slightly beyond the lagenostome.

*Eosperma oxroadense* Barnard (1959) has only two integumentary vascular bundles and lacks integumentary lobes; short spiny trichomes occur on some specimens. Pettitt (1969) demonstrated attachment of three abortive megaspores to a compressed functional megaspore that probably belonged to *Eosperma*.

Bateman (1988) has demonstrated the presence of cf. *Eccroustosperma langtonense* Long from *in situ* strata, the report by Long (1987) having been based on a loose block that was probably transported from nearby Castleton Bay. It has two vascular traces terminating in fairly short integumental lobes that do not extend beyond the lagenostome. Bateman (1988) observed that Oxroad Bay specimens are more strongly curved than those from elsewhere (cf. Long, 1961b, 1975), and may represent the pre-prothallial ontogenetic stage of *Camptosperma berniciense* Long (1961a).

*Deltasperma fouldenense* Long (1961a) was first recognized as fusainized material macerated by Bateman (1988) (Figure 5.20d). Two vascular strands terminate in fairly short integumental lobes that do not exceed the lagenostome. Characteristically large superficial cells cover the whole of the more convex face, and the apical region of the less convex face.

*Pollen-organs*

Three of the four pollen-organs listed by Bateman (1988) and Bateman and Rothwell (1990) appear to be synangiate: cf. *Telangium* sp., cf. *Melissiotheca* sp. and 'pollen-organ C'. The *Telangium*-like synangia terminate repeatedly dichotomous axes. They typically comprise eight, basally-fused, bilaterally-arranged synangia containing finely tuberculate pre-pollen. They therefore share characters with several *Telangium*-like form-genera (e.g. Eggert and Taylor, 1971; Long, 1979a; Millay and Taylor, 1979; Meyer-Berthaud and Galtier, 1986a).

The pedicellate synangia of cf. *Melissiotheca* sp. are shallowly divided into at least two lobes, each consisting of 30–100 basally-fused sporangia that are embedded in a parenchymatous cushion, and contain rugulate pre-pollen (Bateman, 1988; Bateman and Rothwell, 1990). The sporangia are smaller and exhibit less profound lateral fusion than those of *Melissiotheca longiana* Meyer-Berthaud (1986). Numerous compressed synangia from Oxroad Bay are difficult to apportion between cf. *Telangium* sp. and cf. *Melissiotheca* sp. (Bateman, 1988).

Pollen-organ C of Bateman and Rothwell (1990) is a large, hirsute, oviform organ that superficially resembles the ovulate cupule *Calathospermum fimbriatum* Barnard in size and branching pattern. Large sporangia form asymmetrical synangia and enclose dimorphic pre-pollen; the larger pre-pollen morph is considered fertile and the smaller abortive (Bateman, 1988). Similar pre-pollen has frequently been found in the lagenostome of *Salpingostoma dasu* Gordon (Long, 1975; Bateman, 1988). Pollen-organ C was incorrectly referred to cf. *Telangium* by Barnard (1960a) and subsequent workers, and should not be confused with the cf. *Telangium* described above.

Pollen-organ D of Bateman and Rothwell (1990) is a single terete branching axis terminated by a small, thick-walled sporangium. It is associated with abundant foliar organs showing similar histology. Spores found in the sporangium have sparse, blunt-tipped spines, and compare with forms found in the lagenostome of *Deltasperma* (Bateman, 1988).

*General remarks*

At least 11 whole-plant species of tracheophyte grew at Oxroad Bay, and these have produced one of the most diverse assemblages of Early Carboniferous petrifactions. In Britain, it is second in estimated partial-plant diversity only to the Whiteadder (see above, and Scott *et al.*, 1984; Bateman *in* Cleal, 1991). Other taxonomically similar petrifaction assemblages occur at nearby Castleton Bay (Scott and Galtier, 1988); the upper assemblage at Loch Humphrey Burn (see below); Kingswood (see below); and Ballyheighue, Irish Republic (Matten *et al.*, 1980; Matten *et al.*, 1984b). These occurrences vary considerably in age (Strunian to possible early Asbian), sedimentological regime and geographical location, but evidently reflect similar ecological and taphonomic constraints. However, none of the above localities matches Oxroad Bay in whole-plant diversity, or in the potential for correlating petrifaction, adpression and fusain fossils of the same plant organ.

## Conclusion

Oxroad Bay is one of a series of basal Carboniferous sites in southern Scotland that yield anatomically-preserved plant fossils (about 350 million years old). Taken together, they are the most important global source of information on vegetation of this age. This site is especially important for allowing the first rigorous reconstructions for Early Carboniferous representatives of these plant groups. Studies of the club-mosses and horsetails have been completed, but work continues on the more diverse (and thus more problematic) early seed plants and on the ferns. Integration of the reconstructions with palaeoenvironmental data from this unstable volcanic setting has led to recognition of several distinct plant communities. Increasing knowledge of their structure and dynamics will allow comparison with extant vegetation.

## WEAK LAW

## Highlights

Weak Law yields a potentially significant plant petrifaction assemblage of Early Carboniferous age. Stems of the form-genus *Pitus* are particularly well preserved, and show evidence of attachment of young fronds.

## Introduction

Following his work on the plant petrifactions from Pettycur (p. 141), W.T. Gordon extended his investigations to cover similar volcaniclastic deposits in SE Scotland. In 1914, Gordon made his first major discovery, Weak Law, on the coast between Gullane and North Berwick (NT 499858). To investigate further the assemblage, the British Association for the Advancement of Science set up a committee to organize an excavation of the site, employing explosives, but the intervention of the 1914-1918 war prevented Gordon from going ahead (Kidston *et al.*, 1917). The only detailed descriptions of plant fossils from here have been Gordon's (1935a) account of the *Pitus* species, and a description of a *Lyginorachis* by Galtier and Scott (1986a). However, a more detailed investigation is currently in progress by A.C. Scott and his colleagues.

## Description

### Stratigraphy

In the virtual absence of any biostratigraphical data, it is impossible to estimate the chronostratigraphical position of the site, beyond it being Lower Carboniferous. Kidston *et al.* (1917) argued that it was probably homotaxial with the Pettycur deposits (see below), now known to be Asbian, but the evidence on which this was based was limited. According to Gordon (1935a), the plants probably grew on the side of an active volcano, were killed by a violent eruption and buried by the resulting ash flow. The ash may well have had a high moisture content, and the resulting colloidal solution caused the plants to become permineralized.

### Palaeobotany

The assemblage here consists exclusively of petrifactions, including the following:

Lycopsida:
*Lepidodendron* sp.
*Stigmaria* sp.

Filicopsida:
*Botryopteris* sp.
*Bensonites* sp.

Lagenostomopsida:
*Pitus primaeva* Witham
*Pitus dayi* Gordon
*Pitus rotunda* Gordon
*Lyginopteris gordonii* Galtier and Scott

## Interpretation

Gordon (1935a) described three species of *Pitus* from here, although Long (1979a) has questioned the validity of the distinction. They differ in details of wood structure, particularly the width of the medullary rays. *P. dayi* Gordon was found with spirally-attached leaf-bases, which Gordon interpreted as phyllodes, similar to those found in extant araucarias. However, Long (1979a) suggested that they are more likely the petioles of juvenile, unexpanded fronds, of the *Lyginopteris*-type (see also Beck, 1960).

Of passing interest is the discovery of a *P. primaeva* Witham specimen in the hearth of a nearby Neolithic dwelling (Gordon, 1935b). The fossil resembles closely a piece of drift wood, which must have disappointed its original discoverer on attempting to burn it! This must represent one of the earliest examples of a palaeobotanical discovery.

A single specimen was the basis of Galtier and Scott's (1986a) description of the pteridosperm frond *Lyginorachis gordonii*, which shows a trifurcate petiole. The two lateral racheis probably produced the two foliar halves of a bipartite frond, and the median rachis bore fertile structures (either cupules or pollen-organs). It is essentially similar in structure to other *Lyginorachis* species (Calder, 1935; Long, 1963), as well as *Pitus dayi* Gordon and *Calathopteris heterophylla* Long (1976), but differs in the details of the vascular configuration and the relative positions of the racheis.

Most of the assemblage has not been described in detail, and so a full comparison with other localities cannot be given. However, the reported presence of *Bensonites*, which usually occurs in the upper Visean (such as at Pettycur - see p. 148), contrasts with the abundant *Pitus* stems, which are characteristically Tournaisian and lower Visean (as at Lennel Braes, Whiteadder, Kingwater - see elsewhere in this chapter). The site clearly has potential for further significant palaeobotanical discoveries, which will help unravel the history of Early Carboniferous vegetation in this country.

## Conclusion

Weak Law is one of a series of sites showing Lower Carboniferous rocks in southern Scotland, which yield anatomically-preserved plant fossils, about 350 million years old. Nowhere else in the world has so many sites of this age yielding such well preserved fossils. This particular site is so far mainly known for well-preserved stems of early seed plants (known as *Pitus*) in which the position of attachment of young fronds can be recognized. However, the potential of this site has yet to be fully realized.

## KINGWATER

### Highlights

Kingwater has yielded the best example of *in situ* arborescent pteridosperm stumps in the Lower Carboniferous of Britain, and provides important information as to the size, palaeoecology and density of this type of vegetation.

### Introduction

This Visean site lies in the bed of Kingwater Beck, 4 km north-west of Gilsland, near Haltwistle, Cumbria (NY 608697). *In situ* fossil tree stumps were discovered here by Day (1970), and described in detail by Long (1979a).

### Description

#### Stratigraphy

The geology of this site was described by Day (1970). The stumps occur just below the Desoglin Limestone, near the top of the Middle Border Group (Figure 5.21). Fossil marine invertebrates and algae from near this horizon indicate the $S_1$ Zone, which in modern terms places it in the Arundian Stage. The deposits probably represent a fluvial environment, although detailed sedimentological information is not available.

#### Palaeobotany

Ten *in situ* stumps found here have been identified as *Pitus primaeva* Witham. The largest is just under 2.5 metres in diameter, and evidently belonged to a tree of considerable size, perhaps 10 metres or more in height. The spacing between the stumps varies from 4 to 40 metres. However, this variation may be partially the result of some stumps being no longer preserved, due to their removal either immediately after death or by recent erosion.

Such examples of a forest floor with surviving *in situ* stumps dating from the Early Carboniferous are rare worldwide, and this is the only one presently available showing a stand of gymnosperm trees.

### Interpretation

Based on the fossils present, it can be calculated that there must have been at least 2000 trees per $km^2$ in this area, indicating an open woodland environment. According to Long (1979a), they grew at sufficiently low altitudes for them to have become flooded and buried by marine deposits.

# *Pettycur*

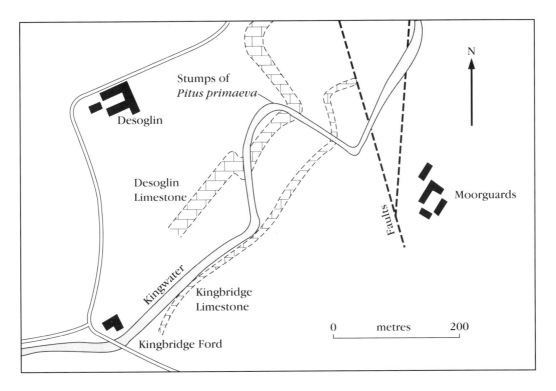

**Figure 5.21** Map of that part of Kingwater where *in situ* tree stumps are preserved in the Lower Carboniferous. Based on Day (1970, figure 20).

## Conclusion

Kingwater has the only known examples of gymnosperm tree-stumps, still preserved in position, from about 340 million years ago. They provide a valuable insight into the nature of these very early forests, that pre-dated by some 30 million years the tropical coal-swamp forests that covered the palaeoequatorial belt (including Britain) in Late Carboniferous times.

## PETTYCUR

### Highlights

Pettycur is one of the classic Lower Carboniferous palaeobotanical sites and has yielded the most diverse British Visean petrifaction assemblage. It is the type locality for a number of species and is particularly important for lycopsids and ferns; it is also the only known locality for the equisete family Cheirostrobaceae. It is one of the most important palaeobotanical sites in Britain.

### Introduction

Prior to the discoveries at Rhynie (see Chapter 4), Pettycur possessed by far the best known assemblage of pre-Late Carboniferous plant petrifactions in the world. Together with the coal-ball petrifactions from Yorkshire and Lancashire, it played a key role in developing an understanding of Palaeozoic plant life. The site (Figure 5.22), on the coast at Pettycur, near Burntisland, Fife (NT 262863), was discovered in 1871 by the local Fifeshire geologist George Grieve. In May of that year, a fossil was exhibited to the Botanical Society of Edinburgh by J.H. Balfour. However, the 'formal unveiling' of the site may be taken as being at the meeting of the British Association for the Advancement of Science held in Edinburgh in August 1871. Details of the site were described during the meeting and afterwards an excursion was conducted there (Gordon, 1909). Numerous publications followed, dealing with parts of this important Lower Carboniferous flora, usually described in tandem with material from the Upper Carboniferous coal balls (Williamson, 1872, 1873, 1874a, b, 1877, 1893; Williamson and Scott, 1894,

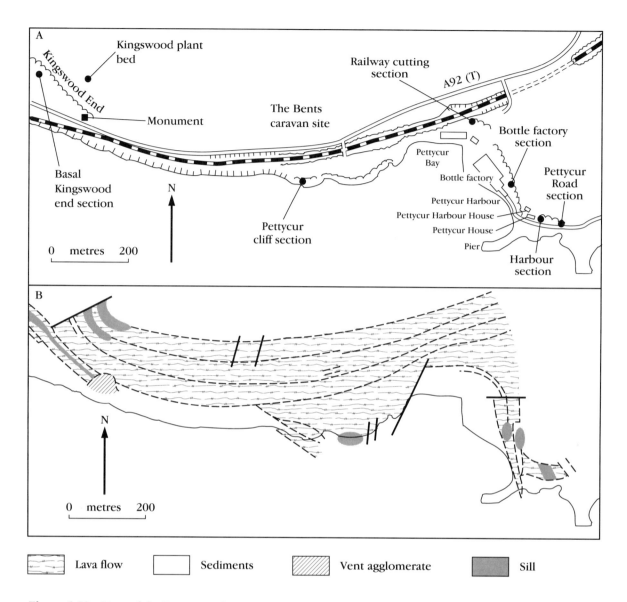

**Figure 5.22** Maps of the Pettycur and Kingswood End GCR palaeobotany sites. The upper illustration is a locality map. The lower illustration shows the distribution of the main lithologies. Based on Rex and Scott (1987, figure 2).

1895; Scott, 1897, 1900, 1901, 1908; R. Scott, 1908).

The most detailed investigation of the site was by Gordon (1908b, 1909, 1910a, b, 1911a, b, 1912). Other significant publications were by Benson on the lycopsids, ferns and pteridosperms (Benson, 1908, 1911, 1914, 1922, 1933), as well as some shorter contributions by Kidston (1907, 1908), Bertrand (1907, 1909), Chodat (1912) and Graham (1935).

As at Rhynie, the development of the acetate peel method of investigating petrifactions (Joy *et al.*, 1956) had a significant impact on the study of the Pettycur fossils, allowing more detailed reconstructions to be achieved. It proved particularly significant for the ferns, whose complex branching patterns could only be properly resolved by very fine serial sectioning. The most important contributions using the method have been by Surange (1952a, b), Lacey *et al.* (1957), Chaloner (1958), Holden (1962), Chaphekar (1963) and Chaphekar and Alvin (1972). In subsequent years, there has been little published work on the palaeobotany of Pettycur, although there has been considerable interest in its stratigraphy and sedimentology (Scott *et al.*, 1984; Scott and Rex, 1987; Rex and Scott, 1987).

## Description

### Stratigraphy

The geology at Pettycur has been extensively discussed by Rex and Scott (1987). The sequence (Figure 5.23) consists of about 25 metres of volcanic igneous rocks (lavas and sills) and mainly volcanogenic sediments of Asbian age (Scott *et al.*, 1984). Petrifactions occur at two main horizons. The most important is the Pettycur Limestone, numerous blocks of which are found on the beach, but which has not yet been located *in situ*. Rex and Scott interpret it as a preserved peat, which developed in a relatively tranquil environment, probably prior to the development of volcanic activity in the area; it is in some ways a Lower Carboniferous equivalent of the habitat represented by the Upper Carboniferous coal-balls (Scott and Rex, 1987). The second plant-bearing bed is the Zygopterid Limestone, which also includes abundant fusain. This does not represent an *in situ* peat, but probably the remains of plants growing in a volcanically disturbed habitat, which were transported into a small lake and buried. Rex and Scott also report less well-preserved petrifactions together with fusain at four other horizons (the Harbour 'Peat' and Ashy Limestones 1–3), and some adpressions in a dolomitic mudstone.

### Palaeobotany

The following form-species preserved as petrifactions and/or fusain have been reported here:

Lycopsida:
    *Paralycopodites brevifolius* (Williamson) DiMichele
    *Lepidophloios scottii* Gordon
    '*Lepidodendron*' *pettycurensis* Kidston
    *Flemingites scottii* (Jongmans) Brack-Hanes and Thomas
    *Lepidostrobus cylindricus* Gordon M.S.
    *Lepidocarpon wildianum* Scott
    *Mazocarpon pettycurense* Benson
    *Stigmaria ficoides* (Sternberg) Brongniart

Equisetopsida:
    *Protocalamites pettycurensis* (Scott) Scott
    *Protocalamostachys pettycurensis* Chaphekar
    *Sphenophyllum insigne* Williamson
    *Cheirostrobus pettycurensis* Scott

Filicopsida:
    *Botryopteris antiqua* Kidston
    *Metadineuron ellipticum* (Kidston) Galtier
    *Diplolabis roemeri* (Solms-Laubach) Bertrand
    *Metaclepsydropsis duplex* (Williamson) Bertrand
    *Musatea duplex* Chaphekar and Alvin
    *Stauropteris burntislandica* Bertrand
    *Bensonites fusiformis* Scott

Lagenostomopsida:
    *Heterangium grievii* Williamson
    *Rhetinangium arberi* Gordon
    *Sphaerostoma ovale* (Williamson) Benson
    *Physostoma* sp.
    *Bensoniotheca grievii* (Benson) Mickle and Rothwell
    *Amyelon* sp.

In addition, Rex and Scott (1987) list as adpressions *Lepidodendron* sp., *Lepidocarpon* cf. *waltonii* Chaloner, *Sphenopteris affinis* Lindley and Hutton, *Sphenopteris* sp., *Sphenopteridium* sp., *Adiantites machanekii* Stur and *Cardiopteridium* sp.

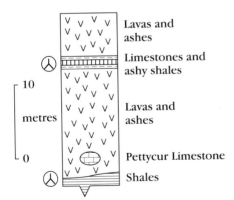

**Figure 5.23** Generalized stratigraphical section at Pettycur. Based on Scott *et al.* (1984, figure 9).

## Interpretation

### Lycopsida

Lycopsid fragments are the most abundant fossils here, particularly in the Pettycur Limestone. The commonest type is *Paralycopodites brevifolius* (Figure 5.24), a lycopsid stem found abundantly in both the Lower and Upper Carboniferous (Williamson, 1872, 1893; DiMichele, 1980). It had a straight, rarely dichotomizing trunk, with a crown of deciduous lateral branches, at the ends of which were small, bisporangiate strobili. The strobili found at Pettycur are known as *Flemingites scottii* (in the Upper Carboniferous, slightly different strobili are associated with *Paralycopodites*, known as *F. diversus* (Felix, 1954) and *F. schopfii* (Brack, 1970)). Megaspores found in *F. scottii* have been reported with endosporal gametophytes (Gordon, 1908a, 1910b; D.H. Scott, 1908-1909), similar to those of the extant *Selaginella* (Gordon, 1908a; Phillips, 1979). The rooting structures are unknown, although Williamson (1872) noted that *Stigmaria ficoides* is often found in close association. Evidence from the Upper Carboniferous suggests that these lycopsids were opportunistic plants, occupying slightly drier parts of the coal-swamps (Phillips and DiMichele, 1992). Its predominance in the Pettycur Limestone (Rex and Scott, 1987), appears to confirm that it favoured peat-accumulating swamps. However, the detailed community structure of the Pettycur swamp has still to be worked out.

Smith (1962c) argued that the stems now called *Paralycopodites* were intermediate between *Lepidodendron* and *Lepidophloios*, but DiMichele (1980) clearly showed they differed from both these lycopsids in having a smooth-surfaced siphonostele, a homogeneous periderm and cortex, and deciduous branches. He further argued that it could be distinguished by the presence of persistent leaves, although some large *Lepidodendron* stems have also now been shown to have been leafy (Leary and Thomas, 1989). A closer analogue might be with the adpressions stems known as *Ulodendron*, as defined by Thomas (1967b). There are differences in the epidermis of the

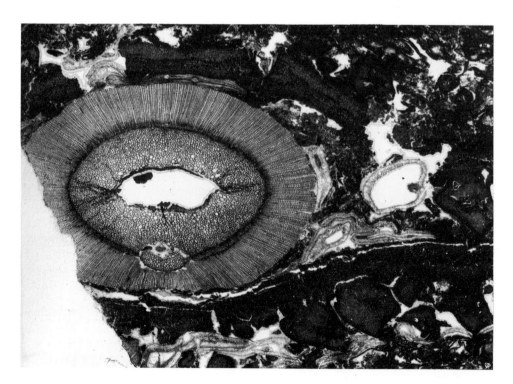

**Figure 5.24** *Paralycopodites brevifolius* (Williamson) DiMichele. Transverse section through lycopsid stem; Natural History Museum, London, specimen WC.502. Pettycur Limestone (Asbian), Pettycur. × 4. (Photo: Photographic Studio, Natural History Museum, London.)

leaves, the stomata being restricted to two bands along each leaf in *Paralycopodites* (Graham, 1935), whereas in *Ulodendron* they are more evenly distributed (Thomas, 1967b); but this may just reflect differences at the species level.

Neither DiMichele (1980) nor Thomas and Brack-Hanes (1984) give any opinion as to the taxonomic position of *Paralycopodites*. However, the likelihood that it bore *Flemingites* strobili must surely place it at least as a satellite form-genus of the Flemingitaceae, as defined by Thomas and Brack-Hanes.

A second, but less abundant, lycopsid in the swamps produced the stem known as *Lepidophloios scottii* Gordon (1908b). These stems have a siphonostele surrounding a mixed pith, a feature which is normally regarded as relatively 'primitive'; other *Lepidophloios* species usually have a fully medullated siphonostele, except in the more distal branches (DiMichele, 1979). Gordon noted that it has leaf cushions similar to those seen in the adpression species *Lepidophloios scoticus* Kidston, which occurs commonly in the Lower Carboniferous of Scotland (see also Galtier and Scott, 1986b). Sections through stomata on the leaf cushions have been described by Thomas (1974).

*Lepidophloios* is thought to have borne megasporangiate strobili known as *Lepidocarpon* (Phillips, 1979; Thomas, 1981b). It is not surprising, therefore, that isolated sporophylls known as *Lepidocarpon wildianum* Scott (1900, 1901) are associated with *Lepidophloios scottii* at Pettycur. Being found as isolated sporophylls supports Phillips' (1979) view that they operated as seed-like disseminules. The Pettycur sporophylls are very similar to *Lepidocarpon lomaxii* Scott from the Upper Carboniferous coal-balls, and Scott distinguished them taxonomically mainly because they came from stratigraphically different horizons. Until complete strobili are found at Pettycur, however, it will be impossible to make any final decision as to the relative taxonomic positions of *L. wildianum* and *L. lomaxii*. The relationship of *L. wildianum* to the adpression from the Pettycur mudstones, described by Rex and Scott (1987) as *Lepidocarpon* cf. *waltonii* Chaloner, is also at present unclear.

A third type of lycopsid stem from the Pettycur Limestone is known as '*Lepidodendron*' *pettycurense*. It often quoted in the literature as a prime example of a primitive *Lepidodendron*, since it has a solid protostele (e.g. Taylor, 1981). However, it is only known from two decorticated axes (Kidston, 1907), and there is little definitive evidence that the fossil really is a *Lepidodendron*. Significantly, the axes are found closely associated with *Mazocarpon pettycurense* Benson (1908) sporophylls, this type of fructification usually being linked with the Sigillariostrobaceae (Schopf, 1941). To determine the taxonomic position of these stems will clearly need better preserved material.

The *Mazocarpon* described by Benson (1908) is only known from incomplete sporophylls, but appears to represent a reproductive strategy similar to that adopted by *Lepidocarpon* (see above). If correctly assigned to the Sigillariostrobaceae, these sporophylls are the oldest evidence of this family in the fossil record.

A fourth type of lycopsid fructification is *Lepidostrobus cylindrica* Gordon M.S. Although mentioned in species lists for the site (Gordon, 1914; Walton and Long, 1964), it has never been effectively published and so for the time being must remain a *nomen nudum*.

### Equisetopsida

The remains of two equisetopsid orders have been found at Pettycur: the Equisetales ('archaeocalamitids') and Bowmanitales ('sphenophylls'). Equisetalean stems are relatively abundant, although Rex and Scott (1987) state that they are often preserved 'in the immature form'. They were originally described as *Calamites pettycurensis* Scott (1902) and then *Protocalamites pettycurensis* D.H. Scott (1908-1909). Chaphekar (1963) argued that the differences from '*Archaeocalamites*' *goeppertii* (*Protocalamites goeppertii* (Solms-Laubach) Bateman) merely reflected different positions within the plant. However, Bateman (1991) has found that the Pettycur stems have consistently fewer primary vascular strands than typical *P. goeppertii*, such as found at Laggan and Loch Humphrey Burn (Walton, 1949b; Chaphekar, 1963), and that they probably represent smaller, shrubby plants.

Associated with these stems are strobili, *Protocalamostachys pettycurensis* Chaphekar (1963). Although never found in organic attachment, the structure of the pedicle is very similar to that of the smaller branches of *Protocalamites pettycurensis*. It differs from *Protocalamostachys arranensis* Walton found at Laggan (p. 155) in being smaller, having more sporangiophores in each whorl, fewer vascular strands in the pedicle, and containing smaller spores (Bateman, 1991).

The Bowmanitales are rare in the Pettycur Limestone. Stems described by Williamson (1874a) and Williamson and Scott (1894) have a primary xylem strand with the characteristic triangular cross-section of this order (cf. Meyen, 1987, fig. 21), and are known as *Sphenophyllum insigne* Williamson. Two specimens have been reported with roots attached to one side (Williamson and Scott, 1894), which supports the evidence from the Upper Carboniferous that *Sphenophyllum* was a creeping or scrambling plant (Batenburg, 1981).

The order is generally rare in the Lower Carboniferous, although there is evidence of its range extending down to the Upper Devonian (Remy and Spassov, 1959). Petrifactions have also been reported from Saalfeld (Solms-Laubach, 1896) and the Montagne Noire (Galtier, 1970), but the Pettycur specimens remain the best documented Lower Carboniferous examples to date.

*Cheirostrobus pettycurensis* (Figure 5.25) is also often included in the Bowmanitales, based partly on its occurrence in a deposit supposedly 'rich with *Sphenophyllum* stems' (e.g. Taylor, 1981); in fact such stems are relatively rare at Pettycur, and Scott (1897) argued that they could not have borne *Cheirostrobus* strobili. Nevertheless, there is an underlying similarity between the peltate sporangiophores of *Cheirostrobus* and those of *Bowmanites* and other bowmanitalean strobili. The most significant difference is that *Cheirostrobus* is a far more complex structure than any other reported bowmanitalean strobilus, and it superficially resembles certain lycopsid strobili such as *Flemingites*. It has whorls of thirty-six sporophylls, each sporophyll bearing four elongate sporangia (Scott, 1897). Its taxonomic position is uncertain, but is usually assigned to a monospecific family (Cheirostrobaceae) within the Bowmanitales. If correct, it suggests that there was a marked reduction in structural complexity of bowmanitalean strobili during the Carboniferous. It remains a considerable palaeobotanical enigma, and is only known from Pettycur.

### Filicopsida

The ferns fall broadly into two groups: (1) species of Zygopteridaceae (Coenopteridales), which occur mainly in the Zygopterid Limestone; and (2) species of Stauropteridaceae (Coenopteridales) and Botryopteridaceae (Botryopteridales), which

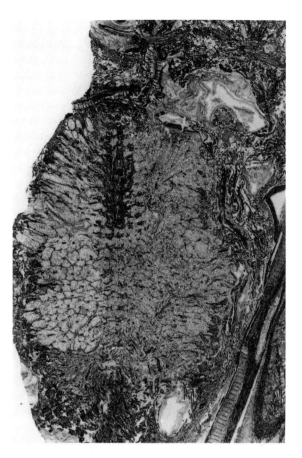

**Figure 5.25** *Cheirostrobus pettycurensis* Scott. Longitudinal section through complex equisetopsid cone showing distribution of sporangia; Natural History Museum, London, SC.3661. Pettycur Limestone (Asbian), Pettycur. × 2. (Photo: Photographic Studio, Natural History Museum, London.)

occur mainly in the Pettycur Limestone (for details of their distribution, see Rex and Scott, 1987, fig. 11).

### Zygopterid community

Pettycur is probably the best known British locality for zygopteridacean ferns, and it has provided valuable information on the evolution of the vascular systems of the stem (Gordon, 1911a, b, 1912; Dennis, 1974) and the phyllophores (Gordon, 1911b; Galtier, 1964). They appear not to have been growing in the main peat-swamp, but were probably a pioneer community which would have invaded the areas nearer the volcanic centres during periods of eruptive quiescence (Rex and Scott, 1987). The best understood of the

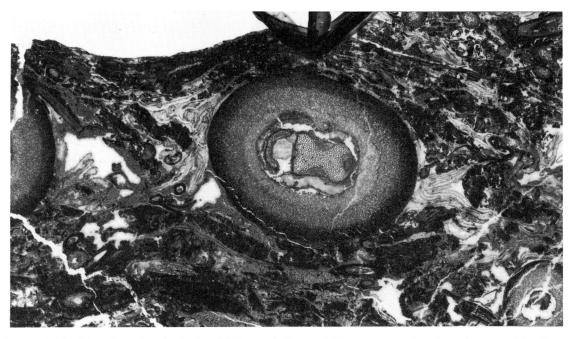

**Figure 5.26** *Metaclepsydropsis duplex* (Williamson) Bertrand. Transverse section through fern rachis; Natural History Museum, London, specimen WC.223. Pettycur Limestone (Asbian), Pettycur. × 4. (Photo: Photographic Studio, Natural History Museum, London.)

Pettycur species is *Metaclepsydropsis duplex* (Figure 5.26; Williamson, 1874b; Bertrand, 1909; Gordon, 1911b; Chaphekar and Alvin, 1972). For many years, it was only known from here, although there are now reliable records also from Glenarbuck (see below), Roannais (Galtier, 1970) and Esnost (Galtier, 1980). It is interpreted as a scrambling plant with little or no secondary wood in the stems and bearing upright fronds. The branching pattern of the vascular traces within the fronds (Gordon, 1911b, text-figs 2–3) is a characteristic feature of the early members of the family. Fertile pinnules associated with these fronds have been named *Musatea duplex* Chaphekar and Alvin (1972); these have sporangial clusters apparently borne superficially at the margins of the abaxial surface (Galtier, 1981). They are similar to fructifications associated with other Lower Carboniferous zygopteridaceans (Galtier, 1968), but contrast markedly with the more complex *Biscalitheca* borne by Upper Carboniferous to Permian zygopteridaceans (Mamay, 1957).

Of similar habit and anatomy was *Diplolabis roemeri* (Figure 5.27) (synonyms *Zygopteris pettycurensis* Gordon, *Diplolabis esnostensis* Renault and *D. forensis* Renault) (Renault, 1896; Gordon, 1909, 1911a). As with *Metaclepsydropsis*, there is little evidence of secondary wood development. The fertile pinnules were also similar to those of *Metaclepsydropsis*, and have been named *Musatea globata* Galtier (1968).

The third of the Pettycur zygopteridacean axes, *Metadineuron ellipticum*, is only known from fragments of phyllophore (Kidston, 1908; Galtier, 1964, 1970). It is presumed to have been of similar habit to the other members of the family from Pettycur.

*Filicopsids of the swamp community*

Within the peat-forming swamp, the Coenopteridales were represented by *Stauropteris burntislandica* (Williamson, 1874b; Bertrand, 1907, 1909). It has also been reported from Roannais in central France (Galtier, 1971), but most work on the species has been based on Pettycur material. Although the overall form of the plant has not been confirmed, it is assumed to have been herbaceous. The fronds were reconstructed by Surange (1952a), and are generally similar to those of the Zygopteridaceae, including the presence of basal aphlebiae (Lacey et al., 1957), but have a less-planated, more primitive aspect.

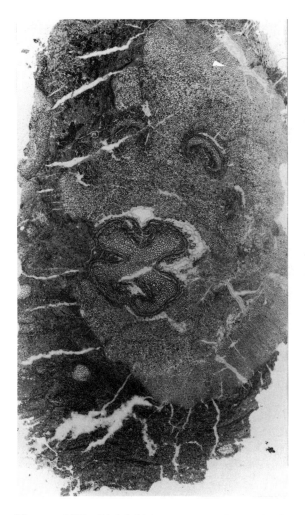

**Figure 5.27** *Diplolabis roemeri* (Solms-Laubach) Bertrand. Transverse section through fern rachis; Natural History Museum, London, specimen GC.789. Pettycur Limestone (Asbian), Pettycur. × 4. (Photo: Photographic Studio, Natural History Museum, London.)

Of particular interest is the fact that *S. burntislandica* was heterosporous. The megasporangia had been initially identified as *Bensonites fusiformis* R. Scott (1908) but the connection with *Stauropteris* was confirmed by Chodat (1912; see also Surange, 1952a). They are spindle-shaped structures, consisting mostly of parenchyma with a longitudinal vascular strand, but with a distal cavity that opens out to the sporangium apex via a narrow tube. The cavity contains a tetrad of two operative and two aborted megaspores, corresponding to *Didymosporites scottii* Chaloner (1958) (see also Hemsley, 1990b). Similar dispersed megaspores are known from various Lower Carboniferous localities in Britain (Chaloner, 1958). For some time, these sporangia were thought of as possible ancestors of gymnosperm seeds (e.g. the nucellar modification concept of Andrews, 1961), but this view has now lost favour (see discussion on the Whiteadder earlier in this chapter). Instead, it would appear that it represents a seed-like reproductive strategy, comparable to that achieved by the lycopsids (*Lepidocarpon*) and equisetes (*Calamocarpon*).

Microsporangia have not been found in organic connection with *S. burntislandica*, but R. Scott (1908) reported that structures very similar to *Stauropteris oldhamia* Binney sporangia occur in close association. Unlike most fern sporangia, they lacked an annulus, and spore-release was achieved via a small stomium.

The Botryopteridales are represented by *Botryopteris antiqua*, for which this is the type locality (Kidston, 1908; Benson, 1911; Surange, 1952b; Holden, 1962). Holden's work, in particular, has clarified the overall habit of the plant, which seems to have had a creeping stem, giving off erect fronds with digitate, three-dimensional pinnules, and a similarity to the adpression form-genus *Rhodeopteridium*. A distinctive character of the fronds is the frequent presence of small plantlets, which were presumably a means of vegetative propagation (see also Galtier, 1969), a feature generally rare in the Filicopsida. The sporangia were borne laterally to the ultimate segments of the frond (Galtier, 1967, 1981). They have a transverse annulus and contain spores of the form-genus *Granulatisporites*.

There has been some disagreement as to the taxonomic position of *B. antiqua*. It differs from typical *Botryopteris* (e.g. *B. forensis* Renault, *B. globosa* Darrah), which has fertile organs consisting of large clusters of several hundred sporangia. Good (1981) argued that it should therefore be transferred to *Psalixochlaena*, but this ignores the differences in vascular structure and position of attachment of the sporangia (Holden, 1960; Holmes, 1977, 1981). It is arguable that *B. antiqua* should be transferred to a different form-genus but, in the absence of a detailed taxonomic analysis of the problem, the traditionally accepted combination is used here.

### Lagenostomopsida

Unlike most other Lower Carboniferous petrifaction sites in Britain, Pettycur does not have a diverse assemblage of pteridosperms; there are just two types each of stem and seed, and one of pollen-bearing organ. The most abundant fossils are the stems *Heterangium grievii*, for which Pettycur is the type and best locality (Williamson, 1873; Williamson and Scott, 1895; Benson, 1933). This is the best known of the 'primitive' group of *Heterangium* species, which are assigned to the subgenus *Heterangium* (synonym *Euheterangium* auct.), and is characterized by a mesarch protostelic vascular system and transverse sclerotic plates in the cortex. The latter feature imparts a transverse bar-pattern to the outer surface of the stems, allowing a correlation with *Sphenopteris elegans* Brongniart and *S. affinis* Lindley and Hutton types of foliage preserved as adpressions. These fronds have deeply dissected pinnules, in contrast to the *Eusphenopteris* fronds borne by *Heterangium* subgenus *Polyangium* from the Upper Carboniferous, which have rounded pinnules (Shadle and Stidd, 1975).

The present distribution of *H. grievii* suggests that it came from a plant that was abundant in the Pettycur Limestone peat-swamp (Rex and Scott, 1987), and was probably a ground-cover creeper, growing below the arborescent lycopsids.

Associated with *H. grievii* are small *Sphaerostoma ovale* seeds (Williamson, 1877; Benson, 1914). Although never found in organic connection, Benson argued strongly that they were parts of the same plant. Such seeds have a number of 'primitive' features, such as the integument and nucellus being only partially fused and an abscission layer between the integument and the cupule. However, the salpinx is considerably reduced and there is a well-developed micropyle. It is thus apparently intermediate between the primitive lagenostomalean seeds from the Tournaisian (e.g. *Stamnostoma* from the Whiteadder – see above) and *Lagenostoma* from the Upper Carboniferous.

Also in association with *H. grievii* are some poorly preserved pollen-bearing organs identified by Benson (1922) as *Heterotheca grievii* (later renamed *Bensoniotheca grievii* by Mickle and Rothwell (1979)). Consisting of an ovoid cluster of completely fused sporangia, they contrast markedly with the loose clusters of free sporangia associated with other Lower Carboniferous *Heterangium* stems, and known as *Telangium* or *Telangiopsis* (e.g. Jennings, 1976). It is possible, however, that they are merely immature forms of *Telangium*.

The second type of pteridosperm stem found at Pettycur is *Rhetinangium arberi* Gordon (1912), for which this is again the type locality. Like *Heterangium*, these stems have a single protostele, but it is exarch and there is no evidence of sclerotic plates in the cortex.

No seeds have been found in organic connection with *Rhetinangium* at Pettycur, but Gordon (1910a) reported a poorly preserved example of what seemed to be *Physostoma* in close association. Gordon's description is only brief and was not accompanied by an illustration. It is perhaps worth noting that in the Tournaisian assemblage at the Whiteadder, there is evidence that *Rhetinangium* bore *Genomosperma* seeds.

Walton and Long (1964) reported *Amyelon* from Pettycur, which may represent pteridosperm roots. However, they have not been described in the literature.

Rex and Scott (1987) list and illustrate a number of pteridosperm foliage fragments preserved as adpressions. However, there is no published taxonomic account of these fossils, and they are thus difficult to judge.

### General remarks

This site has yielded the most diverse Visean plant petrifactions known from Britain, and is of great historical significance, having been central to the work of such leading palaeobotanists as W.C. Williamson, D.H. Scott and W.T. Gordon. One of the plant-bearing horizons (the Pettycur Limestone) is the best known example of an *in situ* peat-swamp from the Lower Carboniferous. It provides a useful comparison with the Late Carboniferous coal-swamps, which were not formed in a volcanic setting, but in a more tranquil, lower delta-plain setting (see Chapter 6). Elsewhere in Britain, Glenarbuck (p. 164) has yielded the most comparable Lower Carboniferous fossil plant assemblage, being also dominated by lycopsids, but the preservation is not so good and it does not have such diverse fern or equisete components. The ferns (especially those from the Zygopterid Limestone) provide a ready comparison with upper Visean petrifactions from Esnost and Roannais in France (Galtier, 1970, 1971, 1980; Scott *et al.*, 1984), but there, in contrast to Pettycur, pteridosperms are virtually absent. Also similar is the Visean assemblage from Glätzisch-Falkenberg, Germany (Solms-Laubach, 1892), but the absence of recent work on that assemblage makes a detailed comparison difficult (Scott *et al.*, 1984). Visean petrifactions from outside Europe are virtually unknown, and this presumably reflects the unusual conditions necessary for the formation and preservation of petrified plant fossils (in this case lake sediments and peats accumulating in a lowland, volcanic environment).

## Conclusion

Pettycur is one of the classic Palaeozoic palaeobotany sites in Britain, yielding anatomically-preserved plant remains about 330 million years old. Together with the peat-floras of the Devonian Rhynie Chert (see Chapter 4) and the Upper Carboniferous coal-balls (see Chapter 6), it was for many years regarded as the best guide to the vegetation of the Palaeozoic. Although many other Lower Carboniferous sites, in this country, France and Germany, are now known to yield anatomically-preserved plant fossils, Pettycur remains of great palaeobotanical significance, particularly for the study of the early fern-like plants and club-mosses. It is also the only known locality for the horsetail family, the Cheirostrobaceae. Despite being one of the oldest known types of equisete, it has by far the most complex reproductive organs ('cones') that have been found in that group of plants, either living or in the fossil record.

## KINGSWOOD END

### Highlights

Kingswood End has yielded a distinctive Visean assemblage of petrifactions and fusain fragments. It differs markedly from the nearby Pettycur assemblage, which is thought to represent a different environmental setting, and is the only known locality for the enigmatic microsporangiate organs *Phacelotheca* and *Melissiotheca*. It has considerable potential for future work on the vegetation of this time.

### Introduction

This site lies on the hillside above Pettycur, near Burntisland, Fife (NS 265864) (Figure 5.22). It was discovered as part of a re-investigation of the Lower Carboniferous plant-bearing rocks near Pettycur, during the mid-1980s. A preliminary report on the assemblage is provided by Scott *et al.* (1986), but to date only two taxa have been subject to detailed systematic analysis (Meyer-Berthaud, 1986; Meyer-Berthaud and Galtier, 1986a).

### Description

#### Stratigraphy

The geology of this site is detailed by Scott *et al.* (1986) and Rex and Scott (1987) (Figure 5.28). The plant fossils occur in the Kingswood Limestone, a brown, fine-grained microsparite that occurs as large (one metre plus), derived blocks incorporated in the upper part of a volcanic agglomerate. Although the limestone bed has not yet been found *in situ* in a sedimentary sequence, it is believed to have been a lacustrine deposit. Palynological evidence given by Scott *et al.* suggests an Asbian age.

#### Palaeobotany

The plant fossils here fall broadly into two groups: petrified lycopsids, and fusainized pteridosperms, ferns and equisetes. Taxa reported to date are as follows:

Lycopsida:
    *Oxroadia* cf. *gracilis* Alvin
    *Achlamydocarpon* sp.

Equisetopsida:
    *Protocalamites pettycurensis* (Scott) Scott

Filicopsida:
    Unidentified fragments

Lagenostomopsida:
    *Lyginorachis* spp.
    cf. *Calathospermum* sp.
    *Dadoxylon* sp.

Uncertain affinities:
    *Phacelotheca pilosa* Meyer-Berthaud and Galtier
    *Melissiotheca longiana* Meyer-Berthaud
    *Amyelon* sp.

In addition, adpressions from a mudstone exposed here have been identified by Rex and Scott (1987) as *Lepidostrobus* sp. and *Lepidostrobophyllum* sp.

# Kingswood End

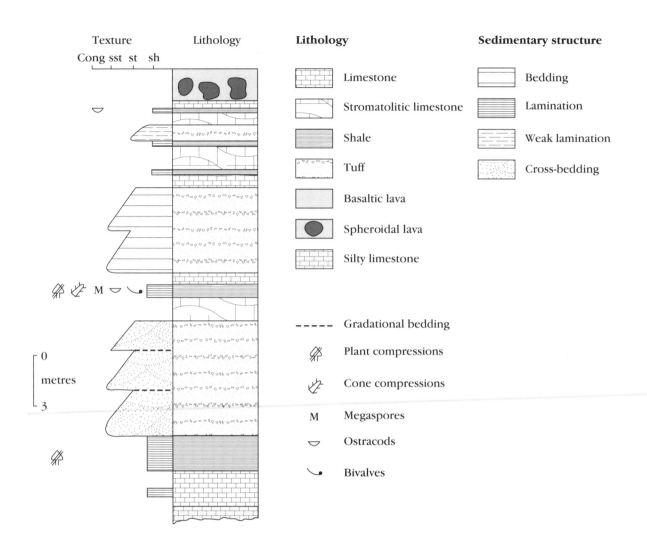

**Figure 5.28** Stratigraphical section at Kingswood End, showing distribution of plant fossils. Based on Rex and Scott (1987, figure 4).

## Interpretation

*Oxroadia* stems are amongst the most abundant and well preserved fossils from here. They have been interpreted as having lived near the margins of the lake in which the limestone was deposited, and had been subjected to little transportation. Their anatomy compares closely with the well-documented *Oxroadia gracilis* Alvin from Oxroad Bay (p. 131), except that the larger stems developed a siphonostele, as opposed to a protostele. As pointed out by Scott *et al.* (1986), however, the Oxroad Bay stems do not reach such a large diameter as the Kingswood End specimens, which may explain the apparently anomalous vascular structure.

Also abundant here are isolated lycopsid megasporophylls. They compare with *Achlamydocarpon varius* (Baxter) Taylor and Brack-Hanes, but have a rather simpler sporangial wall structure (Scott *et al.*, 1986). Phillips (1979) suggests that *Achlamydocarpon* may have been borne by giant lycopsids with *Lepidodendron*-like trunks, but no evidence of the latter has so far been reported from Kingswood End.

The rest of the assemblage consists mainly of fusain fragments. Scott *et al.* (1986) regard them as the transported remains of plants growing some distance away from the lake, and which had been subject to burning induced by volcanic eruption. They report just two decorticated stems of *Protocalamites*, which are principally of interest in that they have secondary wood with pronounced growth rings. Rex and Scott (1987) suggest that this may either reflect local climatic disruption, due to volcanic activity, or variations in the availability of groundwater in a stressful environment.

The rare, fusainized fragments of ferns found to date are too incomplete to allow identification, even at the rank of form-genus. They include a detached fragment of a fertile pinnule with annulate sporangia, and a partly decorticated rachis of possible zygopteridacean affinities (Scott et al., 1986).

The most abundant pteridosperm fragments are rachides probably from lagenostomalean fronds. Scott et al. (1986) compared some of them with *Lyginorachis papilio* Kidston, *L. trinervis* Calder and *L. taitiana* Crookall, but they are too incomplete to allow a definite identification. Also present are fragments of fusainized wood (*Dadoxylon* sp.).

Although no seeds have been found, Scott et al. (1986) illustrate and describe what may be part of a *Calathospermum*-like cupule. Subsequently, however, Rex and Scott (1987) listed it as the calamopityalean foliage *Kalymma*.

Two examples of microsporangiate organs of ?pteridosperms have been described. *Phacelotheca pilosa* Meyer-Berthaud and Galtier (1986a) refers to *Telangium*-like synangia, borne in trusses at the ends of dichotomous axis-systems. Each synangium consists of two to four elongate sporangia, which are fused at the base, and dehisce along a ventral, longitudinal slit. As pointed out by Meyer-Berthaud and Galtier, *Telangium*, as currently interpreted, includes a diverse range of microsporangiate organs. They therefore opted to place the Kingswood End specimens in their own, tightly circumscribed form-genus. Virtually all characters point to it belonging to the Lagenostomales, except that the sporangial trusses are rather more complex than in other described species, and the pre-pollen has an ornamented sexine.

A rather more complex structure was described by Meyer-Berthaud (1986) as *Melissiotheca longiana*. It consists of synangia comprised of 50–150 elongate sporangia, embedded in a lobed, parenchymatous, basal cushion, which in turn was borne distally on a stalk. The pre-pollen are similar to those contained in *Phacelotheca*. Meyer-Berthaud pointed out a number of features which again suggest lagenostomalean affinities, but was unwilling to assign it there because of its complexity and the differences in pre-pollen structure.

Despite its geographical proximity, the Kingswood End assemblage contrasts markedly with that found at Pettycur (p. 141). The latter contains a much higher proportion of ferns and equisetopsids, and relatively few pteridosperms. The lycopsid components are also quite different, consisting at Pettycur mainly of *Paralycopodites* and *Lepidophloios*, and at Kingswood End of *Oxroadia*. The Kingswood End assemblage compares more closely instead with those from some of the older (Tournaisian) Scottish sites, such as Oxroad Bay (p. 127), although it has yet to yield the diversity of pteridosperm seeds that has been found at many of these localities.

## Conclusion

Kingswood End is one of a series of Lower Carboniferous sites in southern Scotland yielding plant fossils with details of their anatomy preserved, and some 330 million years old. It is almost identical in age to the nearby Pettycur site (see above), but represents a significantly drier habitat, which thus supported a different type of vegetation. In particular, there is a much higher proportion of fossils of seed plants than at Pettycur; these are normally taken as good indicators of drier growing conditions. The palaeobotany of this site has still not been completely investigated, although it has already yielded two new and distinctive types of seed plant pollen-bearing organs (*Melissiotheca* and *Phacelotheca*). There is much potential here for new discoveries, which may help in determining aspects of the early evolution of the seed plants.

## LAGGAN

## Highlights

This is a significant site for Lower Carboniferous plant petrifactions, particularly for lycopsids. They include the best known examples of *in situ* petrified stumps of arborescent lycopsids, and some early examples of herbaceous lycopodiaceans. It has also yielded some of the best known examples of anatomically-preserved strobili of the Archaeocalamitaceae (Sphenopsida).

# Laggan

## Introduction

The *in situ* stumps of fossil trees preserved in Lower Carboniferous volcanogenic deposits at Laggan, on the north-east coast of the Isle of Arran (NR 982506), were discovered by E.A. Wunsch in 1865 (the early history of work on the site is summarized by Walton, 1935). Despite early contributions by Carruthers (1869) and Binney (1871), significant progress was not made until W.C. Williamson visited the site in 1877, when a considerable quantity of specimens was collected (a colourful account of the visit is given by Williamson, 1896, pp. 169-177). The lycopsid stumps were described by Williamson (1880, 1883, 1895), and later by Walton (1935). Other elements in the assemblage have been described by Calder (1935), Walton (1949b), Fry (1954), Beck (1958), Pant and Walton (1961) and Chaphekar (1963), mainly based on specimens from Williamson's original collection.

## Description

### Stratigraphy

The most recent detailed accounts of the geology at Laggan appear to be by Tyrrell (1928) and Walton (1935), although no stratigraphical log was given. The fossils occur in a basaltic ash deposit, some distance below the Corrie Limestone, probably in the Upper Sedimentary Group. Their exact chronostratigraphical position is uncertain, but the deposits probably fall in the upper Visean.

### Palaeobotany

The following taxa, preserved as calcite petrifactions, have been described to date:

Lycopsida:
 *Lepidophloios wuenschianus* (Williamson) Walton
 *Levicaulis arranensis* Beck
 *Paurodendron arranensis* Fry
 *Lycostachys protostelicus* Pant and Walton
 *Lepidostrobus* (?) *ambiguus* Binney
 *L. wuenschianus* Binney
 *L. latus* Binney
 *Lepidocarpon* sp.
 *Stigmaria* sp.

Equisetopsida:
 *Protocalamites goeppertii* (Solms-Laubach) Bateman
 *Protocalamostachys arranensis* Walton

Lagenostomopsida:
 *Lyginorachis waltonii* Calder
 *Lyginorachis* sp.
 *Kaloxylon* sp.

## Interpretation

### Lycopsida

The most famous of the Laggan fossils are the *Lepidophloios wuenschianus* stumps (Williamson, 1880, 1883, 1895; Walton, 1935). They are the best known examples of *in situ*, petrified stumps of arborescent lycopsids; the only other reported specimens are *Lepidodendron saalfeldense* Solms-Laubach from Germany, which are smaller and not so well preserved (Solms-Laubach, 1896). They are particularly important as the stumps contain pieces of the stele from different parts of the trunk (e.g. Figure 5.29). This preservational feature appears to have resulted because most of the outer cortex disintegrated shortly after the death of the plants, and the stelar column then collapsed down into the stump, where it became lithified. From these fossils, Walton (1935) was able to deduce that at the base of the trunk there was only a relatively slender protostele, surrounded by a thick layer of secondary wood. In more distal positions, however, the primary xylem formed a wider, medullated siphonostele, but the surrounding secondary wood became concomitantly narrower. These results had important consequences for subsequent ideas about the developmental growth of the arborescent lycopsids (Andrews and Murdy, 1958; Eggert, 1961). Where pieces of the cortex are still preserved, they show that this tissue had clearly developed bands of secretory cells.

Also found in the stumps are fragments of the distal branches of the tree, which are similar to the adpression species *Lepidophloios scoticus* Kidston. They can also be compared with the *Lepidophloios* shoot described from Bearsden by Galtier and Scott (1986b). However, similar shoots are also found associated with *Lepidophloios scottii* Gordon at Pettycur (see above), which differs from the Laggan fossils in not having secretory cells in the cortex. Evidently, different species of

**Figure 5.29** *Lepidophloios wuenschianus* (Williamson) Walton. Transverse section through lycopsid stem; Natural History Museum, London, specimen WC.456a. Visean ash deposits, Laggan. × 2. (Photo: Photographic Studio, Natural History Museum, London.)

lycopsid could produce shoots of essentially identical form.

DiMichele (1979) used *L. wuenschianus* as the 'type' of one of the two groups of *Lepidophloios* species that he recognized, characterized by features of periderm structure, leaf cushion anatomy, and lateral branch architecture. Other members of the group include *L. scottii* Gordon from Pettycur (see above) and *L. johnsonii* from the basal Upper Carboniferous of North America, but *L. wuenschianus* is by far the best known. The differences between the two groups of *Lepidophloios* may reflect palaeoenvironmental differences, the *L. wuenschianus* group occupying more open habitats, whereas the *L. harcourtii* group were forest dwellers.

It has been argued that *Lepidophloios* bore *Lepidocarpon* strobili (e.g. Phillips, 1979). It is not surprising, therefore, that Walton (1935) recorded *Lepidocarpon* in the Laggan stumps, although no description was given.

Binney (1871) described three species of *Lepidostrobus* from Laggan, but they are all incomplete and in need of renewed investigation.

Two other types of lycopsid in the Laggan assemblage were herbaceous, and are both unique to this locality. The stems known as *Levicaulis arranensis* Beck (1958) were less than 40 mm in diameter, with a terete protostele, and apparently without ligules. An associated strobilus known as *Lycostachys protostelicus* Pant and Walton (1961) has a central axis with an almost identical structure and is assumed to have belonged to the same plant. The preserved part of the strobilus was exclusively microsporangiate, although some megaspores were found in association. If this association is merely coincidental and the cone was homosporous, as suggested by Pant and Walton, then the affinities of this plant probably lie with the Lycopodiaceae.

A second type of herbaceous lycopsid is represented by the stems *Paurodendron arranensis* Fry (1954). Unlike *Levicaulis*, *Paurodendron* is ligulate and thus more similar to *Oxroadia* from Oxroad Bay (p. 131), except for details of the vascular anatomy. It was placed in synonymy with *Selaginella fraipontii* (Leclercq) Schlanker and Leisman from the Upper Carboniferous of North America by Schlanker and Leisman (1969). However, in view of their stratigraphical separation, and the fact that details of the strobili are known from *S. fraipontii*, but not the Laggan plant, it seems wiser to keep them separate. Bateman (1988) has argued that *Paurodendron* and *Oxroadia* represent a distinctive order of lycopsids, which is probably a sister group of the Lepidocarpales, although he has subsequently recanted this view (Bateman *et al.*, 1992).

### Equisetopsida

The Laggan equisetopsid stems, known as *Protocalamites goeppertii* (Walton, 1949b) differ from *Protocalamites pettycurensis* from Pettycur in having more primary vascular strands, and probably representing larger plants (Bateman, 1991). They are also associated with larger strobili, known as *Protocalamostachys arranensis* Walton (1949b). Historically, the latter are of interest because they were the first strobili of these primitive equisetopsids to be discovered petrified, which helped clarify the distinctive characters of the Archaeocalamitaceae (i.e. the sporangia were borne on peltate sporangiophores, and there were few intervening sterile bracts).

### Pteridosperms

There has been little work on the pteridosperms at Laggan. Calder (1935) described a distinctive lagenostomalean rachis as *Lyginorachis waltonii* Calder, whilst Walton (1935) recorded *Kaloxylon*, which usually refers to lagenostomalean rooting structures.

### General remarks

Laggan is another of the internationally important Lower Carboniferous petrifaction sites in Britain, which includes a particularly significant lycopsid component. Some of the species have been described from elsewhere, such as *Lepidophloios wuenschianus* from Dalmeny (Seward and Hill, 1900), *Protocalamites goeppertii* and *Protocalamostachys arranensis* from Loch Humphrey Burn, and *Lyginorachis waltonii* from Oxroad Bay and Loch Humphrey Burn. Nevertheless, the balance of taxa at Laggan remains unique, as well as being the only known locality for two herbaceous lycopsids.

## Conclusion

Laggan is one of a series of important sites that show Lower Carboniferous rocks in southern Scotland, and which yield plant fossils with their anatomy preserved; they are probably about 340 million years old. This particular site is especially important for its club-mosses, which include both trees (*Lepidophloios*) and small, herbaceous forms (*Levicaulis* and *Paurodendron*). It has also yielded important specimens of horsetails, including a number of reproductive cones, that have been important for understanding the evolutionary history of this group of plants.

## LOCH HUMPHREY BURN
*R.M. Bateman and C.J. Cleal*

## Highlights

Loch Humphrey Burn has yielded two pivotal assemblages of Lower Carboniferous plant petrifactions, the lower reputedly of Tournaisian and the upper of Visean age (Figure 5.30). The lower is dominated by filicopsids, unlike most Tournaisian petrifaction assemblages; it includes the earliest possible examples of tedeleacean, corynepteridacean and marattialean ferns. The upper assemblage includes the only known fertile specimens of the progymnosperm order Protopityales. It has also provided invaluable information on the anatomy of the early sphenopsid *Protocalamites* and several evolutionarily significant pteridosperms such as *Calathospermum*. Finally, it is one of the few Lower Carboniferous localities that yields both petrifactions and adpressions, thereby offering the potential for correlating features of anatomy and gross morphology. It is a site of outstanding palaeobotanical importance.

**Figure 5.30** Loch Humphrey Burn. Photograph taken during NCC-funded re-excavation of the site in 1985. The coarse dashed lines delimit the plant-rich volcanigenic sediments of Unit 4 (see Table 5.1). The finer dashed lines mark the bases of Bed 17 (the source of Walton's petrified nodules) and Bed 20 (rich in compressions, notably *Pothocites* cones). The overlying Unit 5 includes thin coals and represents a clastic swamp containing giant lycopsids; this correlates with the nearby Glenarbuck site. (Photo: R.M. Bateman.)

# Loch Humphrey Burn

## Introduction

This locality straddles a small tributary of Loch Humphrey Burn (NS 467753) (Figure 5.31). It is one of two classic palaeobotanical localities in the Kilpatrick Hills, near Glasgow (the other, Glenarbuck, is described in the next section). Lower Carboniferous plant fossils were probably first discovered at Loch Humphrey Burn in about 1870 by the Geological Survey, but the earliest documentary evidence is a letter that accompanied specimens sent by James Bennie to Robert Kidston in 1886 (Smith, 1960; Scott *et al.*, 1984). Kidston published many descriptions of the adpressions from here in his classic 1923–1925 monograph, but passed the petrifactions to D.H. Scott for description (Scott, 1899, 1902, 1918, 1920-23, 1924b). The next phase of collecting was undertaken during the early 1930s by Robert Brown, Jessie Wilson and John Walton, who amassed a considerable quantity of material. This formed the basis for a series of studies by Calder (1935), Walton (1940, 1949a, b, c, 1957), Lacey (1953), Smith (1959, 1960, 1962a, b, 1964b) and Chaphekar (1965) (see also Walton *et al.*, 1938). Most recently Scott and Bateman have clarified the stratigraphy and sedimentology at the site. Although Scott *et al.* (1984, 1985) reviewed some of the fossils discovered during their 1982 excavations, there have been only two detailed descriptions published to date (Meyer-Berthaud and Galtier, 1986b; Bateman, 1991).

## Description

### Stratigraphy

The geology of this site is briefly described by Scott *et al.* (1984, 1985), whose work was in turn amended by Bateman (unpublished) following re-excavation of the most important exposure. The sequence is in the lower part of the Clyde Plateau Volcanic Formation (Hall, 1978). The plant-bearing strata immediately underlie the lowermost of several lava flows and represent a fluvio-deltaic complex rich in volcanigenic sediments (Figure 5.32). Detailed logging of the section has resulted in a revised classification of the sequence into five lithostratigraphical units, which will be described fully elsewhere (Bateman, in prep., see summary in Table 5.1). Comparison with palynological evidence summarized by Scott *et al.* (1984, 1985) suggests that Units 1 and 2 are in the CM Zone (upper Tournaisian), Unit 3 in the lower Pu Zone (Chadian, lower Visean) and Units 4 and 5 in the uppermost Pu Zone (Holkerian, middle Visean). Scott *et al.* (1984) postulated a depositional break

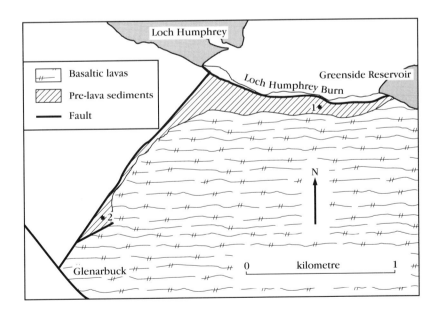

**Figure 5.31** Geological map of area south of Loch Humphrey in the Kilpatrick Hills, showing positions of Loch Humphrey Burn (1) and Glenarbuck (2) sites. Based on Scott *et al.* (1984, figure 5).

## Lower Carboniferous

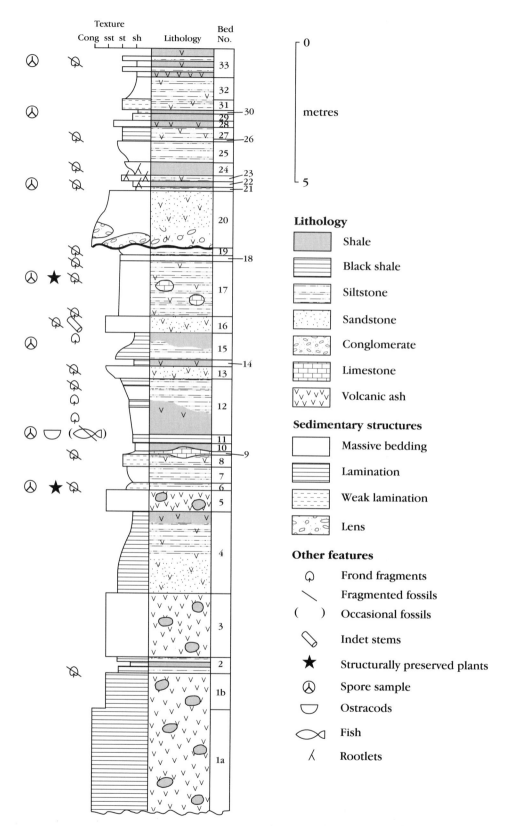

**Figure 5.32** Sedimentological log at Loch Humphrey Burn. Based on Scott *et al.* (1984, figure 8).

# Loch Humphrey Burn

**Table 5.1** Lithostratigraphy of Loch Humphrey Burn (R.M. Bateman, unpublished)

| Lithostratigraphical Unit | Main Lithologies | Environmental Interpretation | Bed Numbers of Scott *et al.* (1984, fig. 8) | Fossil Plant Assemblages |
|---|---|---|---|---|
| Unit 5 | Siltstones and fine sandstones with intercalated coarse sandstones, thin shales, coals and rooted palaeosols | Extensive flood plain | 21–33 | Compressions of rhizomorphic lycopsid rootlets, with rare rhizomorph and aerial axis fragments |
| Unit 4 | Medium to coarse, gritty sandstones, with a few thin shales | Fluvial channel | 13–20 | Nodular petrifactions of many organs representing a wide range of higher taxa (Bed 17); also compressions, especially of pteridosperm and putative progymnosperm foliage |
| Unit 3 | Thinly laminated shales and an impure limestone | Flood plain/ lacustrine | 9–12 | Compressions, especially pteridosperm and putative progymnosperm foliage |
| Unit 2 | Siltstones and fine sandstones | Flood plain | 6–8 | Nodular petrifactions of fragmentary filicopsids, sphenopsids and pteridosperms, especially reproductive organs (Bed 6) |
| Unit 1 | Poorly sorted, coarse sandstones and conglomerates, rich in volcanic ash and clasts | ?Channel/ flood plain | 1–5 | Large, partially petrified pteridosperm axes (Bed 1) |

approximating the Unit 2–3 boundary to explain this apparently long period of time relative to the thickness of the strata.

## Palaeobotany

Petrified, fusainized and adpression plant fossils have been found here. The anatomically-preserved fossils occur at three discrete horizons: (1) towards the bottom of Unit 1, (2) towards the bottom of Unit 2 (both reputedly lower Tournaisian) and (3) in the middle part of Unit 4 (middle Visean). The following list notes the distribution of each species among the numbered lithostratigraphical units, and deviates from the senior author's preferred classification for higher taxa:

Equisetopsida:
  *Protocalamites goeppertii* (Solms-Laubach) Bateman (2,4)
  *Protocalamostachys arranensis* Walton (2,4)
  *P.* cf. *farringtonii* (Bateman) (2)

Filicopsida:
  *Cladoxylon* cf. *taeniatum* Bertrand (2)
  *Hierogramma* sp. (2)
  *Syncardia* sp. (2)
  *Clepsydropsis* sp. (2)

*Metaclepsydropsis* sp. (2)
*Botryopteris* cf. *antiqua* Kidston (2)
cf. *Senftenbergia* sp. (2)
cf. *Musatea* sp. (2)
*Burnitheca pusilla* Meyer-Berthaud and Galtier (2)
*Etapteris tubicaulis* Göppert (4)

Progymnospermopsida:
*Protopitys scotica* Walton (4)

Gymnospermopsida

Lagenostomales:
*Eristophyton fasciculare* (Scott) Zalessky (?4)
*E. waltonii* Lacey (4)
*Bilignea resinosa* Kidston (?4)
*Lyginorachis trinervis* Calder (4)
*Lyginorachis* spp. several (2,4)
*Calathospermum scoticum* Walton (4)
*Geminitheca scotica* Smith (4)
cf. *Pullaritheca* sp. (4)

Calamopityales:
'*Calamopitys*' *radiata* Scott (?1)
*Kalymma* cf. *tuediana* Long (2)
*Kalymma* sp. (4)
*Alcicornopteris hallei* Walton (4)

Gymnosperms (*incertae sedis*):
*Amyelon* sp. (2)

The best preserved adpressions occur in the middle Visean part of the sequence (units 3 and 4). The following taxa have been identified:

Lycopsida:
*Lepidophloios* cf. *kilpatrickensis* Smith
*Stigmaria ficoides* (Sternberg) Brongniart

Equisetopsida:
*Archaeocalamites radiatus* (Brongniart) Stur
*Pothocites grantoni* Paterson

Filicopsida (?):
*Rhodeopteridium* sp.

Progymnospermopsida (?):
*Rhacopteris lindsaeformis* Bunbury
*R. inaequilaterata* Göppert
*R. robusta* Kidston
*R. petiolata* Göppert

Gymnospermopsida

Lagenostomales:
*Sphenopteridium pachyrrachis* (Göppert) Schimper
*S. crassum* (Lindley and Hutton) Kidston
*Sphenopteris affinis* Lindley and Hutton
*S. bifida* Lindley and Hutton
*Calathiops trisperma* Smith

Calamopityales:
*Spathulopteris ettingshausenii* (Feistmantel) Kidston
*S. obovata* (Lindley and Hutton) Kidston
*Staphylotheca kilpatrickensis* Smith
*Alcicornopteris convoluta* Kidston
*A. zeilleri* Kidston

## Interpretation

### Lycopsida

Although lycopsids are considered to be very poorly represented at Loch Humphrey Burn (Smith, 1964b), one of us (RMB) has found in Unit 5 frequent, poorly preserved *Stigmaria* and an axial compression that compares with the outer surface of the petrified *Lepidophloios kilpatrickensis* Smith from nearby Glenarbuck (see below).

### Equisetopsida

*Protocalamites goeppertii* occurs in both petrifaction assemblages here, and its adpression/cast analogue (*Archaeocalamites radiatus*) is frequent in the Visean part of the section. The anatomical details of the stem nodes described by Walton (1949b) prompted Chaphekar (1963) to unify *Archaeocalamites* with *Protocalamites* (the two principal form-genera for Lower Carboniferous sphenophytes), but this action was subject to detailed criticism by Bateman (1991).

Chaphekar (1965) identified adpressions from here as *Pothocites grantonii*, and showed that it was probably the fertile organ of *Archaeocalamites*. Bateman (1991) has demonstrated that a new specimen of petrified *Protocalamostachys arranensis* from Unit 4 is the anatomically-preserved analogue of *Pothocites grantonii*. *Protocalamostachys arranensis*, which was previously only known from a single fragment from Laggan (see above), provides a valuable comparison with the smaller *P. farringtonii* Bateman; fragments from both cone species were recorded in Unit 2 by Scott *et al.* (1985; see also Bateman, 1991; Hemsley *et al.*, in press).

# Loch Humphrey Burn

## *Filicopsida*

Only one petrified fern, the zygopterid phyllophore *Etapteris tubicaulis*, has been reported from the upper assemblage (Walton *et al.*, 1938; Scott *et al.*, 1984). It has not been described in detail, even though Loch Humphrey Burn is the only known British locality for this species.

The Tournaisian assemblage described below is much richer in anatomically-preserved ferns (Scott *et al.*, 1985). The Cladoxylales is represented by the stem *Cladoxylon* cf. *taeniatum* (its only British record) and its branches *Hierogramma* and *Syncardia*. No demonstrably cladoxylalean fructifications have yet been found.

The zygopterids are represented by the phyllophores *Clepsydropsis* and *Metaclepsydropsis*, this being regarded as the lowest recorded stratigraphical occurrence of the latter (Scott and Galtier, 1985). Scott *et al.* (1985) also recorded sporangia similar to *Musatea*, which is generally regarded as the fertile organ of *Metaclepsydropsis*. Another sporangial type recorded by Scott *et al.* (1985) appears to be a precursor of the more familiar genus *Corynepteris*, a relatively common fern in the Upper Carboniferous.

Scott *et al.* (1985) identified axes with a simple xylem anatomy as *Botryopteris* cf. *antiqua*, again the oldest known examples of this species. Associated isolated annulate sporangia are similar in general form to those known from *B. antiqua* (Galtier, 1967) but differ in the details of the annulus. Another coenopteridalean fructification was compared by Scott *et al.* (1985) with the tedeleacean form-genus *Senftenbergia*, which again would represent an oldest known record.

The only sporangium so far described in detail from Unit 2 is *Burnitheca pusilla* Meyer-Berthaud and Galtier (1986b). It consists of a bilaterally symmetrical cluster of eight sporangia, fused basally around a central column. Although it broadly resembles lagenostomalean microsporangiate organs such as *Telangium*, the form of the tapetum and the spores that it contains suggest affinities with the marattialean ferns. If this interpretation is correct, *Burnitheca* is the oldest known example of this extant order. Two further types of sporangia, possibly belonging to ferns, are mentioned by Scott *et al.* (1985) as 'Fructifications G and H'.

The evidence from Loch Humphrey Burn documents the rapid diversification of the ferns in the Early Carboniferous. The supposedly upper Tournaisian Unit 2 contains a remarkable number of first recorded occurrences of fern species, genera and even families, and its taxonomic composition is more consistent with the mid-Visean age attributed to the overlying units at Loch Humphrey Burn. Many of these early ferns lacked fully planated fronds, but possessed fructifications characteristic of recognized fern taxa. By the early Late Carboniferous, more modern-looking ferns had developed, especially in the palaeoequatorial areas. The fossils found at Loch Humphrey Burn, although fragmentary, are thus of considerable value in recording this group of plants at a key stage in its evolutionary history.

## *Progymnospermopsida*

Loch Humphrey Burn is one of the most important localities for Lower Carboniferous progymnosperms. They are thought to have been widespread in the equatorial and northern-temperate palaeolatitudes. However, progymnosperms are difficult to identify with confidence since this requires demonstrating the absence of ovules, and anatomically-preserved fertile material is rare. Loch Humphrey Burn has yielded the only known fertile specimens of the order Protopityales (Walton, 1957; Smith, 1962b). Stems from Loch Humphrey Burn share many anatomical features with *Protopitys buchiana* Solms-Laubach from Yorkshire and Falkenberg (Solms-Laubach, 1893; Walton, 1957, 1969); both have oval siphonosteles emitting distichous leaf-traces, features that characterize the Protopityales. However, the Loch Humphrey specimens have less extensive primary wood, no high biseriate rays, and secondary wood with multiseriate pitting, and so were placed in a new species, *Protopitys scotica* by Walton (1957).

Sporangia attached to these stems were described by Walton (1957) and Smith (1962b). This was the first recorded example of a plant that had both gymnospermous wood and pteridophytic reproductive organs.

Together with the correlation of *Callixylon* wood and *Archaeopteris* foliage by Beck (1960), this was one of the main arguments for recognizing the class Progymnospermopsida as a precursor to the pteridosperms (and arguably the conifers: Beck, 1970, 1971). The spores of *P. scotica* show little variation within sporangia, but considerable variation among adjacent sporangia, and the overall size distribution determined by Smith (1962b) is weakly bimodal. Thus, it is unclear whether the plant was heterosporous (Bateman and DiMichele, in press).

Archangelsky and Arrondo (1966) and Beck (1976) suggested that the adpression foliage *Rhacopteris* is progymnospermous, based on its similarity to *Archaeopteris*. The co-occurrence at Loch Humphrey Burn of adpressed *Rhacopteris* and petrified *Protopitys* lends some support to this argument, though the former is abundant and the latter rare.

## Pteridosperms

Seed plants dominate the assemblages in Units 3 and 4, but Scott *et al.* (1985) reported that they are subordinate in Unit 2. Both the Lagenostomales and Calamopityales are represented.

## Lagenostomopsida

Petrified stems of *Eristophyton* and *Bilignea* were probably found in the Visean part of the section, although their precise origin was not stated. They include the type and one of only two known specimens of *Eristophyton fasciculare* Scott (1899, 1902, 1918), and the only known specimen of *E. waltonii* Lacey (1953). The latter differs from *E. fasciculare* in having a larger parenchymatous pith with sclerotic nests, smaller primary xylem strands emitting undivided leaf traces, and in its ability to develop larger wood rays. *Bilignea resinosa* Scott (1924b) is similarly known from only a single specimen. It resembles *Eristophyton* in many characters, but has a large core of pitted tracheids interspersed with large resin sacs, rather than a parenchymatous pith. Although affinities with the cordaites (Scott, 1899, 1924b; Andrews, 1940; Lacey, 1953) and the calamopityales (Read, 1937) have been suggested for these genera, Long (1987, pers. comm. 1988) and Bateman and Rothwell (1990) have shown that they are lagenostomalean.

Abundant and diverse *Lyginorachis* petioles occur at Loch Humphrey Burn, mainly in the upper part of the section. Of these, only *L. trinervis* Calder (1935) has been formally described. Its unusual tripartite stele and rapid decrease in diameter along its 2 mm length suggest that it is the basal portion of a petiole. Similar petioles have recently been identified from approximately coeval deposits at Kingswood (see above). Another well-preserved petiole from Loch Humphrey Burn resembles *Lyginorachis waltonii*, first described from Laggan on the Isle of Arran (see above). Long (1987) has reported similar axes attached to *Eristophyton beinertianum* stems at Oxroad Bay (see above).

Loch Humphrey Burn is the only known locality for the petrified lagenostomalean cupules *Calathospermum scoticum* Walton (1949a) and *Geminitheca scotica* Smith (1959). *C. scoticum* is the type species of *Calathospermum*, a genus of large, multi-ovular 'megacupules' that also includes *C. fimbriatum* Barnard from Oxroad Bay (see above). These 'megacupules' have been interpreted as entire reduced fronds and are assumed to have been borne directly on the stem (Long, 1977b); in contrast, 'microcupules' are thought to have been only a distal part of a fertile frond that otherwise resembled conspecific sterile fronds. Clusters of microcupules were thus attached to foliage (usually in the fork of a vegetatively bipartite frond) rather than directly to the stem.

Many of the several known specimens of *C. scoticum* are barren, the ovules having been released; the few ovulate cupules are either immature or mature but unfertilized. The ovules resemble *Salpingostoma dasu* Gordon, as found in *C. fimbriatum* Barnard, but are smaller and have more integumental lobes (Walton, 1949a; Barnard, 1960b). *C. scoticum* probably bore a greater number of ovules (c. 48 according to Matten and Lacey, 1981) than *C. fimbriatum*, but had only six undissected lobes. Furthermore, Matten and Lacey (1981) suggested that the *C. fimbriatum* cupule is characterized by an initial dichotomy followed by monopodial or pseudo-monopodial divisions, but that *C. scoticum* has basally-fused cupular lobes. Some enigmatic structures in one *C. scoticum* cupule were interpreted by Walton (1949a) as microsporangia, who thus believed that these cupules were bisexual.

In contrast, *Geminitheca scotica* is a microcupule containing only two ovules. A specimen recently discovered by Scott shows c. 15 cupules borne on a repeatedly dichotomizing fertile frond. Like *Calathospermum scoticum*, it has up to six cupular lobes, but according to Matten and Lacey (1981) shows monopodial branching. Its ovules are also unusual in having an integument that is joined to the nucellus only at the base. This primitive characteristic is shared with *Genomosperma* from the Whiteadder (see earlier in this chapter). However, a salpinx with a central column and a vascular strand extending to the base of the nucellus, characters typical of *Genomosperma*, have not been reported for *Geminitheca*, which has also been correlated with different spores.

Smith (1959, 1960) reported *Telangium*-like sporangia associated with *Geminitheca*. Since they contained similar spores to those found within the lagenostome of *G. scotica*, he argued that they probably belonged to the same plant.

A compressed ovulate cupule was described by Smith (1962b) as *Calathiops trisperma*. It resembles *Geminitheca*, but is smaller and supposedly contains three ovules. The apparently ubiquitous development of ovulate cupules by repeated dichotomies (cf. Matten and Lacey, 1981) renders tri-ovulate cupules uncommon, though the condition has been reported in some specimens of *Stamnostoma oliveri* from Oxroad Bay (see above). *Calathiops trisperma* may be an immature form of *G. scotica*.

A third type of lagenostomalean cupules has been found by one of us (RMB) at Loch Humphrey Burn. Like *Geminitheca*, it is a small cupule pair, but unlike that genus it has numerous cupular lobes and ovules with about ten integumental lobes. It may have affinities with *Pullaritheca longii* from Oxroad Bay (see above).

There is thus evidence of at least three lagenostomalean plants preserved as petrifactions in the upper assemblages of Loch Humphrey Burn. Their foliage presumably occurs in the diverse fern-like foliage preserved here as adpressions. Perhaps the best candidate is *Sphenopteridium*, which has pinnules attached below the main dichotomy, and racheis with transverse bars. Two form-species have been recorded here, *S. pachyrrachis* and *S. crassum*, although there is a morphological continuum between them and they may be conspecific. Other probable lagenostomalean fronds recorded by Walton (1957) and Smith (1964b) were *Sphenopteris affinis*, *S. bifida* and *Rhodeopteridium* sp.

### Calamopityales

The only known specimen of the stem '*Calamopitys*' *radiata* Scott (1924b) originated from an unspecified horizon at Loch Humphrey Burn; its poor state of preservation suggests that it was from Unit 1, recently rediscovered at the base of the sequence. It is characterized by wide rays and probable exarch primary wood, but may be more appropriately segregated from *Calamopitys*. Scott *et al.* (1985) also illustrate, from the lower part of the section (Unit 2), a specimen of *Kalymma* cf. *tuediana*, which is probably part of a calamopityalean frond.

The upper part of the Loch Humphrey Burn section has yielded some of the best preserved examples of calamopityalean reproductive organs, *Alcicornopteris hallei*. These appear to be exclusively microsporangiate (Walton, 1949c; Smith, 1962a), in contrast to the apparently ovulate *Alcicornopteris* described from the Cementstone Group of Berwickshire (Long, 1969). Thus, this genus can encompass both male and female reproductive organs.

The affinities of a second sporangial cluster found at Loch Humphrey Burn are far less certain. The compression *Staphylotheca kilpatrickensis* Smith (1962b) consists of infrequently dichotomizing racheis with clusters of linear organs that bear sporangia. Smith tentatively assigned it to the pteridosperms on the basis of its tracheidal pitting, but its spores are enigmatically dimorphic: Type A is large, round and thin-walled and Type B is small, subtriangular and thick-walled. Smith argued that this morphological discontinuity between the two spore types is rather extreme to represent a contrast between viable and abortive spores. If so, this would imply the presence of heterospory and possible assignment of *Staphylotheca* to the progymnosperms.

The most likely calamopityalean foliage at Loch Humphrey Burn is *Spathulopteris*. Characteristics of these fronds include the thick-limbed, spathulate pinnules, longitudinal striae on the racheis but no transverse markings, and the absence of pinnae attached to the petiole below the initial dichotomy of the petiole. The two form-species described from Loch Humphrey Burn (*S. ettingshausenii* and *S. obovata*) probably represent ontogenetic variations of a single biological species.

### General remarks

Loch Humphrey Burn is of special interest in that it contains two speciose assemblages of anatomically-preserved plant fossils that are taxonomically, stratigraphically and (supposedly) temporally distinct. Fossils in the older assemblages (summarized by Scott *et al.*, 1985) are fragmentary and thus difficult to interpret, both taxonomically and palaeobiologically. Many of its diverse ferns are potential first occurrences of genera or families (e.g. Marattiaceae, Tedeleaceae), but this interpretation requires acceptance of the upper Tournaisian age attributed to the assemblage on palynological evidence; the megafloras and lithostratigraphy are more consistent with the mid-Visean age of the overlying petrifaction assemblages.

The better known upper assemblages are

dominated by pteridosperms and putative progymnosperms, many of the species being endemic. It is quite different in general aspect to the other well-known Visean petrifaction site in Britain at Pettycur (see above), where the assemblage is dominated by ferns and lycopsids. Lycopsids also dominate the more restricted petrifaction assemblage from nearby Glenarbuck (discussed in the next section). Overall, the plant fragments would reconstruct to yield at least eight whole-plant species in the lower assemblage and seven in the upper assemblage. Loch Humphrey Burn is the most significant Visean site in Europe for understanding the evolutionary history of the early gymnosperms and late progymnosperms.

## Conclusion

Loch Humphrey Burn has yielded exceptionally well-preserved plant fossils that reputedly range in age from 340 to 350 Ma. Early ferns predominate in the lower part of the succession and, if correctly dated, include the oldest examples known from anywhere in the world of two families which were common in the Late Carboniferous (Tedeleaceae and Corynepteridaceae), and the earliest known example from anywhere in the world of the extant fern order Marattiales. It has also provided invaluable information on the morphology and anatomy of early sphenophytes ('horsetails') and seed plants, especially reproductive structures such as the 'megacupule' *Calathospermum*. Loch Humphrey Burn is one of the few localities of this age to yield both anatomically-preserved and compressed plant fossils. Although these are not in organic connection, further work may allow correlation by indirect methods, thereby combining features of anatomy and gross morphology. Stratigraphical and palaeoecological investigations have proved fruitful and will continue.

## GLENARBUCK
*R.M. Bateman and C.J. Cleal*

## Highlights

Glenarbuck has yielded an important assemblage of Visean plant petrifactions that is richest in arborescent lycopsids and ferns, including several endemic species. The plants probably grew in a clastic swamp environment.

## Introduction

This Lower Carboniferous locality (Figure 5.31) lies east of a waterfall in Glenarbuck Burn, in the Kilpatrick Hills, near Glasgow (NS 452748). It was probably discovered by the Geological Survey in about 1870 (Smith, 1960). A petrified *Stigmaria* was collected at the locality in 1872, and the presence of this genus was reported by Young (1873) in a general description of the area. Calcareous nodules (possibly pedogenetic) containing a more diverse petrified flora were collected from the upper part of the sequence between 1930 and 1935 by Robert Brown (Smith, 1960), who also redescribed the locality (Brown, 1935). Plants collected by Brown and presented to John Walton were described by Calder (1935), Lacey (1953) and Smith (1960, 1962c, 1964b). A summary of the assemblage is given by Scott *et al.* (1984), who were unable to relocate the now degraded main petrifaction-bearing horizon.

## Description

### Stratigraphy

The 12 m-thick Glenarbuck sequence (Figure 5.33) is part of the 'Green tuffs and agglomerates' (the lowest member of the Clyde Plateau Lava Formation), underlain by the 'Spout of Ballagan Sandstone' of the Cementstone Group and overlain by the basaltic Clyde Plateau Lavas (Hall, 1978). Miospores recovered during a recent attempt to date the plant-bearing rocks (Scott *et al.*, 1984) were poor, but appear to belong to the Pu Biozone of the lower Visean (Chadian to Arundian stages). More recently, one of us (RMB) has correlated the entire Glenarbuck section lithostratigraphically with the mid-Visean (Arundian/Holkerian) Unit 5 of the nearby Loch Humphrey Burn sequence.

The sediments consist of siltstones (including reworked volcanigenic material) and shales with thin, often discontinuous coals and abundant, apparently *in situ* rootlets delimiting several palaeosols. They are probably river channel and flood plain deposits, suggesting that the plants represent a local swamp community.

# Glenarbuck

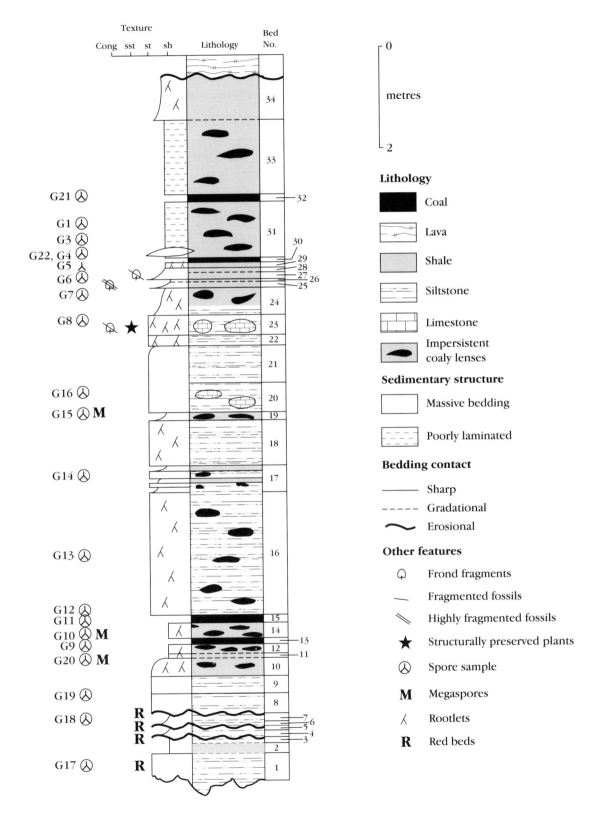

**Figure 5.33** Sedimentological log for Glenarbuck. Based on Scott *et al.* (1984, figure 6).

## Palaeobotany

This locality is significant mainly for the petrified plant fossils, which include the following species:

Lycopsida:
>*Paralycopodites brevifolius* (Williamson) Morey and Morey
>'*Lepidodendron*' *solenofolium* Smith
>*Lepidophloios kilpatrickense* Smith
>*Lepidocarpon wildianum* Scott
>*Stigmaria ficoides* (Sternberg) Brongniart

Equisetopsida:
>*Protocalamites goeppertii* (Solms-Laubach) Bateman

Filicopsida:
>*Metadineuron ellipticum* (Kidston) Galtier
>*Metaclepsydropsis duplex* Williamson
>*Botryopteris antiqua* Kidston

Lagenostomopsida:
>*Heterangium grievii* Williamson
>*Lyginorachis brownii* Calder
>*Endoxylon zonatum* (Kidston) Scott

Uncertain affinities:
>cf. *Mittagia seminiformis* Lignier

In addition, there are adpressions of 'lepidodendrid twigs' (Smith, 1960); possibly *Lepidophloios kilpatrickensis* Smith, *Stigmaria ficoides* (Sternberg) Brongniart, and *Aneimites acadica* Dawson.

## Interpretation

The main interest of the site lies in its apparently endemic lycopsid species (Smith, 1962c). '*Lepidodendron*' *solenofolium* is known from only two specimens, a small twig and a larger branch, both with attached leaves. It is characterized by protostelic axes and leaf cushions with prominent lateral 'wings'. The several known axes of *Lepidophloios kilpatrickensis* are protostelic and range in diameter from 5 to 20 mm. Their leaf cushions are almost equidimensional and approach those of *Lepidodendron sensu lato* in outline. The anatomy of *L. kilpatrickensis* was used by Smith (1962c) to speculate that the primary cortex of these lycopsids remained meristematic for much of the plant's life. This is one of the earliest known diverse assemblages of arborescent lycopsids that show details of the axial anatomy; the ferns too are important. It contrasts with the majority of Lower Carboniferous petrifaction assemblages, which are dominated by seed-ferns and ferns.

This is the only known locality for the lagenostomalean rachis *Lyginorachis brownii*, which is characterized by a corrugated, 'U'-shaped vascular bundle. Another probable pteridosperm is the petrified axis *Endoxylon zonatum*. It has eight endarch protoxylem strands, and widely-separated, undivided leaf-traces; the secondary wood has narrow tracheids and very small, uniseriate rays (Lacey, 1953). It is associated with the adpression foliage *Aneimites acadica*, although they have not been found in organic connection.

Glenarbuck has yielded a small but important amount of palaeobotanical material that is important for showing anatomical details of several early arborescent lycopsids and ferns. It is the type locality for the pteridosperm frond *Lyginorachis brownii* and one of only two known localities for *Endoxylon zonatum*. It is also the only British locality for the enigmatic cf. *Mittagia seminiformis*, a reproductive organ of questionable affinities that was suggested as a possible precursor of the seed plants by Emberger (1968). Palaeoecologically, the site is of interest because it probably represents a swamp community of the same age and depositional environment as the uppermost community found nearby at Loch Humphrey Burn (see previous section).

## Conclusion

Glenarbuck has yielded an important anatomically-preserved assemblage of plant fossils about 340 million years old. They indicate the presence of about ten whole-plant species, which formed a clastic swamp community that was dominated by lycopsid trees with subordinate seed plants. The lycopsid species, which have only ever been found at this locality, are distantly related to small, herbaceous living plants known as club-mosses. The flora provides an interesting contrast with the main plant-bearing units found at the nearby Loch Humphrey Burn (see previous section), which contain fluvially-transported plant fossils. These communities, which were dominated by seed plants and their immediate progymnospermous ancestors, are thought to immediately pre-date those at Glenarbuck.

# PUDDLEBROOK QUARRY

## Highlights

This is the best known locality for plant fossils from the Visean Drybrook Sandstone, and has yielded four apparently endemic species (Figure 5.34). These include *Muscites plumatus*, which may be the oldest known moss, and *Diplopteridium holdenii*, one of the most completely reconstructed Visean pteridosperms.

## Introduction

The Puddlebrook locality consists of a small quarry in the Lower Carboniferous Drybrook Sandstone, just north of Drybrook in the Forest of Dean (SO 647184) (Figure 5.35). Plant fossils from here were first recorded by Allen (1961), and were the subject of a monographic study by Lele and Walton (1962b). Some of the component species have been reviewed by Thomas (1972) and Thomas and Purdy (1982), but the only comprehensive re-assessment of the assemblage is in an unpublished thesis by Rowe (1986), work done partly in conjunction with the Geological Conservation Review Unit. Certain parts of the latter study have been subsequently published (Rowe, 1988a, b, c, 1992).

## Description

### Stratigraphy

The geology is described by Rowe (1986), and briefly summarized by Rowe (1988b, c). The exposed strata belong to the Drybrook Sandstone Formation, and palynological evidence suggests an Asbian age. The plant fossils occur in a 1.2 metre thick lens of shale within the sandstone. The lens is interpreted as the infill of an abandoned river channel.

**Figure 5.34** Puddlebrook. Asbian fluvial deposits of the Drybrook Sandstone, prior to the 1982 excavations at the site. (Photo: C.J. Cleal.)

## Lower Carboniferous

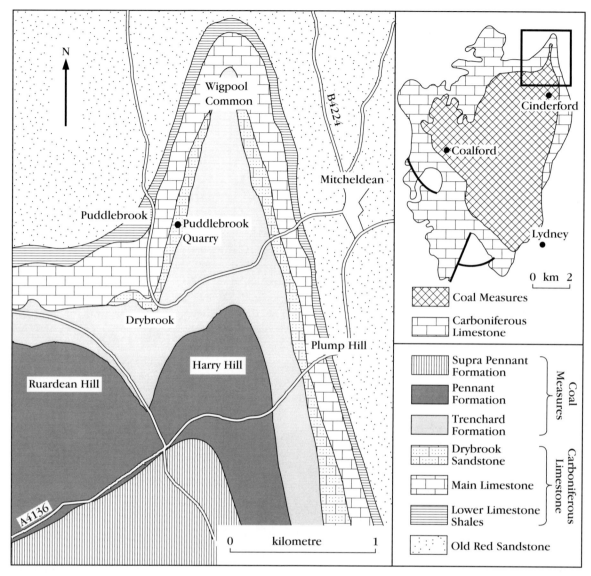

**Figure 5.35** Geological map of the northern part of the Forest of Dean, showing the location of Puddlebrook Quarry. Based on Sullivan (1964, text-figure 1).

### Palaeobotany

The fossils are preserved as adpressions, sometimes with cuticles preserved, or as fusain. The following species have been described to date:

Bryophyta(?):
   *Muscites plumatus* Thomas

Lycopsida:
   *Eskdalia variabilis* (Lele and Walton) Rowe
   *E. fimbriophylla* Rowe
   *Lepidostrobophyllum fimbriatum* (Kidston) Allen
   cf. *Stigmaria* sp.
   *Selaginellites resimus* Rowe

Lagenostomopsida:
   *Diplopteridium holdenii* Lele and Walton
   *Dichotangium quadrothecum* Rowe
   *Sphenopteris obfalcata* Walton
   *S. cuneolata* Lindley and Hutton
   *Archaeopteridium tschermakii* (Stur) Kidston

*Telangiopsis* sp.
*Carpolithus puddlebrookense* Thomas and Purdy

## Interpretation

Thomas (1972) described a small, leafy shoot from here as *Muscites plumatus*. Although there is no evidence of fertile structures, Thomas argued that it could be a bryophyte, possibly a moss. If so, it would be the oldest known moss, the next oldest being from the Stephanian of France (Renault and Zeiller, 1888). No liverworts have so far been described from Puddlebrook, but Sullivan and Hibbert (1964) reported *Tetrapterites visensis* from another outcrop in the Drybrook Sandstone, which Lacey (1969) has argued might be a spore-bearing body of a liverwort.

The most abundant lycopsid remains here belong to the form-genus *Eskdalia*. The stems were initially described by Lele and Walton (1962b) as *Scutelocladus variabilis* Lele and Walton, who interpreted them as having no leaf cushions, ligule pits or parichnos tissue. Thomas and Purdy (1982) subsequently demonstrated that they had expanded leaf cushions with a lateral wing, basal heel and adaxial ligule pit. They also described grooves in the middle of the leaf cushions as being possibly the surface manifestation of infrafoliar bladders (spongy tissue connected to the parichnos, and thought to be part of an aerating system in certain lycopsids). As a result, they transferred the species to *Tomiodendron*, a form-genus previously only reported from Angaran assemblages of Siberia and the north slope of Alaska (Spicer and Thomas, 1987). Most recently, Rowe (1988c) has investigated fusainized fragments, which indicate that an infrafoliar, aerating bladder was not in fact present in the stems, and so he transferred the species to *Eskdalia*, as interpreted by Thomas and Meyen (1984).

Rowe (1988c) also described several specimens of *E. variabilis* stems with small, terminally-borne strobili (Figure 5.36). The strobili have a slender central axis with spirally-arranged sporophylls with entire margins. Although some evidence of sporangia was found, none yielded spores, and so it is still not possible to place *Eskdalia* in any of the families outlined by Thomas and Brack-Hanes (1984). Nevertheless, these Puddlebrook specimens are of considerable interest as the only known evidence of the fructifications of the shrubby lycopsid *Eskdalia*, although Thomas (1992) has argued that not all species of this stem form-genus may have had the same fructifications.

Rooting structures associated with *E. variabilis* were described by Lele and Walton (1962b) as cf. *Stigmaria* sp. Very similar structures are known from Moel Hiraddug (see below), where they are associated with '*Lepidodendron*' *perforatum* Lacey, a species that may belong to *Eskdalia* (Rowe, 1988c).

A second species of *Eskdalia* from Puddlebrook has been described by Rowe (1988c) as *E. fimbriophylla*. Isolated leaves were originally identified by Lele and Walton (1962b) as *Lepidophyllum* cf. *fimbriatum*, but no evidence was found of attached sporangia, as would be expected if they were sporophylls. Rowe demonstrated that they were leaves attached to *Eskdalia*-like stems, although they are wider and have larger leaf cushions than *E. variabilis* stems. *E. fimbriophylla* has not been identified from any other locality; however, there is a possible comparison with the leaf *Lepidophylloides fisheri* Crookall, described by Crookall (1966) from the more or less coeval Scremerston Coal Group of Northumberland.

Allen (1961) described abundant, isolated lycopsid sporophylls from here as *Lepidostrobophyllum fimbriatum* (Kidston) Allen. They have a clearly fimbriate margin, and thus differ from the entire-margined sporophylls of the *Eskdalia variabilis* strobili. It is tempting instead to speculate a connection with *E. fimbriophylla*, whose leaves are also fimbriate, but direct evidence of attachment is so far lacking.

A third type of lycopsid found at Puddlebrook is a small herbaceous plant known as *Selaginellites resimus* Rowe, 1988a. Although only represented by fragmentary material, Rowe was able to reconstruct it as having mainly recumbent, rambling stems, which produced vertical, dichotomous shoots. Some of these vertical shoots bore what Rowe interpreted as terminal strobili, which included sporangia containing megaspores. It thus appears to be a typical example of the Selaginellaceae, one of the most conservative families of vascular plants, which has remained essentially unaltered since the Late Devonian (Fairon-Demaret, 1977).

The most completely understood pteridosperm from Puddlebrook is *Diplopteridium holdenii* (Figure 5.37). The basic features of the foliage were described by Lele and Walton (1962b), but a more comprehensive study by Rowe (1988b) has enabled him to provide a reconstruction of most of the plant, including some details of the

**Figure 5.36** *Eskdalia variabilis* (Lele and Walton) Rowe. Leafy lycopsid shoots bearing terminal fructifications; Natural History Museum, London, specimen V.62432a. Drybrook Sandstone (Asbian), Puddlebrook Quarry. x 1. (Photo: Photographic Studio, Natural History Museum, London.)

fructifications (Figure 5.7). It appears to have been herbaceous, with an upright stem bearing a crown of fronds. As in most pteridosperms, the fronds have an essentially bipartite architecture, the main rachis dichotomizing near the base to produce bipinnate branches. In fertile fronds, however, there is a trichotomy, the third branch bearing the fructifications. Rowe was able to demonstrate organic connection between the fertile branches of the *D. holdenii* frond and cupulate seed/ovule-bearing structures, that Lele and Walton (1962b) had earlier named as *Calathiops* sp. This is only the second known species of *Diplopteridium*; the other being *D. teilianum* Walton from Teilia (p. 178), from which it differs in having narrower pinnule segments.

Rowe (1988b) also described a synangiate organ associated with *D. holdenii*, which Lele and Walton (1962b) had initially identified as *Telangium* sp. Although Rowe presented considerable circumstantial evidence that it was the pollen organ of *D. holdenii*, in the absence of direct evidence of attachment he assigned it a separate name, *Dichotangium quadrothecum*. Rowe (1986) described other sporangial structures from Puddlebrook, including one that he compared with *Telangiopsis*. However, the results of this work have yet to be published.

Other pteridosperm fronds from Puddlebrook, were identified by Lele and Walton (1962b) as *Sphenopteris obfalcata* and *S. cuneolata*. The former is only otherwise known from Teilia (p. 179), and there is as yet inadequate knowledge of its frond architecture or fructifications for its taxonomic position to be firmly established.

The holotype of *S. cuneolata*, from the Oil Shale Group of Scotland, was poorly illustrated and Kidston (1923b, p. 156) reported that it was lost. Kidston (1923b, pl. 214) illustrated a second

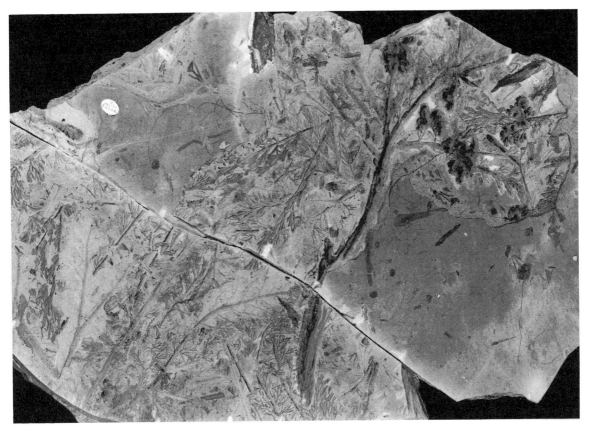

**Figure 5.37** *Diplopteridium holdenii* Lele and Walton. Pteridosperm frond with fructifications; Natural History Museum, London, specimen V.62331a. Drybrook Sandstone (Asbian), Puddlebrook Quarry. x 0.5. (Photo: Photographic Studio, Natural History Museum, London.)

specimen under this name, but Lele and Walton (1962b) note that it differs markedly from Lindley and Hutton's specimen in its nervation and the outline of the pinnules. Lele and Walton therefore nominated one of the Puddlebrook specimens as 'lectotype' (in fact a neotype), but it is far from clear that it is conspecific with Lindley and Hutton's original concept of the species.

Thomas and Purdy (1982) describe some apparently radiospermic (i.e. radially symmetrical) seeds or ovules as *Carpolithus puddlebrookense*. They had an integument fused to the nucellus except at the apex, where it formed four apical lobes. The nearest comparison seems to be with the Tournaisian petrified seeds/ovules *Eurystoma angulare* Long from the Whiteadder (p. 124). However, *Eurystoma* was borne in *Alcicornopteris* cupulate structures, of a type not so far reported from Puddlebrook. Furthermore, *Eurystoma* is thought to have calamopityalean affinities, and none of the foliage found at Puddlebrook is of a type normally associated with that order (e.g. *Triphyllopteris*, *Spathulopteris*).

Two fragmentary Puddlebrook specimens were identified by Lele and Walton (1962b) as *Archaeopteridium tschermakii*. Better specimens have been since described by Rowe (1992), who also found pre-pollen organs which confirm that the species belongs to the pteridospermous class Lagenostomopsida. Although generally rare, this is a widely occurring species, having been reported from several localities in Scotland, Germany and Poland. The Puddlebrook specimens are the only examples known from England or Wales.

Although plant fossils have been reported from elsewhere in the Drybrook Sandstone (e.g. Cleal, 1986a), this is by far the most diverse assemblage from the formation. Five of the taxa listed at the beginning of this section have been reported only from the Drybrook Sandstone, and four of these only from Puddlebrook. There is some overlap in composition with other British Visean assemblages, particularly those from North Wales (Teilia and Moel Hiraddug - see p. 175). However, the overall balance of species separates Puddlebrook from most other palaeobotanical sites of this age. This presumably represents its spatial isolation, being separated from these other, more northerly localities by the Wales-Brabant 'landmass'.

## Conclusion

This is the best site for Lower Carboniferous plant fossils in southern Britain. They represent the vegetation growing here about 330 million years ago, which consisted mainly of shrubby club-mosses and primitive seed plants. They include fossils that demonstrate the connection between vegetative and reproductive structures, which allows some of the species to be viewed more as whole, living plants, rather than just as disarticulated organs. Several of the species are unique to this locality, including *Diplopteridium holdenii* Lele and Walton, one of the most completely reconstructed early seed plants.

## MOEL HIRADDUG

### Highlights

Moel Hiraddug is the best available locality for plant fossils in the Visean Foel Formation of North Wales. A number of taxa are unique to here, including *Lepidodendropsis jonesii*, *Archaeosigillaria stobbsii*, *Rhacopteris weissii* and *Calathiops dyserthensis*. It is also one of the few Visean localities where the plant fossils have cuticles preserved.

### Introduction

This Lower Carboniferous site consists of two small quarries south of Dyserth, North Wales (SJ 061783) (Figure 5.38). Some authors have referred to it as 'Dyserth' (e.g. Lacey, 1962). There is a Dyserth Quarry *c.* 0.5 km further north, however, and so the alternative name Moel Hiraddug Quarries is used here (named after the hill on the side of which they occur). Plant fossils were first recorded by Morton (1871, 1886, 1898), who noted similarities with Bowman's (1837) assemblage from a similar stratigraphical interval on the opposite side of the Vale of Clwyd, at Craig-y-Forwen near St Asaph. Fossils from Moel Hiraddug were also mentioned by Hind and Stobbs (1906) and Hind (1907), although the first detailed description with illustrations was Walton's (1926) analysis of *Rhacopteris weissii* Walton. Walton (1928), Neaverson (1930) and Hirmer (1939) all briefly mentioned the Moel Hiraddug assemblage, but it was not until Lacey made extensive collections here between 1946 and 1960 that its full diversity could be assessed. Provisional taxonomic lists were given by Lacey (1952a, b) who later published a full monographic treatment (Lacey, 1962).

### Description

#### Stratigraphy

There have been numerous geological accounts of this area (e.g. Morton, 1871, 1886; Strahan, 1885a; Hind and Stobbs, 1906; Neaverson, 1930, 1945; Somerville *et al.*, 1989). The strata exposed at Moel Hiraddug Quarries are about 17 metres of brown limestones, with lenticular shale and sandstone bodies, and they belong to the Foel Formation (traditionally called the Lower Brown Limestones). Somerville *et al.* (1989) have placed this formation in the upper Chadian, and suggest that it represents shallow marine deposits.

#### Palaeobotany

The fossils are mainly compressions, sometimes with cuticles still preserved. The assemblage comprises the following taxa:

Algae (division unknown):
    *Koninckopora inflata* (de Koninck) Wood
    *Bythotrepis plumosa* Kidston
    *B. nodosa* Lacey

Lycopsida:
    *Archaeosigillaria stobbsii* Lacey
    *Clwydia decussata* Lacey
    *Lepidodendropsis jonesii* Lacey
    *Lepidostrobophyllum fimbriatum* (Kidston) Allen
    *Lepidodendron* sp.
    *Stigmaria ficoides* (Sternberg) Brongniart
    *Stigmaria* sp.
    *Knorria acicularis* Göppert
    ?*Halonia* sp.

# Moel Hiraddug

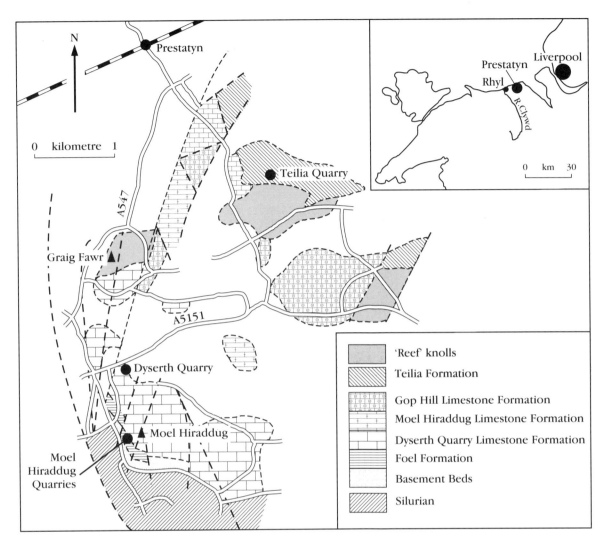

**Figure 5.38** Lower Carboniferous geology of the area south of Prestatyn, showing the position of the quarries at Moel Hiraddug and Teilia. Based on Somerville *et al.* (1989, figure 1).

Equisetopsdia:
 *Archaeocalamites radiatus* (Brongniart) Stur
 *Bowmanites tenerrimus* (Ettingshausen)
  Hoskins and Cross

Progymnospermopsida(?):
 *Rhacopteris weissii* (Walton) Hirmer
 *R. subcuneata* Kidston

Lagenostomopsida:
 *Calathiops dyserthensis* Lacey

## Interpretation

This is the type locality for *Archaeosigillaria stobbsii* (Figure 5.40), and has yielded information on both gross morphology and epidermal structure. It is very similar in gross morphology to *Archaeosigillaria kidstonii* Krausel and Weyland from the Carboniferous Limestone of Cumbria (Kidston, 1885a, 1901b; Crookall, 1966; Chaloner *in* Boureau *et al.*, 1967). However, since the epidermal structure of *A. kidstonii* is unknown, Lacey (1962) opted to place the better-preserved Moel Hiraddug specimens in a new species.

Associated with the *Archaeosigillaria* stems are slender leafy shoots. Lacey (1962) noted that they had a different phyllotaxis and epidermal structure to *A. stobbsii*, and so assigned them to a different species, *Clwydia decussata* (Figure 5.39). However, similar leafy shoots have been

**Figure 5.39** *Clwydia decussata* Lacey. Lower Carboniferous lycopsid leafy shoot; National Museum of Wales, specimen 84.27G125. Foel Formation (Chadian), Moel Hiraddug. x 2. (Photo: Photographic Studio, National Museum of Wales.)

**Figure 5.40** *Archaeosigillaria stobbsii* Lacey. Lower Carboniferous lycopsid leafy shoots; Natural History Museum, London, specimen V.16012. Foel Formation (Chadian), Moel Hiraddug. x 1. (Photo: Photographic Studio, Natural History Museum, London.)

reported in association with *Archaeosigillaria* at other Visean localities (e.g. Kidston, 1885a), and many workers consider them to be parts of the same plant. Central to the debate is the interpretation of two of Kidston's specimens (Chaloner *in* Boureau *et al.*, 1967, fig. 340 c, d) as showing the attachment of such leafy shoots to *Archaeosigillaria* stems. Rowe (1988a) has recently concluded that these specimens are too poorly preserved for the case to be proved, and that the taxonomic separation of the stems and shoots is best retained, at least until the group as a whole can be revised in detail.

Lacey (1952a) identified a single stem fragment as *Lepidophloios* cf. *laricinus* Sternberg, but, in a footnote to the paper, Jongmans suggests that it might be a *Lepidodendropsis*. Lacey (1962) later demonstrated that the stem was eligulate, thus tending to confirm Jongmans' view. It has a very similar gross morphology to *Lepidodendropsis vandergrachtii* Jongmans and Gothan (e.g. Jongmans *et al.*, 1937, figs 33-39, 49), except that the leaf cushions are more elongate and the leaves more obliquely attached. These differences, and the absence of information on the *L. vandergrachtii* epidermal structure, caused Lacey to refer the Moel Hiraddug specimen to *Lepidodendropsis jonesii*.

The lycopsid sporophylls *Lepidostrobophyllum fimbriatum* occur commonly here. Elsewhere, they are often associated with *Eskdalia* stems (e.g. Puddlebrook - see p. 169), but there is no evidence of such stems at Moel Hiraddug. However, at another Foel Formation exposure, at Graig Quarry, stems were identified by Lacey (1962) as *Lepidodendron perforatum* Lacey, which Rowe (1988c) considers to belong to *Eskdalia* (although he makes no formal transfer of the species).

Equisetopsids are mainly represented by *Archaeocalamites radiatus* stems, without any evidence of the foliage. In addition, Lacey (1962) described a single sphenophyllalean strobilus with incised sporophylls, under the name *Bowmanites*

**Figure 5.41** *Lepidodendron* sp. Lower Carboniferous lycopsid stem; National Museum of Wales, specimen 84.27G122. Foel Formation (Chadian), Moel Hiraddug. x 1. (Photo: Photographic Studio, National Museum of Wales.)

*tenerrimus*; but again there is no evidence of the foliage.

The only unequivocal evidence of pteridosperms is the fructification *Calathiops dyserthensis*. This form-genus is usually regarded as having been attached to lagenostomalean fronds such as *Diplopteridium* and *Diplothmema*, but no such fronds have so far been reported from Moel Hiraddug.

Two species of the ?progymnosperm *Rhacopteris* have been reported, *R. weissii* and *R. subcuneata* (Walton, 1926; Lacey, 1962). The former is only known from this locality and is of interest in being one of the few rhacopterid species from which cuticles have been prepared (Lacey, 1962). This may eventually help resolve the taxonomic position of these leaves.

This is easily the best available locality for plant fossils in the Foel Formation. Only Graig Quarry near Denbigh has yielded an assemblage of comparable quality; but, although this site is still extant, the plant bed is now obscured by a new roadway leading to the upper part of the quarry. Many of the species found in the Lower Brown Limestone are unique to the formation, including *Lepidodendropsis jonesii*, *Archaeosigillaria stobbsii*, *Rhacopteris weissii* and *Calathiops dyserthensis*; also '*Lepidodendron*' *perforatum* (?*Eskdalia* sp. – Rowe, 1988c), although this has not so far been reported from Moel Hiraddug. It is also of particular significance as one of the few British Visean palaeobotanical sites where cuticles are still preserved, allowing details of the epidermal structure to be observed.

## Conclusion

Moel Hiraddug has yielded a diverse assemblage of plant fossils, dating from the early part of the Carboniferous period, about 330 Ma. It represents the remains of vegetation growing on the northern margins of St George's Land (an island that extended over parts of central Britain), probably in a lowland, swampy habitat, and which drifted into the nearby sea. Seventeen species have been identified so far, with the remains of giant club-mosses and progymnosperms (the immediate ancestors of the seed plants) predominant. A rare reproductive organ, thought to be *Bowmanites*, has also been reported, that is the oldest known example from Britain of this particular group of horsetails, which, in the Late Carboniferous, were important members of the tropical swamp vegetation. Of special interest is that many of the fossils still have their outer protective 'skin' (known as the cuticle). Plant cuticles have been widely studied in younger fossil floras and have proved of considerable value, as they are often the only source of anatomical detail still preserved. For reasons that are not clear, however, cuticles are extremely rare in the Lower Carboniferous, which makes their presence at Moel Hiraddug of considerable potential importance.

## TEILIA QUARRY

## Highlights

Teilia Quarry is the only locality to yield plant fossils from the Gronant Group of North Wales, and includes a number of endemic taxa. It is particularly important for the putative progymnosperm *Rhacopteris*, yielding the only known British example with fructifications. It has also yielded some of the best known examples of lagenostomalean fronds from Britain, including *Diplopteridium* and *Sphenopteridium*, and what may be the earliest evidence of the order Trigonocarpales. It is a site of outstanding palaeobotanical significance.

## Introduction

Lower Carboniferous plant fossils were first collected from this disused quarry north-east of Gwaenysgor, near Prestatyn, North Wales (SJ 080814), by the amateur geologist J.B. Shone, during the mid-nineteenth century. Morton (1886) published the earliest species list based on Shone's material, although many of his identifications were wrong. The first reliable account of the assemblage was by Kidston (1889a), who also later included a number of Teilia specimens in his classic monograph on Carboniferous plant fossils (Kidston, 1923-1925). Further material was collected during the 1920s, and resulted in a revision of the assemblage by Walton (1926, 1928, 1931); Walton's identifications were also quoted by Neaverson (1930), Crookall (1932) and Hirmer (1939). Most recently, Benson (1935a, b) has given brief accounts on fructifications.

## Description

### Stratigraphy

Morton (1886) and Walton (1931) have provided accounts of the geology at Teilia. The sequence exposed consists of about 8 metres of thinly-bedded, dark grey limestones belonging to the Gronant Group (previously the Upper Black Limestone Group). In these sedimentary rocks, land-plant fossils are closely associated with fossils of marine animals, and Walton regarded this as evidence that the strata were lagoonal or shallow marine. The fauna mentioned by Morton (1886), Hind and Stobbs (1906), Jackson *in* Walton (1928) and Neaverson (1930) all clearly point to an early Brigantian age (George *et al.*, 1976).

### Palaeobotany

The plant fossils are preserved here mainly as adpressions, but no evidence of cuticles has so far been reported. The following taxa have been described:

Lycopsida:
    cf. *Lepidodendron* spp.

Equisetopsida:
    *Archaeocalamites radiatus* (Brongniart) Stur

Filicopsida(?):
    cf. *Rhodeopteridium tenue* (Gothan) Purkynová

Progymnospermopsida(?):
    *Rhacopteris circularis* Walton
    *R. robusta* Kidston
    *R. petiolata* (Göppert) Schimper
    *R. fertilis* Walton

Lagenostomopsida:
    *Diplopteridium teilianum* (Kidston) Walton
    *Sphenopteridium capillare* Walton
    *S. pachyrrachis* (Göppert) Schimper
    *Lyginopteris bermudensiformis* (Sternberg) Patteisky
    *Adiantites antiquus* (Ettingshausen) Kidston
    *A. machanekii* Stur
    *Sphenopteris obfalcata* Walton
    *Spathulopteris ettingshausenii* (Feistmantel) Kidston
    *S. clavigera* (Kidston) Walton
    *Calathiops acicularis* Göppert
    *C. glomerata* Walton
    *C. renieri* Walton

Cycadopsida:
    *Neuropteris antecedens* Stur
    *Holcospermum ellipsoideum* (Göppert) Walton
    *Carpolithus* sp.

## Interpretation

### Lycopsida

The lycopsids are represented at Teilia by a few stem fragments, described by Walton (1931). Most belong to a *Lepidodendron*, which Walton compared with *L. obovatum* Sternberg based on a suggestion by Jongmans. However, *L. obovatum* is a confused taxon whose types probably fall in synonymy with *Lepidodendron aculeatum* Sternberg (Thomas, 1970), a species quite different from the Teilia specimens. A single specimen was compared by Walton with *L. calamitoides* Nathorst, but the leaf cushions are poorly preserved and the leaf scar not distinguishable. Considering the poor preservation of this material, it is best recorded for the time being merely as cf. *Lepidodendron* spp.

A single small fragment of a third type of lycopsid was illustrated by Walton in a rather sketchy line diagram. He referred it to *Sigillaria* sp., but the available evidence is inadequate to justify this assignment. Crookall (1966) does not refer to it in his monograph on British *Sigillaria* species.

## Equisetopsida

At Teilia, as in most Visean assemblages, equisetopsids are represented by *Archaeocalamites radiatus*. Walton (1931, pl. 26, fig. 39) shows a stem with the characteristic forked leaves of this species.

## Filicopsida(?)

The frond fragments assigned to *Rhodeopteridium* have much smaller pinnules with very slender lobes. It is possible that they are parts of small pteridosperm fronds, but their small size rather suggests that they are filicopsid fronds. Walton (1931) assigned the Teilia specimens of this type to *Rhodeopteridium tenue* (syn. *Rhodea tenuis* Gothan), a species first described from the lower Namurian of Silesia (Gothan, 1913). However, Gothan's diagnosis mentions that the rachises are 'winged', a feature not yet demonstrated in the Teilia specimens.

## Progymnospermopsida(?)

Among the commoner plant fossils found at Teilia is foliage belonging to the form-genus *Rhacopteris*. Although there is some evidence available on the fructifications of this form-genus, including one specimen described by Walton (1926) from Teilia, its affinities have never been firmly established. On balance, however, it seems most likely that it belongs to the progymnosperms (Archangelsky and Arrondo, 1966; Beck, 1976). It was initially established for Upper Carboniferous species (Schimper, 1869-74), but a number of Lower Carboniferous species were later included (Stur, 1875). Oberste-Brink (1913) placed these Lower Carboniferous species in a separate section, *Rhacopteris (Anisopteris)*, based partly on their stratigraphical position, and partly on the greater asymmetry of the pinnules (some authors have gone as far as raising it in rank to form-genus – Hirmer and Guthörl, 1940; Boureau and Doubinger, 1975). However, this division is probably artificial (Walton, 1926) and the Lower Carboniferous species are best referred to simply as *Rhacopteris*.

The commonest *Rhacopteris* species found at Teilia, *R. circularis*, has round, fairly symmetrical pinnules and a radiating nervation. There have also been tentative records of this species from France (Corsin and Dubois, 1932), the Sahara (Boureau, 1953), Nova Scotia (Bell, 1960) and the Himalayas (Høeg et al., 1955; Pal and Chaloner, 1979), but they are all based on poorly preserved fragments. Somewhat better material has been described from Argentina by Frenguelli (1944) and Peru by Doubinger and Alvarez Ramis (1980), but it is closely associated with *Pseudorhacopteris ovata* (McCoy) Archangelsky, an Upper Carboniferous species from Gondwanaland with very similar shaped pinnules to *R. circularis*. Teilia is the only known locality to yield undoubted examples of *R. circularis*.

The specimens from Teilia identified by Kidston (1889a) as *Rhacopteris inaequilaterata* (Göppert) Stur were included within the synonymy of *R. circularis* by Walton (1926).

A second species from Teilia is *Rhacopteris robusta* (syn. *R.* cf. *petiolata* Walton, 1926, non Göppert). It has asymmetrical pinnules with an incised distal margin, and is thus readily distinguishable from the commoner *R. circularis*. The Teilia specimens have rather smaller pinnules than the types of *R. robusta* from Scotland (Kidston, 1923c, pl. 51, figs 5-7) but compare closely with specimens from Silesia figured by Patteisky (1929, pl. 6, fig. 1).

Walton (1931, pl. 25, fig. 20) figured a single specimen as *Rhacopteris machanekii*, a species based on just one specimen from Moravia (Stur, 1875). Walton implied that it might be a small form of *R. robusta*, a view which is accepted here.

A third species present at this site is *R. petiolata*, which has deeply digitate pinnules. Only one undoubted specimen has been documented (Kidston, 1923c, pl. 53, fig. 5), although some other more fragmentary examples may also belong here (Kidston, 1923c, pl. 53, fig. 3; Walton, 1926, pl. 16, fig. 11).

The fertile specimen of a *Rhacopteris* described by Walton (1926) was essentially similar to the only other example then known (*R. paniculifera* Stur, 1875, pl. 8, fig. 3), in that it showed a dichotomous axis bearing clusters of sporangia. However, unlike Stur's specimen, the one from Teilia has no sterile pinnules attached and so Walton assigned it to a separate species, *R. fertilis*. This remains the only known British example of a *Rhacopteris* bearing fructifications, and has an important bearing on determining the taxonomic position of this group of leaves, which form such an important component of Lower Carboniferous adpression floras. It suggests that they may be progymnosperms, as they appear similar to *Archaeopteris* fructifications (cf. Phillips *et al.*, 1972). However, the fertile rhacopterids were described before the progymnosperms had been recognized as a major taxon, and they have not been subsequently re-assessed. Teilia should thus play a significant role in the much needed revision of this widely occurring fossil foliage.

### Lagenostomopsida

Dominating the assemblage are pteridosperm frond fragments, most probably belonging to the Lagenostomales. The most completely known is *Diplopteridium teilianum*, following the work of Walton (1926, 1931). It has an essentially bipartite architecture, with a dichotomy of the main rachis near the base of the frond; it is thus similar in many ways to *Sphenopteridium*. However, it shows no evidence of the transverse bars that typically occur on *Sphenopteridium* racheis, and so the species was for some time referred to the generalized form-genus *Sphenopteris* (Kidston, 1889a, 1923a). However, Walton demonstrated that at least some fronds produced a third branch in the angle of the 'dichotomy', which probably bore the fructifications. Consequently, he assigned it to a new form-genus, *Diplopteridium*. Seeds or sporangia have not been found attached to this third branch, but a number of *Telangium*-like structures were found in close association (that it was a fertile branch has been confirmed in the related species *Diplopteridium holdenii* Lele and Walton from Puddlebrook - see p. 169). Although these *Telangium*-like structures are not well preserved, Benson (1935a) concluded that they were cupules, which originally bore seeds/ovules.

A number of other bipartite fronds found at Teilia have rachises with transverse bars and so are assigned to *Sphenopteridium*. To date, no evidence has been found of their fructifications. Walton (1931) described three forms. *S. capillare* is quite distinctive, having small, deeply incised pinnules packed closely to one another along the rachis; to date it has only been reported from Teilia. The other specimens have larger, more widely spaced pinnules with more extensive lamina. They mostly belong to *S. pachyrrachis*, a well-known species with a wide distribution, known from Scotland (Kidston, 1923b), Bavaria (Lutz, 1933), the Loire (Bureau, 1913-1914) and Moravia (Patteisky, 1929). The third type recognized by Walton is only known from Teilia by one small pinna fragment, and was identified as *S. crassum*. It only differs from *S. pachyrrachis* in having slightly smaller, broader pinnules and Walton argued that it might merely be an extreme form of that species. It has not therefore been included here in the species list for Teilia.

Walton (1931, pl. 24, figs 16, 17) figures two specimens of *Lyginopteris bermudensiformis*. Two forms have been recognized within this species, *Lyginopteris bermudensiformis* fa. *bermudensiformis* (synonyms *Lyginopteris bermudensiformis* fa. *schlotheimii* Stur, 1875 and *Lyginopteris bermudensiformis* fa. *typica* Kidston, 1923c) and *Lyginopteris bermudensiformis* fa. *geinitzii* Stur, 1875. The Teilia specimens are too fragmentary to be certain, but Walton (1931) suggested that they have most in common with *Lyginopteris bermudensiformis* fa. *bermudensiformis*. The specimen figured by Walton (1931, pl. 24, fig. 16) is of particular interest in that it seems to be a 'miniature' version of a *Lyginopteris* frond, complete with dichotomy. There may be a correlation with the bipartite 'mini-fronds' described by Boersma (1972) for *Mariopteris*, thought to represent the small fronds from the most distal part of the plant.

Kidston (1889a, p. 425-6) recorded some poorly preserved fragments as ?*Sphenopteris schlehanii* (Stur) Gothan. This species is now generally included within *Lyginopteris* (Patteisky, 1957) and it is thus likely that Kidston's reported specimens belong to *L. bermudensiformis*.

Associated with these lagenostomalean fronds are a number of fructifications of the *Calathiops*-type (Walton, 1931). None of them show any internal structure, nor are they attached to foliage, and so their affinities must be speculative. However, *C. acicularis* and *C. glomerata* can be compared with the so-called 'micro-cupule' type of lagenostomalean fructification, such as *Telangiopsis* (Kidston, 1924; Long, 1979b). A third type described by Walton as *C. renieri* is nearer to the 'mega-cupule' type of fructification, such as *Megatheca* from the Oil Shale Group of Scotland (Andrews, 1940). Yet another fructification from Teilia, *Calathiops gothanii* Benson, 1935a, seems indistinguishable from *C. glomerata*, to which Benson does not refer.

Other forking fronds in this assemblage belong to *Spathulopteris*, distinguished from *Sphenopteridium* and *Diplopteridium* by the swollen shape of the pinnules, and the absence of pinnae or pinnules attached to the main rachis below the dichotomy. There is no evidence available of fructifications attached to these fronds or of the rachial anatomy (except that there are no transverse bars across the rachises), but it is likely that they belong to the Calamopityales. The commonest species at Teilia is *S. ettingshausenii*, which has relatively slender pinnule-lobes. It is a widely distributed species, having been recorded from the Visean of Scotland (Kidston, 1923b), Moravia (Patteisky, 1929) and Silesia (Gothan, 1937). A single small fragment was also described by Walton (1931) as *S. clavigera*, which has more swollen lobes than *S. ettingshausenii*. However, this specimen shows little evidence as to the frond architecture and is thus included in the species list with a '?'.

A single seed identified by Walton (1931) as *Carpolithes* sp. gives the impression of being platyspermic, similar to *Samaropsis* described from the Tournaisian of Scotland. As remarked in the discussion on the Whiteadder (p. 124), such seed compressions might be correlatives of the petrified *Lyrasperma* seeds, and these in turn may have calamopityalean affinities.

In contrast to most of the other bipartite fronds in the Teilia assemblage, a number have wedge-shaped pinnules with entire margins, and belong to *Adiantites*. They are also distinguished by the more complex branching of the sterile foliage (Walton, 1931, text-fig. 2). Walton (1931) pointed out, however, that there is often a scar at the dichotomy of the main rachis, which might be the origin of a fertile branch, similar to that seen in *Diplopteridium*. If correct, this suggests lagenostomalean affinities. Two species of this type have been recognized from Teilia: *A. machanekii* (Figure 5.42), which is the most abundant, and has slender straight-sided pinnules (Walton, 1931, pl. 23, fig. 6, pl. 24, fig. 12); and *A. antiquus*, with more rounded pinnules (Kidston, 1889a, pl. 1, fig. 1; Kidston, 1923b, pl. 45, fig. 1; Walton, 1931, text-fig. 2). The specimens described by Kidston (1889a) and Walton (1931) as *A. tenuifolius* and *A.* sp. are too small for a positive identification.

Walton (1931) recorded, but did not figure, *Sphenopteris* cf. *filiformis* Kidston from Teilia. He suggested that it was perhaps only a young *Adiantites* frond. It has thus not been included here in the list of species for this locality.

Less completely known types of pteridosperm(?) frond from Teilia have been referred to *Sphenopteris obfalcata* Walton, 1931. It has rather swollen pinnule lobes, similar to the type normally found in *Spathulopteris*, but its frond architecture is too imperfectly known for it to be placed there. To date, this species has only been recorded from Teilia.

### Cycadopsida

Two small fragments of *Neuropteris antecedens* were described by Walton (1931) and Crookall (1959), and are the oldest evidence of the cycadopsid order Trigonocarpales (Medullosales of some authors) in the fossil record. This pteridosperm order is of considerable interest, being one of the major components of Late Carboniferous and Early Permian equatorial vegetation. It is also of some evolutionary interest, as it is thought to have been ancestral to the extant group, the cycads. The Teilia specimens are too small to confirm the trigonocarpalean affinities of this species, but the more complete material from the former Czechoslovakia figured by Stur (1875) has a distinctly trigonocarpalean aspect. Also of interest is that the Teilia frond fragments are associated with the seeds *Holcospermum ellipsoideum*, which look remarkably like the trigonocarpalean seeds found in the Upper Carboniferous; a similar association has been reported from slightly younger strata in France by Bureau (1913-1914).

### General remarks

Teilia Quarry is the only known locality to yield this distinctive assemblage of plant fossils from the Gronant Group. Many of the other assemblages of similar age, such as from south-east Spain (Jongmans, 1956) and the Loire (Bureau, 1913-1914) are dominated by lycopsids and are thus quite different. More comparable is the assemblage from the Posidonienschiefer of Moravia (Stur, 1875; Oberste-Brink, 1913; Patteisky, 1929; Hartung and Patteisky, 1960). In Britain, the closest

**Figure 5.42** *Adiantites machanekii* Stur. Almost complete pteridosperm frond; Natural History Museum, specimen V.2755. Gronant Group (Brigantian), Teilia Quarry. x 0.75. (Photo: Photographic Studio, Natural History Museum, London.)

comparison is found in the upper Oil Shale Group of Scotland, such as at Wardie Shore (see below). There are, however, several of the Teilia species which have not been reported from any of these other areas or localities, including *Rhacopteris circularis*, *Diplopteridium teilianum* and *Sphenopteridium capillare*; also, the Welsh assemblage does not appear to contain certain elements which characterize many other upper Visean assemblages, such as *Archaeopteridium* and *Fryopteris*. Whether this reflects the lagoonal setting for the Gronant Group or some other palaeoecological factor, or is merely a function of collecting and/or taxonomic bias, is as yet unclear.

## Conclusion

Teilia Quarry is one of the best known sites in Britain for plant fossils from the Lower Carboniferous, about 330 million years old, and has been studied now for over a century. It is particularly well-known for a type of leaf known as *Rhacopteris*, which occurs very commonly in fossil floras of this age. The Teilia material shows an unusually wide range of shapes of these leaves, as well as including one of only two known specimens bearing reproductive organs. The fossil flora here is also very rich in early seed plant remains, including a number of complete fronds, and several types of fructification. Of especial interest are fragments of the frond known as *Neuropteris*. These are the oldest known remains of an order of plants known as the Trigonocarpales, which later in the Carboniferous and Early Permian became very common in the tropical forests, and which is thought to be ancestral to the living group, the cycads. It is also possibly ancestral to the Bennettitales, a group of plants that became important in the later Mesozoic vegetation, and which may be ancestral to the flowering plants. The only other similar fossil flora has been found in Moravia. Other assemblages of the same age tend either to be dominated by lycopsids ('club-mosses'), such as those found in Spain and France, or, as in the Scottish assemblages, to contain quite different species of seed plants.

## WARDIE SHORE

## Highlights

Wardie Shore is the best available site for plant fossils from the Visean Wardie Shales (Figure 5.43). It is the type locality for the lycopsid stem *Bothrodendron wardiense* Crookall and the equisete strobilus *Pothocites grantoni* Paterson.

## Introduction

Records of Visean plant fossils from this shoreline Oil Shale exposure east of Granton Harbour, on the Firth of Forth (NT 245771), date back to the mid-nineteenth century, the earliest being that of Paterson (1841). Subsequent records include those by Kidston (1883b, c, 1889b, c), but the most comprehensive accounts of the fossils are in the monographs by Kidston (1923b, 1924) and Crookall (1964, 1969). There has been no recent work on the palaeobotany of this site.

## Description

### Stratigraphy

The geology of this site is described by Peach and Horne (1910). The exposed sequence (Figure 5.44) belongs to the Wardie Shales Formation in the Lower Oil Shale Group. According to Carruthers (1927), these shales were probably laid down under alternating lagoonal and estuarine conditions. Scott (1985) regards them as probably late Holkerian in age.

### Palaeobotany

The plant fossils found here are mainly compressions, which are readily separated from the matrix. Their appearance suggests that cuticles should be preserved but, in practice, they have proved impossible to separate from the bituminous carbonized tissue (C.H. Shute, pers. comm., 1989). The following species have been described to date:

Lycopsida:
 *Lepidodendron veltheimianum* Sternberg
 *Bothrodendron wardiense* Crookall

Equisetopsida:
 *Archaeocalamites radiatus* (Brongniart) Stur
 *Pothocites grantoni* Paterson

Lagenostomopsida:
 *Sphenopteris affine* Lindley and Hutton
 *Spathulopteris dunsii* Kidston

## Lower Carboniferous

**Figure 5.43** Wardie Shore. Foreshore exposures of the upper Holkerian Wardie Shales Formation. (Photo: C.J. Cleal.)

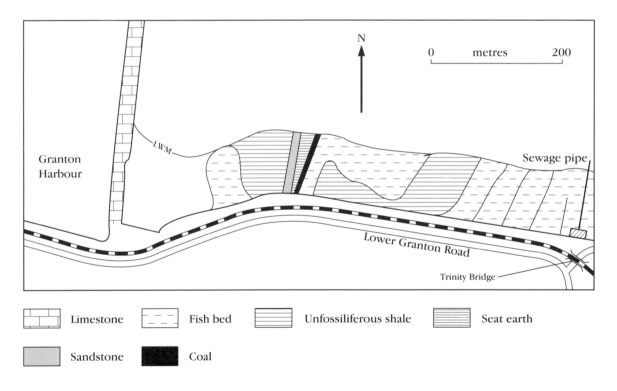

**Figure 5.44** Map showing the main lithologies exposed at Wardie Shore. Based on McAdam and Clarkson (1986, Map 7).

# Wardie Shore

## Interpretation

Crookall (1964, pl. 64, fig. 1, pl. 71, fig. 1) shows a large specimen of *Lepidodendron veltheimianum* with a row of so-called 'ulodendroid-scars'. Such scars are thought to be the result of branch abscission, and occur commonly in many arborescent lycopsids (Jonker, 1976). A similar specimen from the Burdiehouse Limestone, identified by Carruthers (1870) as *Ulodendron ovale* Carruthers, was transferred to *L. veltheimianum* by Kidston (1885b). *Ulodendron* has been used by many authors as an artificial form-genus for lycopsid axes with such abscission scars; Thomas (1967b) has shown that it in fact represents leafy lycopsid twigs with a distinctive epidermal structure.

This is the only known locality for another species of lycopsid stem, *Bothrodendron wardiense*. It was originally described by Kidston (1889b, c) as *Bothrodendron wiikianum* Heer, but he later recognized that the ligule pit aperture is more distantly placed above the leaf scar (see also Crookall (1964) and Thomas (1980)). Although he proposed an alternative name in manuscript, it was not validly published until after his death, by Crookall (1932). According to Thomas (1980), *Bothrodendron* differs from *Eskdalia*, another lycopsid stem frequently found in the Visean of Britain, in having a clearly marked parichnos mark in the leaf scar; and the Wardie Shore specimen indeed seems to show such a mark. Thomas also notes differences in the epidermal structure, but such information is unavailable at present for *B. wardiense*.

This is also the type locality for the equisete strobilus *Pothocites grantoni* (Paterson, 1841; Kidston, 1883b, c). Unfortunately, however, the type specimens are reported by Chaphekar (1965) as lost, and her re-assessment of the species had therefore to be based on material from Loch Humphrey Burn and Glencartholm (both sites discussed elsewhere in this chapter). Stur (1875) reported similar strobili attached to *Archaeocalamites* stems (see also Chaphekar, 1963), and so it is not surprising to find this type of stem associated with *Pothocites* at Wardie Shore.

*Spathulopteris dunsii* has only been reported to date from the Oil Shale Group of Scotland (Kidston, 1923b). Kidston lists it as occurring at Wardie Shore, but no specimens from here were illustrated. It is similar to another spathulopterid from the Oil Shale Group of Scotland, *S. decomposita* Kidston, 1923b, but the latter has more closely spaced pinnae and larger pinnules. Clearly, however, it is possible that they are just small and large fronds of a single biological species. The affinities of *Spathulopteris* are at present unproved, but may be with the Calamopityales, one of the two main orders of seed plants found in the Lower Carboniferous.

Kidston (1924) figured a number of specimens of the lagenostomalean frond *Sphenopteris affine* from the Oil Shale Group of Scotland. He mentions Wardie Shore as one of the many sites to yield the species, although none of the figured specimens originated from here. Kidston assigned the species to the form-genus *Telangium*, since he was able to show that the fronds bore *Telangium* fructifications. However, such fructifications are only rarely found attached to such fronds, and it is unwise to use a fructification form-genus for fronds which are normally found in a sterile condition (compare comments by Cleal, 1986b on the combination of form-genera representing different organs of a plant). Since *S. affine* has an essentially similar frond architecture and bore the same fructifications as the type-species of *Sphenopteris* (*Sphenopteris elegans* (Brongniart) Sternberg; synonym *Diplothmema adiantoides* (Schlotheim) Gothan – see Kidston, 1923c), it is best if it is retained in that form-genus.

The Oil Shale Group of Scotland has yielded some of the best preserved Visean plant compressions in Britain. The Burdiehouse Limestone was for many years one of the best stratigraphical horizons for such fossils (e.g. Lindley and Hutton, 1831-37; Miller, 1857), but there are now no suitable exposures of this bed. The oil shales themselves, however, particularly the Wardie Shales, have also produced some remarkably well-preserved specimens. When the matrix dries, the fossil sometimes peels off more or less intact (Kidston, 1923b, c, 1924), and almost has the appearance of a compressed piece of modern plant, albeit blackened (some excellent examples of such isolated fossils are figured by Kidston, 1924, pl. 101 figs 7-10). Plant fossils have been recorded from a number of exposures of the Wardie Shales in Edinburgh and environs, in particular Hailes Quarry and at a number of localities along the Water of Leith (Kidston, 1923a, 1925), but only Wardie Shore now still yields them. The site has only so far yielded the limited assemblage listed above, but this may just reflect the limited amount of collecting that has been done here, especially in recent years. Kidston also lists from other nearby exposures of the Wardie Shales the following taxa, which may also eventually prove to be present at Wardie Shore:

>*Sphenopteris cuneolata* Lindley and Hutton
>*Sphenopteridium pachyrrachis* (Göppert) Potonié
>*S. crassum* (Lindley and Hutton) Kidston
>*S. speciosum* Kidston
>*Spathulopteris obovata* (Lindley and Hutton) Kidston
>*S. decomposita* Kidston
>*Rhacopteris lindsaeformis* (Bunbury) Kidston
>*R. inaequilaterata* (Göppert) Stur
>*R. geikiei* Kidston
>*Schuetzia bennieana* Kidston

Similar British assemblages to that found in the Wardie Shales have been reported from Glencartholm and to a lesser extent Teilia Quarry (see comments elsewhere in this chapter on these sites for details of the comparison). Outside Britain, comparable assemblages are known from Upper Silesia (Patteisky, 1929; Hartung and Patteisky, 1960). Most other Visean assemblages from Europe differ from that of the Wardie Shales in both overall balance and species composition (Hirmer, 1939; Vakhrameev *et al.*, 1978). In particular, the Wardie assemblage has a much higher proportion of lycopsids and equisetes, suggesting a wetter, more swampy environment. According to Raymond and Parrish (1985), a rain-shadow caused by newly formed mountains to the west resulted in much of western Europe having a relatively dry climate during the Early Carboniferous. However, the Scottish sites such as Wardie seem to have avoided its influence, perhaps due to local topographic factors.

## Conclusion

Wardie Shore is the only site that still yields abundant plant fossils from the Scottish Oil Shale Group, about 335 million years old. These shales were once a famous source of plant fossils. However, as the shales are no longer a commercial hydrocarbon source, there are now no quarries or mines from which the fossils can be collected. This particular site has yielded a number of club-mosses and early seed plants. One of its main interests, though, is as the type locality for a cone known as *Pothocites grantoni*, which is now known to be an early type of horsetail fructification, and has proved important for establishing the early evolutionary history of this extant class of plants. The general balance of the assemblage, with its predominance of club-mosses and horsetails, suggests a much wetter environment than is represented by the vegetation of this age found elsewhere in Europe.

## GLENCARTHOLM

### Highlights

Glencartholm has yielded one of the best examples of a *Neuropteris antecedens* Zone fossil flora from Britain. Many of the species found here are widely distributed, but there are a number for which this is the only or best locality, such as *Rhacopteris geikiei*, *Sphenopteridium macconochiei* and *Rhodeopteridium machanekii*. Glencartholm has provided particularly fine examples of *Spathulopteris*. It is a site of considerable palaeobotanical significance.

### Introduction

Lower Carboniferous rocks in the bed of River Esk, about 4 km north-west of Canonbie, Dumfries and Galloway (NY 377795), are best known as a site for fish and crustacean fossils. However, they have also yielded a diverse plant fossil assemblage. They were first discovered in 1876 by A. Macconochie, during the mapping of the area by the Geological Survey. Kidston (1883a) made a preliminary report on the assemblage (see also Kidston, 1886), and later published a systematic account (Kidston, 1903b). Specimens from here were extensively used in the Geological Survey monographs on Carboniferous plant fossils (Kidston, 1923–1925; Crookall, 1955–1975). In addition, there have been accounts of individual taxa from here, including *Pothocites* (Kidston, 1882, 1883b, c; Chaphekar, 1965), *Eskdalia* (Thomas, 1968) and *Lycopodites* (Kidston, 1901b). Thomas's *Eskdalia* study is particularly important since it shows that cuticles are preserved in at least some of the Glencartholm fossils.

### Description

#### Stratigraphy

The plant fossils occur in the Glencartholm Volcanic Group, a sequence of tuffs, shales and

# Glencartholm

sandstones near the base of the Upper Border Group (Lumsden *et al.*, 1967). They are interpreted as shallow marine or lagoonal deposits, which were subject to periodic burial from volcanic outfall. Foraminifera from the sequence suggest a position near the Holkerian–Asbian boundary (George *et al.*, 1976).

## Palaeobotany

The following taxa have been reported to date:

Algae (divisions unknown):
  *Bythotrepis acicularis* (Göppert) Kidston
  *B. plumosa* Kidston
  *B. simplex* Kidston
  *B. scotica* Kidston

Lycopsida:
  *Eskdalia minuta* Kidston
  *Lepidodendron veltheimianum* Sternberg
  *Lepidostrobus ornatus* Brongniart
  *Lepidostrobophyllum lanceolatum* (Lindley and Hutton) Bell
  *Stigmaria ficoides* (Sternberg) Brongniart
  *L. fimbriatum* (Kidston) Allen
  *Lycopodites stockii* Kidston

Equisetopsida:
  *Archaeocalamites radiatus* (Brongniart) Stur
  *Pothocites grantoni* Paterson

Filicopsida(?):
  *Rhodeopteridium machanekii* (Ettingshausen) Purkyňová

Progymnospermopsida(?):
  *Rhacopteris lindsaeformis* (Bunbury) Kidston
  *R. geikiei* Kidston

Lagenostomopsida:
  *Sphenopteris bifida* Lindley and Hutton
  *Sphenopteridium pachyrrachis* (Göppert) Schimper
  *S. macconochiei* Kidston
  ?*S. crassum* (Lindley and Hutton) Kidston
  *Spathulopteris obovata* (Lindley and Hutton) Kidston
  *S. decomposita* Kidston
  *Cardiopteridium nanum* (Eichwald) Walton
  *Carpolithes* sp.

## Interpretation

### Algae

Kidston (1883a, 1903b) described a variety of enigmatic fossils from here, which he interpreted as algal remains. He assigned them to four species of the 'artificial' form-genus *Bythotrepis*, but could draw no conclusions about their affinities.

### Lycopsida

These are relatively rare at Glencartholm, but the most abundant are the stems *Lepidodendron veltheimianum* and rooting structures *Stigmaria ficoides*. In association are strobili described (Kidston, 1903b) as *Lepidostrobus variabilis* Lindley and Hutton, but which Crookall (1966) later transferred to *Lepidostrobus ornatus* (it should be noted that this is a true *Lepidostrobus* as interpreted by Brack-Hanes and Thomas (1983), i.e. a microsporangiate strobilus). Most *Lepidodendron* species, however, are thought to have borne *Flemingites* strobili, and so there must be some doubt as to whether *L. ornatus* was attached to *L. veltheimianum*.

A second lycopsid plant represented at Glencartholm had stems belonging to the form-genus *Eskdalia*. Kidston (1883a, 1903b) interpreted it as a fern stem, but Thomas (1968) demonstrated that it was a ligulate lycopsid (see also Chaloner *in* Boureau *et al.*, 1967; Thomas and Meyen, 1984; Rowe, 1988c). A key aspect of Thomas's study was the preparation of cuticles, which yielded evidence of the ligule, stem stomata and 'resistant excrescences'. However, no evidence of a parichnos was found. It appears to be a wide-ranging form-genus, having been reported from Puddlebrook (see above), and from several localities in Siberia (Thomas and Meyen, 1984). It is also very similar to the '*Lepidodendron*' *perforatum* Lacey, 1962 from North Wales (Rowe, 1988c). Nevertheless, Glencartholm remains one of the most important localities for this form-genus.

There seems to be a consistent association between *Eskdalia* and the isolated sporophylls *Lepidostrobophyllum fimbriatum* (cf. also Puddlebrook Quarry - see above). However, the only species of *Eskdalia* to which strobili have been found attached have sporophylls of a different type (*E. variabilis* (Lele and Walton) Rowe, 1988c). This problem has been discussed under Puddlebrook Quarry (p. 169).

A poorly preserved specimen of an herbaceous lycopsid was described by Kidston (1884, 1901b) as *Lycopodites stockii*. It shows a leafy shoot with sporangia arranged both in a terminal strobilus and amongst the leaves. Bower (1908) argued that this arrangement of sporangia points to it belonging to the subsection *Phlegmaria*, in the classification established for extant species. It is one of the earliest known examples of a fossilized lycopodiacean with fructifications preserved.

## Equisetopsida

The equisetes are represented here by the ubiquitous (in the Lower Carboniferous) *Archaeocalamites radiatus*. Strobili found in close association were identified by Kidston (1882) as *Pothocites calamopsoides* Kidston (Figure 5.45), but he later recognized that they were indistinguishable from *P. grantoni* (Kidston, 1883b, c).

## Filicopsida(?)

Glencartholm is the best known locality for *Rhodeopteridium machanekii*. It has yielded several large specimens, showing the distinctive large, deeply incised pinnules. Relatively little is known of the branching architecture or pinnule variation in these fern-like plants, and so large specimens such as those found here have considerable significance.

## Progymnospermopsida(?)

The commonest type of foliage which probably belongs to this group is *Rhacopteris lindsaeformis*. Kidston (1903b) initially identified it as *Rhacopteris inaequilaterata* (Göppert) Stur. However, he later revised this identification (Kidston, 1923c), although he still listed Glencartholm as a locality yielding *R. inaequilaterata*; this was probably just a clerical error.

Glencartholm is the type and best known locality for a second species of ?progymnosperm foliage, *R. geikiei*. It has very distinctive pinnules, which are deeply incised with slender lobes, but not markedly asymmetrical, as in most other deeply incised rhacopterid species.

## Lagenostomopsida

The most abundant pteridosperm fossils found here to date are fronds. The most abundant belong to the Lagenostomales. Three species of

**Figure 5.45** *Pothocites grantoni* Paterson. Cone of archaeocalamitid equisetopsid; Natural History Museum, London, specimen V.195. Glencartholm Volcanic Group (Holkerian–Asbian), Glencartholm. × 1. (Photo: Photographic Studio, Natural History Museum, London.)

*Sphenopteridium* have been recorded. *S. pachyrrachis* is by far the most common (Figure 5.46). ?*S. crassum* has never been illustrated, but there must be a strong possibility that the ?*S. crassum* is merely a small form of *S. pachyrrachis*. The third, *S. macconochiei*, is only known from Glencartholm. Details of the venation and rachis ornamentation appear to support its inclusion in *Sphenopteridium*, but, compared with the other known species of the genus, it has more robust, subrhomboidal, entire-margined pinnules.

Kidston (1924) figured a fructification from here as *Sphenopteris bifida*, but details of the

## Glencartholm

**Figure 5.46** *Sphenopteridium pachyrrachis* (Göppert) Schimper. Almost complete pteridosperm frond; Natural History Museum, London, specimen V.186. Glencartholm Volcanic Group (Holkerian-Asbian), Glencartholm. × 1. (Photo: Photographic Studio, Natural History Museum, London.)

sterile part of the frond were not given to support the identification.

A number of relatively large portions of *Spathulopteris* fronds has been illustrated from here (Kidston, 1924). Neither of the recorded species (*S. obovata* and *S. decomposita*) is rare or particularly distinctive. Nevertheless, the Glencartholm material is of interest as providing some of the best examples of this form-genus to be recorded in the literature. Fructifications or stem/rachis anatomy are unknown, but the fronds are believed to belong to the Calamopityales (see discussion on Loch Humphrey Burn, above).

Two small specimens of *Cardiopteridium nanum* have been found here. The taxonomy of this species has had a complex history (Walton, 1941), but it is now generally assumed that a range of species names used in the past in fact just reflects a marked degree of infra-specific variation in this taxon. Walton retained a distinction at the rank of forma between the large and small types of pinnules, but this seems to be an artificial division of doubtful utility.

### General remarks

The occurrence together at Glencartholm of *Spathulopteris* and *Cardiopteridum* clearly indicates that the assemblage belongs to the lower part of the *Neuropteris antecedens* Zone (*Diplopteridium* Subzone) of Cleal (1991). There is some comparison with the Wardie Shales assemblage (p. 181), which has a number of species in common, such as *Rhacopteris lindsaeformis*, *Pothocites grantoni* and *Lepidodendron veltheimianum*. The Wardie assemblage, however, has a higher proportion of lycopsids. There are also differences in the pteridosperm composition, such as the presence at Wardie of *Sphenopteris affinis* instead of the closely related *S. bifida*. Also comparable among the British Visean assemblages is that from Teilia Quarry (p. 175), but the similarity here seems to be mainly at the rank of form-genus, the only species in common being *Archaeocalamites radiatus* and *Sphenopteridium pachyrrachis*.

Outside of Britain, the nearest comparison is with assemblages reported from the Pollak Stollen Formation of Upper Silesia (Patteisky, 1929; Hartung and Patteisky, 1960), which includes among others *Lepidodendron velthiemianum*, *Rhodeopteridium machanekii* and *Sphenopteridium pachyrrachis*. However, most of the other Lower Carboniferous assemblages from central Europe differ markedly from that found at Glencartholm, both in overall balance and in species composition (for reviews of these assemblages see Hirmer, 1939 and Vakhrameev *et al.*, 1978). The reason for these differences has not been properly investigated, and it is not clear whether it reflects palaeoecological or biostratigraphical variation, or even just taxonomic and/or collecting bias. However, as one of the best documented fossil floras of the Upper Visean of Europe, Glencartholm will clearly play a central role in unravelling the patterns of vegetational distribution within the palaeoequatorial belt.

## Conclusion

Glencartholm has yielded one of the best documented fossil floras from the upper part of the Lower Carboniferous of Britain, representing vegetation growing some 330 million years ago. Only two species have been found here, the possible fern *Rhodeopteridium machanekii* and the seed plant *Sphenopteridium macconochiei*. It has also yielded some exceptionally complete examples of leaves of the ?progymnosperm (i.e. probably belonging to the immediate ancestors of seed plants) *Rhacopteris*, and of the early seed plant *Spathulopteris*, which have proved important for understanding the affinities of these plants. The club-mosses here include a small, leafy form (*Lycopodites*) very similar to the living *Lycopodium*. In contrast to many other fossil floras of this age from northern Britain, however, club-mosses are relatively rare and probably represent vegetation growing in drier habitats. A much closer comparison can be made with fossil floras found in continental Europe, in particular from Upper Silesia (Poland).

## VICTORIA PARK

## Highlights

Victoria Park has the best known examples of *in situ* stumps of arborescent lycopsids in the Lower Carboniferous, providing a unique insight into the forests which were starting to dominate the palaeoequatorial regions at that time (Figure 5.47).

# Victoria Park

**Figure 5.47** Victoria Park. *In situ* fossilized tree stumps in the Limestone Coal Group (lower Pendleian). (Photo: B.A. Thomas.)

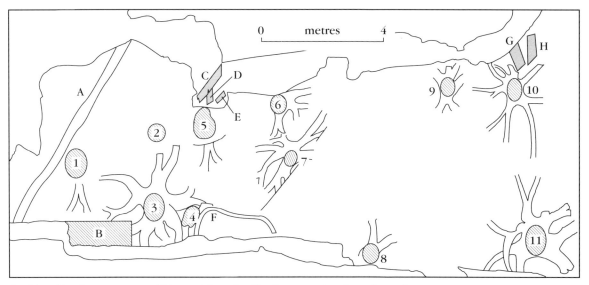

1 – 11  Fossil tree stumps (*Stigmaria*)   A – H  Prostrate trunks or branches

**Figure 5.48** Plan of the Fossil Grove at Victoria Park, showing distribution of *in situ* lycopsid stumps. Based on MacGregor and Walton (1972, figure 1).

# Lower Carboniferous

## Introduction

This famous locality is under a covered enclosure in Victoria Park, Glasgow (NS 541673), and shows a cluster of *in situ* Namurian lycopsid stumps, sometimes called the 'Fossil Grove' (Figure 5.48). In addition to its palaeobotanical interest, it is significant as one of the earliest examples of a conserved earth science site (Black, 1988). The stumps were discovered in 1887 (Kidston, 1888), but were not described in detail until MacGregor and Walton (1948, 1972; see also Mclean, 1973; Lawson and Lawson, 1976). The most comprehensive account to date is by Gastaldo (1986).

## Description

### Stratigraphy

The stumps lie within the Limestone Coal Group, of early Pendleian age (Figure 5.49). They are rooted in a silty mudstone, which Gastaldo (1986) interpreted as a palaeosol, and the aerial parts were originally entombed in sandstones, which were probably crevasse-splay deposits.

### Palaeobotany

Eleven stumps are preserved *in situ* here (Figure 5.49), and are of an average height of *c.* 0.4 m. The rooting structures are clearly of a *Stigmaria*-type, and so the trees must have been arborescent lycopsids. There would appear to be the equivalent of about 4500 trees per km$^2$.

## Interpretation

Such stands of *in situ* lycopsid stumps are relatively common in the Upper Carboniferous (see Williamson, 1887 for a review of some of the early evidence), where they represent the remains of extensive forests that came to dominate the equatorial regions. However, these are the only conserved examples of such lycopsid stumps in the Lower Carboniferous.

A curious feature of the stumps is that they have an elliptical transverse section and are all aligned in about the same direction. MacGregor and Walton (1972) interpreted this as due to tectonic distortion, but Gastaldo (1986) has pointed out that there is little other evidence of tectonic deformation here. Instead, Gastaldo argued that

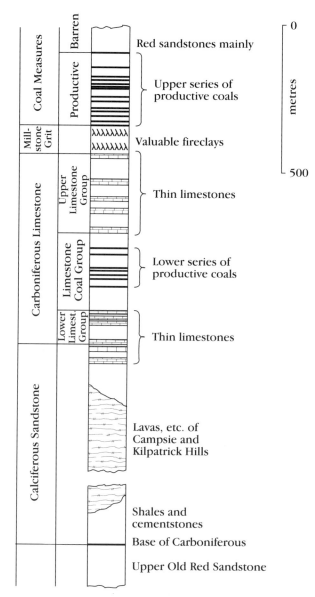

**Figure 5.49** Generalized section through the Carboniferous of the Glasgow area, showing the position of the Limestone Coal Group. Based on MacGregor and Walton (1972, figure 8).

the stumps were at least partially hollowed-out prior to being fully engulfed in sediment, and that the distortion was a result of 'streamlining' by the entombing sediment.

## Conclusion

Victoria Park has the only preserved examples of *in situ* stumps of giant club-mosses in the Lower

## Victoria Park

Carboniferous, some 325 million years old. They allow us to estimate the tree density in the forests growing at this time in Britain as about 4500 per km$^2$. Most localities containing plant fossils only yield fragments of stems, leaves or reproductive organs, that were washed from the site where they grew. This makes localities such as Victoria Park all the more remarkable.

*Chapter 6*

# Upper Carboniferous

# Palaeogeographical setting

The productivity of terrestrial vegetation during the Late Carboniferous was greater than at any other time during the Palaeozoic. This is most clearly seen in the development of thick, coal-forming peats in the tropical lowland habitats – the result of dense forests, dominated by lycopsids, ferns and in some cases cordaites (see Figure 6.3). However, there is also evidence of extensive vegetation in the tropical extra-basinal areas, as well as in parts of the high latitudes, in Angara and Gondwana (Figure 6.1).

Conditions at this time were clearly favourable for plant life in the tropics (e.g. high levels of precipitation). Probably because of this, there were relatively few major evolutionary innovations in these floras; most major events seem to have occurred in other habitats, such as the tropical extra-basinal habitats (e.g. the appearance of conifers) and at higher latitudes (e.g. the appearance of glossopterids). Nevertheless, the plant fossils found in the Upper Carboniferous tropical deposits are of considerable interest, as they represent the acme of palaeophytic vegetation. They also have a major economic importance, through the hydrocarbon fuels (coals, oil and gas) that they generated. Consequently, the palaeobotany of these strata has been more intensely studied over the years than any other part of the geological column.

## PALAEOGEOGRAPHICAL SETTING

At the start of the Late Carboniferous, Britain was on the southern margins of the Laurussian continent, positioned near the equator (Figure 6.1). Probably sometime in the late Westphalian or early Stephanian, however, the Gondwanan continent collided with Laurussia to form the Pangaean 'super-continent' that stretched from the south pole to high northern latitudes. During much of the subperiod, a series of major fluvial deltas prograded over large areas of present-day northern Europe and eastern North America. The resulting delta-plains provided the substrate for the growth of extensive forests – the very first major tropical rain-forests.

The forests were at their peak during the Westphalian Epoch, and the peat deposits that they produced have resulted in extensive coals. Economically significant Upper Carboniferous coalfields occur throughout the palaeoequatorial belt, from central North America (Arkansas, Kansas), through much of Europe to the Ukraine, as well as in China. The British coalfields belong to what is often called the Paralic Belt, which was a trough or 'valley' that extended from Poland in the east, through northern Germany, the Netherlands, Belgium, northern France to the British Isles and the Maritime Provinces of Canada.

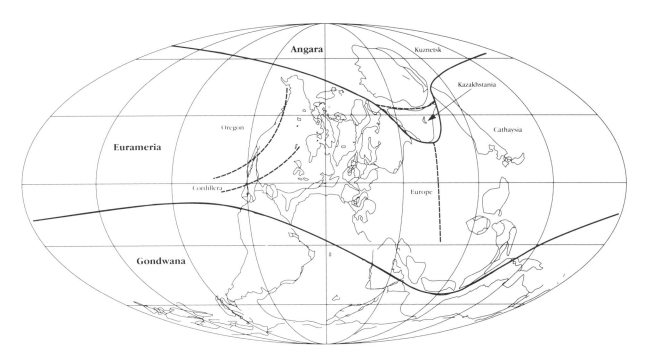

**Figure 6.1** The palaeogeography of the Late Carboniferous, showing the distribution of the major floristic zones (phytochoria). Based on Scotese and McKerrow (1990) and Cleal and Thomas *in* Cleal (1991).

Although largely formed under non-marine conditions, the trough was periodically flooded by eustatic rises in sea level. Coal formation in the Paralic Belt eventually ceased towards the end of the Westphalian or very early Stephanian; the youngest British coals are in the basal Cantabrian (Cleal, 1978, 1984b, 1986c).

During the Namurian and most of the Westphalian, Britain was subject to a hot and wet tropical climate. However, the formation of Pangaea disrupted the oceanic and atmospheric currents, and was at least partially responsible for a change to a significantly arid climate in the tropical regions towards the end of the Carboniferous (Rowley *et al.*, 1985). The collision between Laurussia and Gondwana also caused a degree of topographical uplift of the tropical regions. This in turn made the area less suitable for the growth of the lycopsid-dominated forests, which were physiologically constrained to lowland swamps. The lycopsids in the tropical forests of Pangaea were initially replaced by tree-ferns and pteridosperms, which were still capable of generating considerable peat accumulations. Eventually, in the Permian, the latter were in turn largely replaced by tracts of conifers, better adapted to the drier habitats. However, it is unlikely that these were anywhere near as aerially extensive as the lycopsid forests, and certainly never resulted in any significant peat accumulations. It is interesting to postulate that the reduction in the tropical plant biomass towards the end of the Carboniferous may itself have been a factor causing climatic change, through a 'greenhouse effect'.

There is a marked provincialism in plant fossil distribution in the Upper Carboniferous (Chaloner and Lacey, 1973; Chaloner and Meyen, 1973; Vakhrameev *et al.*, 1978; Meyen, 1987; Chaloner and Creber, 1988; Allen and Dineley, 1988). Four palaeokingdoms are usually recognized (Figure 6.1), each having its own discrete set of plant fossil assemblages. The tropical part of Pangaea, including Britain, belongs to the Euramerian Palaeokingdom, which extended from central North America, through Europe and North Africa, to the Caucasus.

## STRATIGRAPHICAL BACKGROUND

As pointed out in the introduction to the previous chapter, it is convenient to divide this part of the discussion on British palaeobotanical sites into Upper and Lower Carboniferous chapters. This corresponds both to a natural division of the plant fossil assemblages, and to the sub-systemal division currently accepted by the IUGS Subcommission on Carboniferous Stratigraphy.

The marked provincialism of both fossil faunas and floras in the Upper Carboniferous has made it impossible to establish a chronostratigraphy for use throughout the world. Even within the equatorial belt, problems of detailed correlation have prevented agreement on a unified classification; the outline scheme proposed by Bouroz *et al.* (1978) still requires considerable refinement before it becomes a practical stratigraphical tool.

The British Upper Carboniferous is classified according to a modified form of the 'Heerlen Classification', named after the conferences on Carboniferous stratigraphy held at Heerlen between 1928 and 1958 (Wagner, 1974). The part of this chronostratigraphy relevant to the following discussion is summarized in Figure 6.11, including the revised stage nomenclature outlined by Engel (1989).

Wagner (1984) proposed eleven plant fossil zones for the Upper Carboniferous Eurameria Palaeokingdom, and a twelfth was added by Cleal (1984b). Of these, the top four (Barruelian to Stephanian C in age) are not found in Britain. The remaining eight are shown in Figure 6.11, with the GCR palaeobotanical sites plotted against them.

## LATE CARBONIFEROUS VEGETATION

There was a significant change in tropical vegetation between the early and late Carboniferous (the latter represented in Figure 6.3), probably due to climatic changes triggered by the growth in the southern polar ice-cap. A number of groups declined or became extinct, such as the archaeocalamites and the callamopityalean pteridosperms, whilst others underwent a major radiation, such as the true ferns (especially the marattialeans), the trigonocarpalean pteridosperms, and the cordaites (Figure 6.2).

As in the Early Carboniferous, lycopsids were the most important component of this vegetation. If the adpression record is viewed in isolation, this is not immediately apparent, as the dominant fossils preserved there tend to be pteridosperms, ferns and equisetes. However, if the coal-forming peat deposits are examined, either by looking at coal-balls (e.g. Phillips, 1980; Phillips and DiMichele, 1992) or palynology (A.H.V. Smith,

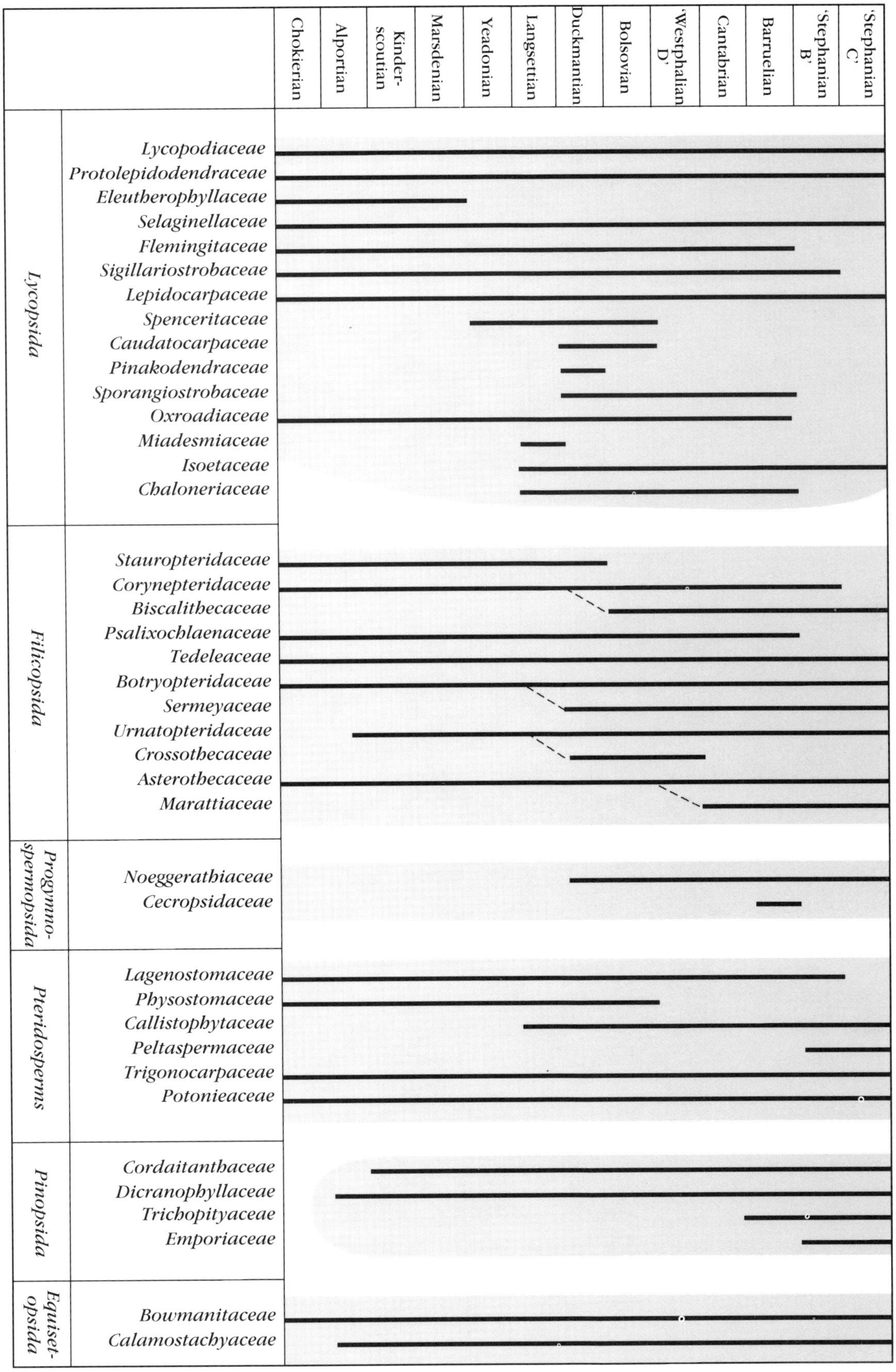

**Figure 6.2** The distribution of the principal families of vascular plants in the Late Carboniferous. Based on data from Cleal (1993).

**Figure 6.3** Diorama of a Late Carboniferous (Coal Measures) tropical swamp forest. Taken from the 'Evolution of Wales' exhibition, National Museum of Wales, Cardiff.

## Late Carboniferous vegetation

1962), it is found that (at least through most of the Westphalian) they are dominated by lycopsid remains. A number of herbaceous forms have been found, which are strikingly similar to the Recent *Selaginella* (Thomas, 1992). However, the best documented Late Carboniferous lycopsids were the arborescent forms (e.g. Flemingitaceae, Lepidocarpaceae, Sigillariostrobaceae - Thomas, 1978a; Figure 6.4). They dominated many of the tropical forests and, although they suffered a major decline in Pangaea towards the end of the Westphalian, they continued to be important in Cathaysia through to the end of the Permian Period (Li, 1980). Much of the primary work on the Upper Carboniferous lycopsids was done on fossils from British coal-balls (reviewed by Scott, 1920-1923), and later supplemented by work on adpressions (e.g. Thomas, 1967a, b, 1970, 1974, 1977, 1978a, b). In more recent years, however, work on American fossils has come to the fore, resulting in significant progress in understanding the diversity, reproductive strategies and population dynamics of these arborescent lycopsids (e.g. Phillips, 1979; DiMichele and DeMaris, 1987). It is now clear that the forests were extremely heterogeneous, their composition controlled largely by substrate conditions.

Two orders of equisetopsids are known from the forests: the arborescent Calamostachyales and the herbaceous Bowmanitales. Fossils of the former are particularly common, especially pith casts of the stems (*Calamites*) and foliage (*Annularia*, *Asterophyllites*). They represent plants that were superficially very similar to the Recent Equisetales, except in size, reaching heights of ten metres or more (Figure 6.5). However, there are marked differences in their reproductive organs, the calamostachyalean strobili having bracts separating the sporangiophores. Also, the larger stems of the Calamostachyales developed a zone of secondary wood, not seen in the living forms. They are generally thought to have grown along the margins of standing water or on sand bars within streams (Scott, 1977, 1978).

Fossilized foliage of the Bowmanitales (*Sphenophyllum*) is also extremely common. They were exclusively herbaceous, and were probably creeping, ground-cover plants (Batenburg, 1977; Figure 6.6) that were early colonizers of disturbed land within the swamp.

The Coenopteridales persisted into the Late Carboniferous, represented mainly by the form-genus *Corynepteris*, but they were never common (Scott and Galtier, 1985). True ferns, however, became much more common, especially in the tropical forests. Kidston's (1923-25) studies on the British Upper Carboniferous ferns have proved of fundamental importance in understanding their palaeobiology, and have proved the foundation for the more recent studies, such as by Danzé (1956) and Brousmiche (1983). Most of the herbaceous ferns belong to three orders, the Botryopteridales, Urnatopteridales and Crossothecales (Brousmiche, 1983; Meyen, 1987). It was once though that the extant family Gleicheniaceae was present at this time, in the

**Figure 6.4** Reconstruction of a Late Carboniferous giant lycopsid, *Lepidodendron*. From Thomas and Spicer (1987, figure 7.5).

## Upper Carboniferous

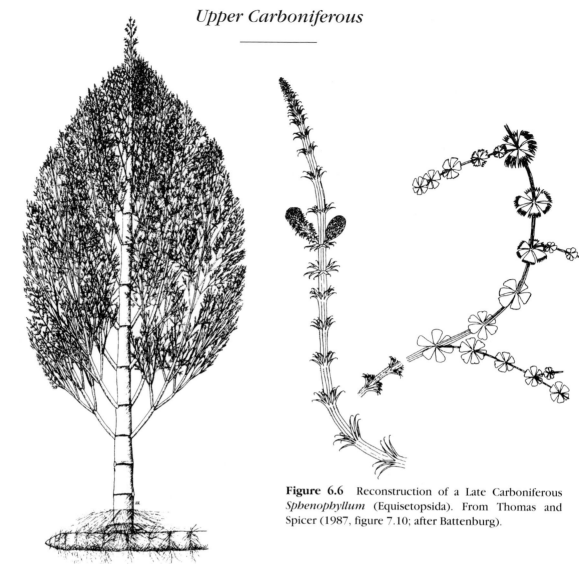

**Figure 6.5** Reconstruction of a Late Carboniferous giant equisetopsid *Calamites*. From Thomas and Spicer (1987, figure 7.11; after Hirmer).

**Figure 6.6** Reconstruction of a Late Carboniferous *Sphenophyllum* (Equisetopsida). From Thomas and Spicer (1987, figure 7.10; after Battenburg).

form of *Oligocarpia*, but these fossils are now assigned to the extinct family Sermeyaceae within the Botryopteridales. Evidence as to the habitat of these ferns is still very limited but they probably grew in open areas or in some cases as understorey within the main parts of the forests.

Tree-ferns also became common at this time (Figure 6.7); particularly after the decline of the lycopsids in the Stephanian, they were dominant components of the tropical lowland forests (DiMichele *et al.*, 1985). The most abundant belong to the order Marattiales, which appears to have had a global distribution. Their foliage (*Pecopteris sensu lato* - Kidston, 1923-1925; Dalinval, 1960) and trunks (*Psaronius, Caulopteris, Megaphyton, Artisophyton* - Crookall, 1955; Morgan, 1959; Pfefferkorn, 1976; Mickle, 1984) are extremely common fossils, especially in the upper part of the Upper Carboniferous. A second group known from the tropical forests (e.g. *Senftenbergia*), was originally thought to belong to the extant family Schizaeaceae but is now assigned to the extinct Tedelaceae (Eggert and Taylor, 1966).

The progymnosperms had declined significantly by the Upper Carboniferous. In the equatorial belt, foliage assigned to the form-genus *Noeggerathia* occurs sporadically, and bears a close similarity to the Lower Carboniferous *Rhacopteris* and *Archaeopteris* fronds. However, their fructifications (known as *Discinites*) are in the form of discrete cones (Hirmer, 1941; Danzé, 1957), quite different from the loose clusters of sporangia of the more typical Early Carboniferous progymnosperms.

Of the gymnosperms, the pteridospermous groups were the commonest in the tropical forests. The traditional concept of pteridosperms was of a group of plants with complex, fern-like fronds, but which reproduced by seeds. However,

## Late Carboniferous vegetation

**Figure 6.7** Reconstruction of a Late Carboniferous marattialean tree-fern *Psaronius*. From Thomas and Spicer (1987, figure 6.6; after Morgan).

the commonest (e.g. Figure 6.8). Some (e.g. *Alethopteris*) had very large fronds (up to seven metres long according to Laveine, 1986), although others (e.g. *Callipteridium*, *Odontopteris*) were much smaller. The seeds, which were attached directly to the fronds, also tended to be large. The Trigonocarpales seem to have been mainly restricted to the equatorial belt; records of *Neuropteris* foliage from the northern high latitudes (e.g. Gorelova *et al.*, 1973) are all extremely doubtful.

The Lagenostomales were the second major group of pteridosperms of the tropical forests (Taylor and Millay, 1981). Unlike most of the Early Carboniferous members of the order, they appear to have been relatively small plants. The *Lyginopteris*-bearing plants were probably shrubs, that favoured the wetter habitats in Namurian and early Westphalian swamps. Most of the others,

the current view is that the pteridosperms were a heterogeneous group of only distantly related plants, which independently developed such complex fronds (e.g. Crane, 1985).

The trigonocarpalean pteridosperms (Medullosales *auct.*) originated in the Early Carboniferous and persisted into the very early Permian, but they are regarded as a characteristically Late Carboniferous group. Adpressions of their foliage (e.g. *Neuropteris*, *Alethopteris* - Laveine, 1967; Wagner, 1968; Cleal and Shute, 1991, 1992; Cleal *et al.*, 1991) are extremely common in the Upper Carboniferous, and their stems are major components of some of the coals (e.g. Delevoryas, 1955), particularly those representing drier habitats. The trigonocarpaleans developed a variety of growth habits, including lianas, ground-creepers, shrubs and trees, although the latter two were probably

**Figure 6.8** Reconstruction of a Late Carboniferous trigonocarpalean pteridosperm, *Alethopteris*. From Thomas and Spicer (1987, figure 10.2; after Stewart and Delevoryas).

however, such as the *Eusphenopteris/ Heterangium* and the *Mariopteris/Schopfiastrum* plants, had a creeping or vinose habit. Most lagenostomalean fronds were consequently rather smaller than those of the Trigonocarpales. The only unequivocal records of this order are from the equatorial belt.

A third pteridosperm order known from the tropical forests was the Callistophytales (Rothwell, 1975, 1981). It first appears both as adpressions and coal-ball petrifactions in the middle Westphalian D, and occurs consistently through the rest of the Carboniferous. The few known species assigned to the order probably represent creeping plants (Figure 6.9), with fronds bearing lobed pinnules (*Dicksonites*), and there is therefore some superficial comparison with the Lagenostomales. However, in just about every other character (e.g. stem and rachial anatomy, seed structure, pollen) they are quite different, and suggest possible affinities with the peltasperms (e.g. *Callipteris auct.*) that occur commonly in the Permian.

In addition to the pteridosperms, the other major group of Late Carboniferous gymnosperms was the cordaites (Rothwell, 1988). They were mostly trees (although some herbaceous forms are also known) with long, strap-like leaves with a parallel nervation (Figure 6.10). There were separate male and female fructifications, each consisting of clusters of cones attached to a central rachis. In the tropical areas

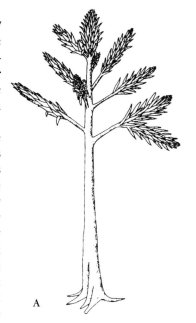

**Figure 6.10** Reconstruction of Late Carboniferous cordaitaleans: (A) an arborescent form found in the palaeoequatorial swamp-forests; (B) a smaller form. From Thomas and Spicer (1987, figure 11.1; after D.H. Scott, and Rothwell and Warner).

**Figure 6.9** Reconstruction of a Late Carboniferous callistophyte liana. From Thomas and Spicer (1987, figure 10.6; after Rothwell).

such as Britain, cordaites were probably most abundant in the drier, extra-basinal habitats, although some also seem to have grown within the swamp-forests, perhaps on the raised levée banks. During the somewhat drier interval in the mid-Westphalian, they even formed major components of the forest, particularly in coastal areas where they were similar to Recent mangroves (DiMichele *et al.*, 1985).

From the detailed structure of the fructifications, the cordaites seem to have been closely related to the early conifers. The conifers themselves were already in existence in the Late Carboniferous (records from North America and Britain), although they mainly grew in the extra-basinal habitats and are thus rarely found in the fossil record (A.C. Scott and Chaloner, 1983; Lyons and Darrah, 1989).

Outside the tropical belt, vegetation was not particularly lush during the Late Carboniferous. In the southern continents of Gondwana, the polar ice-cap had a severely limiting effect on the vegetation, restricting it largely to herbaceous lycopsids and shrubby progymnosperms. This has been described by Retallack (1980) as the *Botrychiopsis* tundra. Only at the very end of the Carboniferous did the ice-cap contract, allowing forests of arberialean ('glossopterid') trees to develop. In the northern continent of Angara there was no significant ice-cover, but vegetation was still very restricted, consisting mainly of shrubby lycopsids; Meyen (1972) described it as a 'cheerless and monotonous "brush" of fairly short straight sticks'. As in Gondwana, conditions seem to have become more favourable to vegetation towards the end of the Carboniferous, and the variety of plants present started to increase, probably as a result of migration from the tropical forests. However, it was not until the Permian that the Angaran vegetation became as lush and diverse as that seen in the Late Carboniferous tropical forests of Eurameria.

## UPPER CARBONIFEROUS PLANT FOSSILS IN BRITAIN

During the Namurian, extensive fluvial deltas in Britain produced sandstone bodies, belonging to the Millstone Grit. There is some evidence of the vegetation growing on these deltas (e.g. Lacey, 1952c), but the preservation of the fossils is often poor and they have not been studied to the same extent as elsewhere in Europe (e.g. Stockmans and Willière, 1952-1953; Josten, 1983). British Namurian vegetation is thus somewhat of an unknown quantity and no GCR sites have been selected for the palaeobotany of this series.

In the Westphalian, the sedimentary regime in Britain changed to mainly middle and then upper delta-plain deposits. The resulting strata, known as the Coal Measures (Figure 6.3), are particularly suited to the preservation of plant fossils. In the South Wales and Bristol–Somerset coalfields, plant fossils can be found throughout the sequence, from the basal Langsettian to the basal Cantabrian (Dix, 1933, 1934; Moore and Trueman, 1942; Cleal, 1978; Cleal and Thomas, 1991). In the Forest of Dean, they are restricted to the upper Westphalian D and Cantabrian (Wagner and Spinner, 1972). In the English Midlands and Pennines coalfields, plant fossils are mainly restricted to the Langsettian and Duckmantian parts of the sequence (e.g. Arber, 1914, 1916; Kidston 1923-1925), as much of the Bolsovian and Westphalian D is represented by red-beds (the Etruria Formation – Besly and Turner, 1983) in which the plant fragments have been mostly removed by oxidation. There is a short interval in the upper Westphalian D of these coalfields where plants are found (the Halesowen Formation and its lateral equivalents – e.g. Arber, 1914). This in turn is succeeded by more red-beds (the Keele Formation – probably upper Westphalian D and Cantabrian) in which they are extremely uncommon and consist mainly of conifer fragments. In northernmost England and Scotland, plant fossils are known mainly from the upper Langsettian to lower Bolsovian, but few sites have been studied in any detail (for an exception see Thomas and Cleal, 1993).

Britain has a number of 'classic' assemblages, such as the Duckmantian Barnsley Seam 'flora' of Yorkshire and the Westphalian D Radstock 'flora' of Somerset (Kidston, 1923-1925; Crookall, 1955-1975; Thomas and Cleal, 1994). Unfortunately, in neither case has it proved possible to find conservable sites to represent these assemblages. Attempts have been made to conserve parts of the tips at Kilmersdon and Writhlington to represent the Radstock assemblage (Cleal, 1985; Jarzembowski, 1989), but the life expectancy of these sites was considered too short to justify selection as GCR sites. In fact, this was a problem found throughout this part of the GCR; despite the extremely widespread distribution of the Upper Carboniferous plant fossils, the majority of work that has been done on them was based on material from underground workings or spoil-tips, which was unsuitable for long-term conservation. Those sites which were eventually selected are shown in Figure 6.11.

Other than some poorly preserved examples from the upper Westphalian Pennant Measures of the South Wales and Bristol–Somerset coalfields, Upper Carboniferous petrifactions are known from the only coal-ball horizon in Britain, known as the Halifax Hard Bed and Union Seam amongst other names (Phillips, 1980). It proved impossible to select a GCR site for coal-balls, however, as they nearly all originated from defunct and now

# Upper Carboniferous

| Chronostratigraphy | | Biostratigraphy | GCR palaeobotany Sites |
|---|---|---|---|
| Series | Stage | | |
| Stephanian | Cantabrian | *Odontopteris cantabrica* | |
| Westphalian | 'Westphalian D' | *Lobatopteris vestita* | |
| | | *Lobatopteris micromiltoni* | Jockie's Syke |
| | | *Linopteris bunburii* | Llanbradach Quarry |
| | Bolsovian ('Westphalian C') | *Paripteris linguaefolia* | Nostell Priory |
| | Duckmantian ('Westphalian B') | *Lonchopteris rugosa/ Alethopteris urophylla* | |
| | Langsettian ('Westphalian A') | *Lyginopteris hoeninghausii/ Neuralethopteris schlehanii* | Cattybrook Claypit Wadsley Fossil Forest Nant Llech |
| Namurian | Yeadonian — Chokierian | *Neuralethopteris larischii/Pecopteris aspera* | |

**Figure 6.11** Chronostratigraphical and biostratigraphical classification of the Upper Carboniferous, and the positions of the GCR palaeobotany sites.

sealed mine workings. One spoil tip was, until recently, still available for collecting coal-balls (Rowley Tip, Burnley) although the quality of the petrifactions was rather poor; this has now been landscaped for recreational use.

Stephanian plant fossils are very poorly represented in Britain. There are records of the basal *Odontopteris cantabrica* Zone from the Forest of Dean (Wagner and Spinner, 1972; Cleal, 1986c) and South Wales (Cleal, 1978), although no conservable outcrops yielding the fossils are known. The zone may also occur in the Keele Formation of the English Midlands. The species list given by Dix (1935) from these strata refers to *Odontopteris* cf. *schlotheimii* Brongniart, a name which has often been given in error for specimens of *O. cantabrica* Wagner (see Wagner *in* Wagner *et al.*, 1969). She also mentions *Pecopteris miltoni* (Artis) Brongniart, which could in fact refer to *Lobatopteris vestita* (Lesquereux) Wagner. Dix's material is in clear need of taxonomic revision.

The best available review of the literature pertaining to British Upper Carboniferous plant fossils was that provided by Jongmans (1940).

## NANT LLECH

### Highlights

Nant Llech has yielded the best known example of a lower *Lyginopteris hoeninghausii* Zone plant adpression assemblage from Britain. It is of particular interest as being coeval with the coal-ball horizons from northern England, which have played such a significant role in developing a better understanding of Late Carboniferous palaeoequatorial floras.

# Nant Llech

## Introduction

This locality exposes Westphalian Coal Measures in the bed of Nant Llech, near Abercrave, Powys (SN 836123). Plant fossils were first discovered here by a local geologist, W.D. Ware, probably in the 1920s. They were briefly mentioned by Robertson (1932), but a more complete account was provided by Dix (1933) including some illustrations. Her identifications were partly modified in a later published list (Dix, 1934). Specimens from here were also figured by Crookall (1959).

In passing, it is worth mentioning the *in situ* lycopsid stumps, discovered here in the 1830s by W.E. Logan. They were the first examples of such stumps in Europe to be properly excavated, and are of some historical interest (see account in North, 1931). However, as the trunks were removed from the site shortly after their discovery, to be placed in the grounds of the Royal Institution at Swansea, they can no longer be regarded as an integral part of the interest of the site.

## Description

### Stratigraphy

The geology of the basal Coal Measures on the north crop of the South Wales Coalfield has been described by Leitch *et al.* (1958), who included a stratigraphical log for this locality (summarized in Figure 6.12). Three shale beds yielding plant fossils are known and were named by Dix (1933): Bed B, immediately above the Farewell Rock; Bed C, 15 metres higher; and Bed D, 9 metres above that. The palaeontology indicates that they are early Langsettian in age. They are almost certainly interdistributary bay deposits in a lower delta-plain setting.

### Palaeobotany

The plant fossils are preserved here as impressions. The following species are known, with their occurrence in the three plant beds (B, C and D) given in parentheses:

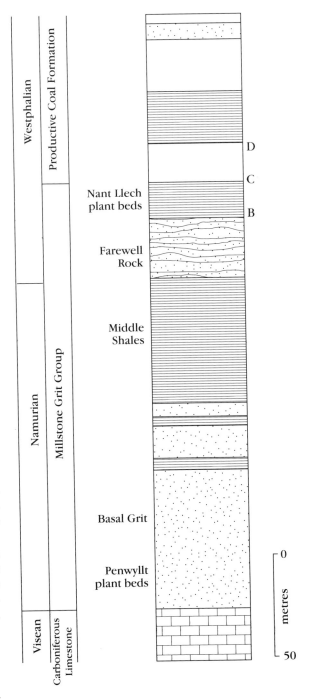

**Figure 6.12** Stratigraphical section at Nant Llech, showing position of plant beds. Based on Dix (1933).

Lycopsida:
    *Lepidodendron aculeatum* Sternberg (B,C)
    *Sigillaria elegans* (Sternberg) Brongniart (C,?D)
    *Lepidophloios acerosus* Lindley and Hutton (B,D)
    *Bothrodendron minutifolium* Boulay (B)
    *B. punctatum* Lindley and Hutton (B,C)
    *Lepidostrobophyllum lanceolatum* (L. and H.) Bell (B,D)
    *Lepidostrobus ornatus* Brongniart (B)

# Upper Carboniferous

*Flemingites olryi* (Zeiller) Brack-Hanes and Thomas (B)

Equisetopsida:
*Asterophyllites equisetiformis* Brongniart (B)
*Calamites* sp. (B,C,D)
*Calamostachys* sp. (C,D)

Filicopsida:
*Pecopteris plumosa* (Artis) Brongniart (B,C)
*P. volkmannii* Sauveur (D)
*P. minor* Kidston (C,D)
*Renaultia gracilis* (Brongniart) Zeiller (B,C)
*R. crepinii* (Stur) Kidston (B)
*Sphenopteris warei* Dix (C)

Cycadopsida:
*Alethopteris lonchitica* Sternberg (B,C)
*A. decurrens* (Artis) Zeiller (B,D)
*A. valida* Boulay (D)
*Neuralethopteris jongmansii* Laveine (B,C,D)
*N. rectinervis* (Kidston) Laveine (C,D)
*Paripteris gigantea* (Sternberg) Gothan (C)
*Trigonocarpus* sp. (B)

Lagenostomopsida:
*Karinopteris acuta* (Brongniart) Boersma (B,C,?D)
*Eusphenopteris hollandica* (Jongmans and Gothan) Novik (D)
*Lyginopteris hoeninghausii* (Brongniart) Gothan (B,C)
*L. baeumleri* (Andrae) Gothan (B,C)
*Adiantites tenellus* Kidston (B)

Pinopsida:
*Cordaites principalis* (Germar) Geinitz (B,C,D)

## Interpretation

The lycopsids, as listed in Dix's papers, are represented by a variety of stems and fructifications. Some of the identifications mentioned by Dix have been generally accepted in the literature (e.g. *Sigillaria elegans* by Crookall, 1966). There are problems with others, however, which are in clear need of critical revision. For instance, she lists both *Lepidodendron obovatum* and *L.* cf. *aculeatum*, two species which have frequently been confused, and which Thomas (1970) places in synonymy. The list given above only mentions *L. aculeatum*, the earliest published of these two names; but it is possible that the specimens might instead belong to *Lepidodendron mannebachensis* Presl (compare comments by Thomas, 1970, pp. 149–51). As pointed out by Crookall (1966), similar problems arise with the species of strobili *Lepidostrobus ornatus* and *L. variabilis*, both of which were again mentioned from Nant Llech. Crookall's interpretation of *L. ornatus* is very flexible, and means little more than compressed 'cigar-shaped' strobili; but Brack-Hanes and Thomas (1983) give a much narrower circumscription, including details of the spores that it contained. The Nant Llech strobili need to be re-examined in the light of the observations by Brack-Hanes and Thomas.

Among the equisetopsid remains, Dix mentioned from Nant Llech stem pith casts as *Calamites* cf. *cistiiformis* Stur. However, *C. cistiiformis* Stur does not normally occur as high as the Westphalian (Leggewie and Schonenfeld *in* Gothan *et al.*, 1959), and so the Nant Llech specimens are listed here merely as *Calamites* sp.

The ferns are represented by a number of species, which can be essentially divided into two groups: the *Pecopteris* species with *Senftenbergia*-type sporangia and belonging to the family Tedeleaceae (Botryopteridales), and the *Renaultia* species, which belong to the Urnatopteridaceae (Urnatopteridales). Dix recorded *Sphenopteris lanarkiana* Kidston from here, but Brousmiche (1983) has transferred this species to *Renaultia crepinii*, although she made no reference to the Nant Llech material (Figure 6.13). Dix's *Sphenopteris warei* has been retained in the list, although it is based on very inadequate material and has been virtually ignored in the literature.

Dix (1933) documented the pteridosperm foliage fragments in rather more detail than the other elements in the assemblage. A number of her identifications have been reviewed and in some cases revised (e.g. *Lyginopteris* by Patteisky, 1957; *Neuralethopteris* by Laveine, 1967; *Eusphenopteris* by van Amerom, 1975) and the species list quoted above has taken these amendments into account.

In addition to the listed pteridosperm taxa, Dix recorded *Neuropteris* cf. *heterophylla* (Brongniart) Sternberg and *Mariopteris* cf. *sphenopteroides* (Lesquereux) Zeiller. *N. heterophylla* has been substantially revised since Dix's work (Laveine, 1967; Cleal and Shute, 1991) and is now normally thought to occur at higher stratigraphical levels. It is possible that Dix's record refers to distal fragments of *Neuralethopteris*.

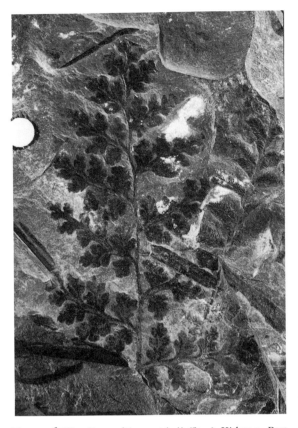

**Figure 6.13** *Renaultia crepinii* (Stur) Kidston. Part of frond from a small herbaceous fern; Natural History Museum, London, specimen V.23353. Lower Productive Coal Formation (Langsettian), Nant Llech. × 2. (Photo: Photographic Studio, Natural History Museum, London.)

As pointed out by Boersma (1972), *M. sphenopteroides* is a taxon of doubtful validity. He does not refer directly to the Nant Llech material, but argued that the only record of the species from Europe prior to Dix was by Zeiller (1886–1888, pl. 19, figs 3 and 4), and that this belongs to *Fortopteris latifolia* (Brongniart) Boersma. Dix's identification needs to be reviewed in the light of Boersma's comments.

This assemblage belongs to the lower part of the *Lyginopteris hoeninghausii* Zone of Wagner (1984), now referred to as the *Neuralethopteris jongmansii* Subzone of Cleal (1991), and is the best example of its type in Britain (Figure 6.14). Similar assemblages are known from the Tenby–Saundersfoot coast in Pembrokeshire (Goode, 1913; Jenkins, 1962) but the fossils from there are not as well preserved, due largely to tectonic deformation. They are also known from the Bideford Formation in North Devon (Arber, 1904b), mainly from 'culm workings' which are no longer in existence. Despite extensive outcrop of the formation along the Devon coast near Westward Ho!, few plant fossils have been found there. From northern Britain, similar assemblages have been reported from the basal Coal Measures, such as above the Black Coal (Walton, 1932) and the Kilburn Coal (Dix, 1934), but there are no sites now yielding the fossils.

Almost identical assemblages of species occur throughout the palaeoequatorial belt, from eastern North America (e.g. New River Formation of West Virginia – Gillespie and Pfefferkorn, 1979) through the Ruhr (Josten, 1962), Silesia (Stopa, 1962), the Ukraine (Novik, 1952) to the Caucasus (Anisimova, 1979) (for a more complete list of occurrences and references, see Wagner, 1984).

The Nant Llech assemblage is of particular interest as being coeval with the coal-ball horizon of northern England (the Halifax Hard or Union coals; Phillips, 1980). This was the first discovered and remains the best studied of all the Carboniferous coal-ball seams in the world. It is of considerable value to have a coeval adpression assemblage, with which to compare preservational variations and environmental settings.

## Conclusion

Nant Llech yields the best fossil flora from the basal Coal Measures in Britain, representing plants that lived about 310 million years ago. The assemblage of plant fossils is dominated by seed plants (so-called pteridosperms) and ferns, and is typical of the vegetation growing on the topographically raised levée-banks of the rivers which flowed through the tropical swamps at this time. Similar fossil floras of this age occur in many of the coalfields of North America and Europe, including West Virginia, the Ruhr, Silesia, Ukraine and the Caucasus. They represent the start of the growth of the large-scale swamp-forests that spread throughout the tropical belt at this time and lasted for some 10–15 million years. The peats generated by these forests were converted by geological processes into the economically important coal deposits of Europe, North America and China.

**Figure 6.14** *Neuralethopteris jongmansii* Laveine. Part of trigonocarpalean frond; Natural History Museum, London, specimen V.23359. Lower Productive Coal Formation (Langsettian), Nant Llech. × 1. (Photo: Photographic Studio, Natural History Museum, London.)

## WADSLEY FOSSIL FOREST

### Highlights

Wadsley Fossil Forest is the only conserved example of *in situ* fossilized tree stumps in the Coal Measures of Britain (Figure 6.15).

### Introduction

This group of *in situ* Westphalian fossil tree stumps was discovered in 1873, during excavations for new buildings, in the grounds of what was then the Wadsley Lunatic Asylum, now the Middlewood Hospital, Sheffield (SK 318913; Thorpe, 1972). Formerly, out-buildings were constructed to protect the fossils but over the years they have become seriously degraded and covered by debris. However, recent work by Gaynor Boon and other staff of the Sheffield City Museum has uncovered them again, and has allowed the stumps to be re-examined. This account includes information supplied by Ms Boon, which is very gratefully acknowledged. The only published scientific account of the stumps was by Sorby (1875).

### Description

#### Stratigraphy

The stumps occur in sandstones of the Middle Rock Formation, between the Coking and Clay coal seams, and are thus early Langsettian in age. The sandstone is likely to be a crevasse-channel deposit.

#### Palaeobotany

Sorby (1875) stated that ten stumps were originally present, spread over a distance of 35–45 metres, but the recent re-excavations have only revealed four of these. They are up to 1.5 metres in diameter and show the remains of the stigmarian rooting structures penetrating the underlying sea earth. Although surface details are poorly preserved, Sorby maintained that *Sigillaria*-like markings could just be made out.

# Wadsley Fossil Forest

**Figure 6.15** Wadsley Fossil Forest. *In situ* fossilized tree stumps preserved in the lower Langsettian Middle Rock Formation. (Photo: G. Boon, Sheffield City Museum.)

## Interpretation

A point emphasized by Sorby was that the stumps appear to show a regular distortion. He argued that this reflected a prevailing westerly wind at the time and was apparently impressed by the fact that this was very similar to present-day wind directions in Sheffield! However, Gastaldo (1986) has suggested that similar distortion in the Lower Carboniferous stumps at Victoria Park (see Chapter 6) was a 'streamlining' effect, caused by movement of the entombing sediment.

This is the only place where such stumps are conserved *in situ* in the British Coal Measures. Their significance lies in establishing a palaeoecological model for these strata. There have been a number of attempts to establish such a model using the transported fragments that form the majority of the plant macrofossils found in the Coal Measures (e.g. Scott, 1977, 1978, 1979). Establishing a robust model using such transported fragments is extremely difficult, however, even when there is evidence of the hydrological regime operating at the time; and Gastaldo (1985, 1987) has argued that *in situ* stumps such as those found at Wadsley are the only reliable means of determining plant distribution (see also Cleal, 1987a). The Wadsley stumps appear to confirm that at least parts of the Late Carboniferous palaeoequatorial floodplains were covered by lycopsid forests, rather than pteridosperm-dominated assemblages as suggested in the Scott model.

## Conclusion

Wadsley Fossil Forest is the only conserved example in Britain of stumps from the Coal Measures swamp-forests, about 310 million years old. They are the remains of giant club-mosses, probably originally up to 40 metres high, embedded in the fossilized soil in which they grew. Although historically well-known, for many years they had become inaccessible, and were only recently

## Upper Carboniferous

re-excavated to allow their study to be resumed. The site is a rarity, and provides a graphic impression of these ancient forests.

## CATTYBROOK CLAYPIT

### Highlights

Cattybrook Claypit has yielded the best plant fossil assemblage from the upper *Lyginopteris hoeninghausii* Zone in Britain. It is of particular interest as having yielded exceptionally large examples of *Karinopteris acuta* and *Sphenophyllum cuneifolium*.

### Introduction

This Westphalian site is in part of the brickworks at Almondsbury near Bristol (ST 592833). Brick-making clays have been worked here since the 1860s and the geology of the site was discussed by Smith and Reynolds (1929). Plant fossils from the Upper Carboniferous were first recorded by Moore and Trueman (1942), whose list of identifications were repeated by Welch and Trotter (1961). The only systematic account of the assemblage, however, is by Cleal and Thomas (1988).

There are two discrete pits at Cattybrook (Figure 6.16): the Golden Quarry, from which much of the earlier information was derived, and the Red Quarry, which was the basis of the Cleal and Thomas (1988) study (the names derive from the colour of the bricks after firing). The Golden Quarry described by Smith and Reynolds (1929) is now seriously overgrown.

### Description

#### Stratigraphy

The Coal Measures exposed near Almondsbury are part of the Ridgeway Thrust Zone, a Variscan structural feature which has caused considerable distortion to the strata. They are often referred to as part of the Severn or Avonmouth Coalfield (e.g. Welch and Trotter, 1961), but they are in fact more properly regarded as part of the Bristol–Somerset Coalfield (Cleal and Thomas, 1988).

The most recent account of the geology of the Red Quarry, including a lithological log, is by Cleal and Thomas (1988). The 85 metre thick sequence consists of coals, seat earths and mudstones (Figure 6.17), which probably represent 'floodplain' deposits formed in the middle or upper

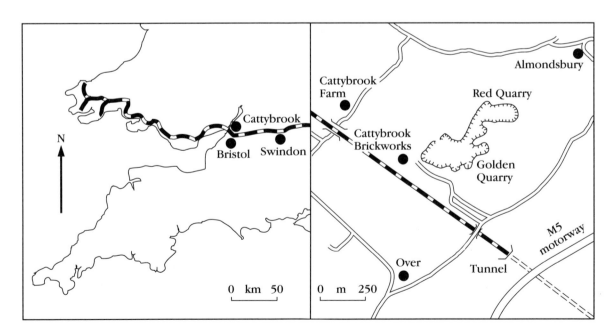

**Figure 6.16** Locality maps for Cattybrook Claypit. Based on Cleal and Thomas (1988, figure 1).

# Cattybrook Claypit

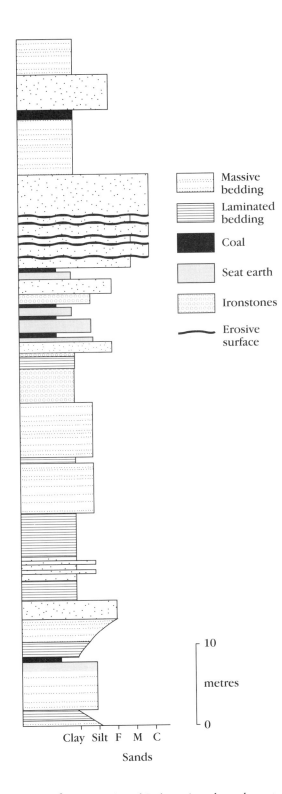

Figure 6.17 Stratigraphical section through part of the lower Productive Coal Formation (upper Langsettian) at Cattybrook Claypit. Based on Cleal and Thomas (1988, figure 2).

regions of a fluvial delta. There are also sandstone bodies, at least one of which contains large pieces of plant fossil, and which may be crevasse-channel deposits. The biostratigraphical evidence (mainly from the plant fossils) discussed by Cleal and Thomas suggests that the sequence belongs to the topmost Langsettian.

## Palaeobotany

The plant fossils here are preserved as adpressions. Those from the mudstones are fragmentary, but often retain carbonized tissue (i.e. compressions). The fossils from the sandstone bodies are impressions. The following taxa have been described:

Lycopsida:
    *Lepidodendron aculeatum* Sternberg
    *L. lycopodioides* Sternberg
    *Lepidostrobophyllum* sp.
    *Sigillaria scutellata* Brongniart

Equisetopsida:
    *Calamites carinatus* Sternberg (Figure 6.18)
    *Asterophyllites equisetiformis* Brongniart
    *Annularia* cf. *radiata* (Brongniart) Sternberg
    *Calamostachys paniculata* Weiss
        (Figure 6.18)
    *Sphenophyllum cuneifolium* (Sternberg)
        Zeiller

Filicopsida:
    *Pecopteris plumosa* (Artis) Brongniart
    *Corynepteris angustissima* (Sternberg)
        Němejc
    *Renaultia* cf. *schatzlarensis* (Stur) *sensu*
        Brousmiche
    *R.* cf. *crepinii* (Stur) Kidston

Cycadopsida:
    *Laveineopteris loshii* (Brongniart) Cleal,
        Shute and Zodrow
    *L. tenuifolia* (Sternberg) Cleal, Shute and
        Zodrow
    *Cyclopteris orbicularis* Brongniart
    *Paripteris pseudogigantea* (Potonié) Gothan
    *Alethopteris decurrens* (Artis) Zeiller
    *A.* cf. *lancifolia* Wagner
    *Lonchopteris rugosa* Brongniart

Lagenostomopsida:
    *Karinopteris acuta* (Brongniart) Boersma
        (Figure 6.19)

**Figure 6.18** *Calamites carinatus* Sternberg and *Calamostachys paniculata* Weiss. Stem and cones of giant equisetopsids; National Museum of Wales, specimen 86.101G54. Productive Coal Formation (Langsettian), Cattybrook Claypit. × 1. (Photo: Photographic Studio, National Museum of Wales.)

*Eusphenopteris* cf. *neuropteroides* (Boulay) Novik
*Palmatopteris geniculata* (Germar and Kaulfuss) Potonié

Pinopsida:
*Cordaites principalis* (Germar) Geinitz
*Cordaitanthus* sp.
*Cordaicarpon* cf. *cordai* Geinitz

## Interpretation

The Cattybrook assemblage is typical for the *L. loshii* Subzone, being dominated by *Laveineopteris loshii*, *Karinopteris acuta* and *Sphenophyllum cuneifolium*. It differs from assemblages from the underlying *Neuralethopteris jongmansii* Subzone (e.g. from Nant Llech – see p. 204) in the absence of *Lyginopteris* and *Neuralethopteris*. Both these form-genera tend to be rare in the *L. loshii* Subzone, a point which was recognized by Dix (1934) and was one of the arguments used by Cleal (1991) to justify separating the two subzones of the *L. hoeninghausii* Zone.

Crookall (1955) states that *Lonchopteris* is usually rare in Britain, although it can be locally abundant. Cattybrook is one of those few localities where it is relatively abundant. Its distribution is generally very uneven, being abundant in certain coalfields in Europe (e.g. Upper Silesia), but absent from North America except for the Maritime Provinces of Canada. It is perhaps significant that the *Lonchopteris* specimens from Cattybrook were found in the crevasse-channel sandstone, rather than the mudstones that yielded most of the other taxa.

Another Cattybrook taxon that is otherwise rare to Britain is *Alethopteris lancifolia*. Only one other specimen is known from Britain, from the Duckmantian of Yorkshire (Crookall, 1955, pl. 5, fig. 1 – see Wagner, 1961).

Perhaps of most interest at Cattybrook, however, is the presence of large specimens of *Karinopteris acuta* and *Sphenophyllum cuneifolium* (up to

## Cattybrook Claypit

**Figure 6.19** *Karinopteris actua* (Brongniart) Boersma. Pteridosperm frond; National Museum of Wales, specimen 86.101G25. Productive Coal Formation (Langsettian), Cattybrook Claypit. x 0.5. (Photo: Photographic Studio, National Museum of Wales.)

0.5 metres across) in the crevasse-channel sandstones. Being preserved in sandstone, they do not show fine surface details (e.g. nervation) particularly well. Nevertheless, being probably preserved not far from their position of growth, they provide valuable information as to the growth habit of these plants. Cleal and Thomas (1988) argued that the specimens from here suggest that *K. acuta* was a liana-like plant, and that *S. cuneifolium* was a ground-creeper, adding support to similar suggestions based on material from elsewhere by Batenburg (1977) and Scott (1978).

This is the best available site in Britain for plant fossils of the upper *Lyginopteris hoeninghausii* Zone of Wagner (1984), and referred to as the *Laveineopteris loshii* Subzone of Cleal (1991). Similar upper Langsettian assemblages are known throughout the palaeoequatorial belt, from eastern North America to the Caucasus (evidence reviewed by Wagner, 1984), but mainly from underground or temporary workings. In Britain, the best documented assemblage of similar age is the so-called Ravenhead 'flora' collected from Thatto Heath railway cutting in the Lancashire Coalfield (Kidston 1889d; Cleal, 1979). This has many taxa in common with the Cattybrook assemblage, although it has a rather higher preponderance of ferns. However, the Thatto Heath exposure no longer yields plant fossils. Dix (1934) lists species from a number of similar assemblages from South Wales, and which she assigns to her 'Flora' D, but they are nearly all from underground mines. No published systematic account exists for these Welsh assemblages. According to the evidence reviewed by Jongmans (1940) comparable assemblages have been reported from the North Wales, South Staffordshire, Yorkshire and Central Scottish coalfields, but these are again nearly all from underground mines, and taxonomic analyses have never been published.

## Conclusion

Cattybrook Claypit has yielded the best plant fossils in Britain representing the vegetation from a time known as the late Langsettian, about 310 million years ago (part of the Carboniferous Period). It was the time when the tropical swamp-forests, of which these plants were a part, were starting to reach their maximum development. It marks a clear change from the plant fossils found lower in the Langsettian (such as at Nant Llech - see p. 204), which probably represented a less densely forested setting, more like that evinced by the plant beds in the underlying Millstone Grit. Of particular interest at Cattybrook is the existence of a sandstone, thought to have formed as a river deposit near a levée-bank on which many of these plants lived, and into which large pieces of the plants fell and were buried. This has provided a useful insight into the growth form of some of them, such as the horsetail *Sphenophyllum*, which seems to have been a ground-creeper, and the seed plant *Karinopteris*, which was a liana. Also, Cattybrook has yielded some of the largest known examples of the seed plant frond *Lonchopteris*, which is one of the very earliest examples of a leaf with mesh-veining, as now developed in most dicotyledonous flowering plants.

## NOSTELL PRIORY BRICKPIT

## Highlights

Nostell Priory Brickpit has yielded one of the best documented plant fossil assemblages in Britain belonging to the *Paripteris linguaefolia* Zone (indicating the Bolsovian). It is particularly significant for cordaites and is the type locality for two species of cordaite cone (Figure 6.20).

## Introduction

This brickpit (SE 430170), between Ackworth Moor Top and Crofton, about 8 km south-east of Wakefield, is cut in Westphalian shales. Plant fossils were first discovered in these rocks by James Wright in the early 1930s and were described in detail by Barker (*in* Barker and Whittle, 1944). Cordaite cones were subsequently re-described by Crookall (1970).

## Description

### Stratigraphy

Barker and Whittle (1944) describe the bed containing the plant fossils as 5.5 metres of 'laminated mudstones', overlying the Shafton Marine Band. They are thus of early Bolsovian age. Barker and Whittle interpret these beds as 'estuarine'.

### Palaeobotany

The fossils are preserved as adpressions. The following species have been described:

Lycopsida:
*Lepidodendron* cf. *simile* Kidston
*Lepidodendron* sp.
*Bothrodendron punctatum* Lindley and Hutton
*Stigmaria ficoides* (Sternberg) Brongniart
*Flemingites* sp.

Equisetopsida:
*Calamites suckowii* Brongniart
*Annularia radiata* Brongniart
*A. sphenophylloides* (Zenker) Gutbier
*Asterophyllites equisetiformis* Brongniart
*A. grandis* (Sternberg) Geinitz
*Palaeostachya ettingshausenii* Kidston
*Calamostachys* sp.
*Myriophyllites gracilis* Artis
*Pinnularia columnaris* (Artis) Zeiller
*Sphenophyllum emarginatum* Brongniart
*S. majus* Bronn

Filicopsida:
*Lobatopteris miltoni* (Artis) Wagner
*Crossotheca* cf. *crepinii* Zeiller
cf. *Renaultia* sp.

# Nostell Priory Brickpit

**Figure 6.20** Nostell Priory Brickpit. Working quarry, as seen in 1985. The beds are associated with the lower Bolsovian Shafton Marine Band. (Photo: C.J. Cleal.)

Cycadopsida:
    *Alethopteris lonchitica* Sternberg
    *Laveineopteris loshii* (Brongniart) Cleal,
       Shute and Zodrow
    *Cyclopteris* sp.

Lagenostomopsida:
    *Mariopteris sauveurii* (Brongniart) Zeiller
    cf. *Karinopteris robusta* (Danzé-Corsin)
       Boersma
    *Rhodea wrightii* Barker
    *Carpolithus reticulatus* Sternberg
    *C. minimus* Sternberg

Cordaites:
    *Cordaites borassifolius* (Sternberg) Unger
    *Cordaitanthus flagellibracteatus* Barker
    *C. nostellensis* Barker
    *Cordaicarpus ventricosus* Grand'Eury
    *Samaropsis orbicularis* Ettingshausen
    *S. pyriformis* Barker

## Interpretation

The presence together of *Asterophyllites grandis*, *Annularia sphenophylloides* and *Sphenophyllum emarginatum* clearly points to this assemblage belonging to the middle *Paripteris linguaefolia* Zone of Wagner (1984), or the *Laveineopteris rarinervis* Subzone of Cleal (1991). This is compatible with its position above the Shafton Marine Band, which is in about the middle Bolsovian.

### *Lycopsida*

Those uncovered here to date are generally unexceptional, comprising mainly leafy shoots and stigmarian rooting structures. However, Barker reported an incomplete strobilus and some associated megaspores, which might reward further study. Barker identified it as *Lepidostrobus variabilis* Lindley and Hutton, and Crookall (1966) transferred it to *Lepidostrobus ornatus* Brongniart. In view of the reported association of megaspores, however, the strobilus is more likely to be a species of *Flemingites* (Brack-Hanes and Thomas, 1983).

### Equisetopsida

There is rather more variety in the calamitid equisetopsids. At least one species of *Annularia* and two of *Asterophyllites* have been reported. Barker has also described specimens of *Annularia radiata* Brongniart. However, the illustrated specimen shows leaves which are too slender and parallel-sided and may instead belong to *Asterophyllites equisetiformis*.

Also, a number of calamitid strobili have been reported from here. The most abundant was described by Barker as *Palaeostachya ettingshausenii*, and the figured specimen seems to confirm the identification (compare with similar specimens figured by Crookall, 1969). A single specimen of a much more slender strobilus was described by Barker as *Calamostachys* ?sp. nov. Since he did not illustrate this specimen, it is difficult to judge, but the dimensions given in his description suggest a comparison with *Calamostachys ramosa* Weiss.

A number of species of sphenophyll foliage were listed by Barker, but were only poorly described and illustrated. In view of the more recent work on the variation of leaf form in different parts of the sphenophyll plant (e.g. Storch, 1966; Batenburg, 1977), the identity of the Nostell Priory specimens must be regarded as tentative.

### Filicopsida

At Nostell Priory, ferns are relatively uncommon, which is typical of middle Westphalian palaeoequatorial assemblages. The most abundant is the marattialean *Lobatopteris miltoni*, a species recently reviewed by Shute and Cleal (1989). In addition, Barker reported a single example of *Crossotheca*. He compared it with *C. boulayi* Zeiller, but Brousmiche (1982) has demonstrated that this is merely part of the range of morphological variability within *C. crepinii*. Although many palaeobotanists (e.g. Taylor and Millay, 1981) still regard *Crossotheca* as a lagenostomalean pteridosperm fructification (following Kidston, 1923d), Danzé (1955, 1956) and Brousmiche (1982) have shown that it was a fern.

### Cycadopsida

The dominant pteridosperm is an alethopterid. Its identification as *A. lonchitica* is based on the authority of Barker, but he did not illustrate any specimens. As pointed out by Wagner (1968, 1984), however, this species is widely misidentified, and records from the middle Westphalian often refer to *Alethopteris urophylla* (Brongniart) von Roehl.

Neuropteroid foliage is represented here by *Laveineopteris loshii*. Barker identified it as *Neuropteris heterophylla* (Brongniart) Sternberg, a species which has frequently been confused with *L. loshii* (see Laveine, 1967; Cleal and Shute, 1991, 1992). It is perhaps significant that it is associated here with cyclopterid pinnules, which occur in the lower part of the *L. loshii* frond but not of the *N. heterophylla* frond. There is no direct evidence of *N. heterophylla* in the Nostell Priory assemblage. A single fragment was also identified by Barker as *Neuropteris* cf. *obliqua* (Brongniart) Zeiller, although there seems little reason for separating it from the *L. loshii*.

### Lagenostomopsida

Two types of mariopteroid frond were reported (Barker *in* Barker and Whittle, 1944). One was identified as *Mariopteris sauveurii*, which, from the illustrations, appears to be correctly identified. The second was stated to be specifically identical with *Mariopteris* sp. D of Kidston (1925), which Danzé-Corsin (1953) formally named *Mariopteris robusta*, and which Boersma (1972) assigned to the form-genus *Karinopteris*. There is indeed an apparent comparison with Kidston's figured specimens, especially with the one selected by Boersma as the lectotype of *K. robusta*, but additional material from Nostell Priory will be needed before this rare species (otherwise only known from a few specimens from South Wales, Nord-Pas-de-Calais, the Ruhr and the Donets) can be unequivocally recorded from here.

The specimen figured by Barker under the new name *Rhodea wrightii* appears to be a small type of *Palmatopteris*. In the absence of more complete material, however, it is difficult to assess this species.

### Pinopsida

Barker reports a number of excellently preserved cordaite fossils, including some large leaf fragments and isolated seeds. Of most interest, however, was the discovery of a male and a number of female cones, which were made the types of *Cordaitanthus nostellensis* and *C. flagellibracteatus*, respectively. The latter is particularly distinctive, having very slender, elongate bracts, quite unlike those of any other described species. Both species are known only from this locality.

*General remarks*

Adpression floras dominated by pteridosperms, cordaites and equisetopsids are relatively uncommon in the upper Duckmantian and lower Bolsovian of Britain. In South Wales, for instance, conditions seem to have become rather wetter and less favourable for the development of this type of vegetation (Davies, 1929; Dix, 1934); lycopsid-dominated vegetation instead seems to have been the norm. In much of the English Midlands, most strata of this age are in the Etruria red-bed facies, which seem to have been unfavourable for the preservation of plant fossils (Besly and Turner, 1983). There is some evidence of a similar assemblage from the Bradford Four Feet Seam in the Lancashire Coalfield (Kidston, 1892, 1894b), but illustrations of the fossils have never been published and the identifications have not been revised in nearly a hundred years. There is very little evidence of plant fossils from coeval strata in northernmost England or Scotland (Jongmans, 1940).

This seems to follow the same pattern seen in most of the paralic coalfields of the palaeoequatorial belt, where assemblages of plant fossils, similar to those at Nostell, are only sporadically found (see Wagner, 1984 for a review of the available evidence). This may reflect the greater marine influence on the delta at this time (Guion and Fielding, 1988), which would allow relatively few river levées to develop. In those relatively few situations where the levées did develop, however, they supported a pteridosperm/cordaite/equisete-dominated type of vegetation, such as found at Nostell.

In addition to their relative scarcity value, sites in the English Pennines, such as Nostell, have the advantage over most of these other areas because the fossils often still retain their cuticles, which can provide important information about the affinities of the plants (e.g. Cleal and Shute, 1991, 1992). The best comparison from this point of view is with the intra-montane basins of central Europe, such as with the Sulzbach Formation in Saar-Lorraine (Laveine, 1989) and the Radnice Member of Central Bohemia (Wagner, 1977), where cuticles are often preserved. As pointed out by Gothan (1954), however, the composition of species found in these intra-montane basins is different from that of the paralic belt. Nostell Priory Brickpit is thus of considerable importance for studying the vegetation of the paralic belt of coalfields.

## Conclusion

Nostell Priory Brickpit has yielded some of the best documented plant fossils from middle Westphalian rocks in Britain, about 305 million years old. The assemblage consists mainly of primitive and now extinct seed plants with fern-like fronds (pteridosperms) and horsetails, that were typical of the river, levée-bank vegetation growing within the swamp-forests of the time. The site is also important for another group of primitive and now extinct seed plants, the cordaites, which were related to the conifers, but had large, palm-like leaves. The flora here is a typical example of the so-called Coal Measures flora, representing the height of development of the tropical swamp-forests in Late Palaeozoic times, and which generated the thick, economically important coals of the northern and central European coalfields.

## LLANBRADACH QUARRY

### Highlights

Llanbradach Quarry yields the best examples of plant petrifactions from the South Wales Pennant Measures. The only other Upper Carboniferous petrifactions known from Britain are the rather older (early Langsettian) coal-balls of northern England.

### Introduction

This site is a disused quarry in upper Westphalian sandstones (Figure 6.21), south of the village of Llanbradach, 3 km north of Caerphilly (SO 146894). The only detailed account of the plant fossils from here is that given by Crookall (1931b). They appear to have been first discovered by J. Storrie, at some time in the second half of the nineteenth century, and Storrie's collection

# Upper Carboniferous

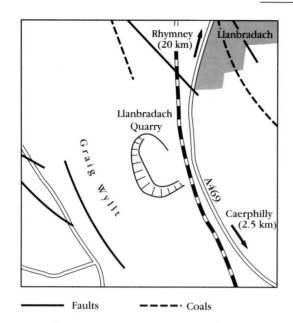

**Figure 6.21** Locality map for Llanbradach Quarry.

formed the basis of Crookall's work. Additional material was, however, also collected during the 1920s by W.D. Ware.

## Description

### Stratigraphy

The plant fossils here occur in the Brithdir Beds, the topmost member of the Lower Pennant Measures. They are therefore early Westphalian D in age (Cleal, 1978). Some ten metres of mainly massive Pennant sandstones are exposed here. The best preserved specimens were found in conglomeratic bands within the coarse sandstones. They clearly underwent considerable transportation before being entombed in the sediment, and so any palaeoecological interpretation is difficult. They are, however, unlikely to represent plants from the swampy, peat-forming parts of the forests, but rather levée-bank or extra-basinal assemblages.

### Palaeobotany

The fossils are calcareous petrifactions, with only the more robust, woody tissue being preserved. Crookall mentioned the following taxa:

*Calamites* sp.
*Psaronius* sp.
cf. *Mesoxylon* spp.
*Dadoxylon* sp.

## Interpretation

The Llanbradach specimen of *Calamites* only shows secondary wood. It was assigned to this form-genus because the wood has mainly scalariform pitting on the radial xylem walls, the rays consist of vertically elongate cells, and the presence of 'infra-nodal canals'.

Llanbradach has yielded only the second known species of the marattialean tree-fern *Psaronius* from Britain, the other being the coal-ball species *P. renaultii* Williamson. The specimens described by Crookall show the characteristic peripheral 'root-zone', that forms the outer part of the *Psaronius* stems. Although some evidence of the vascular and cortical structure was found by Crookall, better preserved material will be needed before a specific identification can be made.

The remaining specimens from Llanbradach are probably cordaite wood. Two of them described by Crookall show a siphonostele and were identified as cf. *Mesoxylon* sp. They are unusual for cordaite stems in that they have very poorly developed rays in the secondary wood. However, neither specimen was well enough preserved to demonstrate the maturation of the cauline stele or the structure of the leaf traces, which is essential for a reliable identification of such stems. The situation is further complicated by the proposals by Rothwell and Warner (1984) and Trivett and Rothwell (1985) to change the concept of the two main form-genera of petrified cordaite stems, *Cordaixylon* Grand'Eury (synonyms *Cordaioxylon* Felix and *Pennsylvanioxylon* Vogellehner) and *Mesoxylon* Scott and Maslen, to include fertile structures and leaves, thus making them in effect whole-plant genera.

The specimens assigned by Crookall to *Dadoxylon* sp. are mainly just fragments of secondary wood. One might have a siphonostele, but the primary region in the centre of the stem is not preserved. The identification of such secondary wood in isolation is difficult but, in the Upper Carboniferous, it usually belongs to the cordaites. The tracheid pitting and form of the rays would seem to support this.

This is the best site for plant petrifactions in the Pennant Measures. Lillie (1910) and Crookall (1927) describe similar specimens from Staplehill near Bristol, although this site no longer exists. Elsewhere in Britain, Upper Carboniferous plant petrifactions are known only from the coal-balls of the basal Coal Measures (lower Langsettian) of northern England (Phillips, 1980). These are not only significantly older than the Pennant fossils, but represent quite a different, swampy habitat, dominated by lycopsids. Abroad, the nearest comparisons are with coal-ball assemblages from the Upper Moscovian of the Ukraine (Snigirevskaya, 1972) and the middle of the Carbondale Formation of Indiana (Canwright, 1959), but again these contain predominantly the lycopsids of the swamp habitats, rather than the equisetes, ferns and pteridosperms of the levée vegetation. Llanbradach thus yields the only known anatomically-preserved plant fossils representing the levée-bank vegetation of the early Westphalian D.

## Conclusion

Llanbradach Quarry is the only site in Britain to yield plant fossils from the upper Coal Measures with anatomical features preserved. The fossils include the remains of giant horsetails, tree-ferns and seed plants, which were growing on raised levée-banks of rivers flowing through the tropical swamp-forests of the time. The fossils nearest in age showing details of anatomy are some five million years older, found in the Coal Measures of northern England, and these are dominated by quite different types of plant, mainly giant club-moss trees that grew within the wetter parts of the swamp-forests. Although the Llanbradach fossils were subject to a considerable degree of transport and erosion, having drifted some distance down rivers, they give a good impression of the type of vegetation growing in South Wales in the late Westphalian times, about 305 million years before the present.

## JOCKIE'S SYKE

### Highlights

Jockie's Syke is the best available locality for *Lobatopteris micromiltoni* Zone plant fossils in Britain.

### Introduction

Late Westphalian plant fossils can be found in this small stream (Figure 6.22), a tributary of Liddel Water, near Longtown, Cumbria (NY 424756). They were first collected from here in 1879 by A. Macconochie of the Geological Survey and were briefly mentioned by Kidston (1894b). Further collecting was done by Macconochie and Kidston in 1902, and their results were published by Kidston (1903a, b). There has been no subsequent work on the palaeobotany here, apart from an illustration of one of Kidston's alethopterids by Crookall (1955), and some passing comments made by Crookall (1969) and Wagner (1968).

## Description

### Stratigraphy

The plant fossils occur in red mudstones within the Red Sandstone Group of the Canonbie Coalfield (*sensu* Peach and Horne, 1903) (Figure 6.23). According to Ramsbottom *et al.* (1978), this part of the sequence contains an *Anthraconauta tenuis* Zone fauna, which indicates an age no older than latest Bolsovian (Cleal, 1984a). The plant fossils appear to belong to the *Lobatopteris micromiltoni* Zone, indicating a middle Westphalian D age (Cleal, 1978, 1984b). The beds probably represent floodplain deposits within a fluvial plain environment.

### Palaeobotany

The plants are preserved as fossil impressions in a red mudstone. They were reported by Kidston

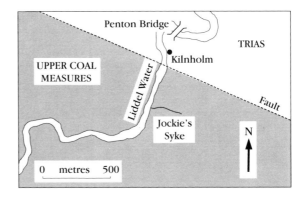

**Figure 6.22** Geological map showing the area near Jockie's Syke. Based on Peach and Horne (1903, plate 1).

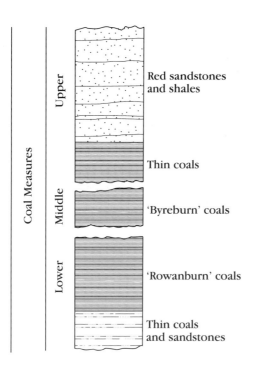

**Figure 6.23** Stratigraphical succession in the Canonbie Coalfield. Based on Peach and Horne (1903, plate 4).

(1903a, b) from four distinct locations within the site. In the absence of details as to their relative stratigraphical distributions, however, the following is a combined list of species from all four:

Lycopsida:
*Lepidodendron fusiforme* (Corda) Corda
*Lepidostrobophyllum* sp.
*Stigmaria ficoides* (Sternberg) Brongniart

Equisetopsida:
*Calamites undulatus* Sternberg
*C. carinatus* Sternberg
*Calamites* sp.
*Asterophyllites equisetiformis* Brongniart
*Annularia stellata* (Sternberg) Wood

Filicopsida:
*Cyathocarpus* aff. *arborescens* (Brongniart) Weiss
*Cyathocarpus* sp.

Cycadopsida:
*Neuropteris ovata* Hoffmann
*N. flexuosa* Sternberg
*Macroneuropteris scheuchzeri* (Hoffmann)

Cleal *et al.*
*Alethopteris ambigua* Lesquereux
?*A. grandinioides* Kessler
*Alethopteris* sp.

## Interpretation

The above species list points strongly to the assemblage belonging to the *Lobatopteris micromiltoni* Subzone of Cleal (1991) (previously regarded as a separate zone by Cleal, 1984b); *Cyathocarpus* aff. *arborescens* does not extend below the base of this zone, and there is no evidence of the species that characterize the base of the overlying *Lobatopteris vestita* Zone: *Lobatopteris vestita* (Lesquereux) Wagner, *Polymorphopteris polymorpha* (Brongniart) Wagner and *Dicksonites plueckenetii* (Sternberg) Sterzel. This indicates that the strata exposed at Jockie's Syke are mid-Westphalian D in age.

Kidston (1903b) recorded specimens from here as *Calamites ramosus* Artis, but Kidston and Jongmans (1917) showed that this was a later synonym of *C. carinatus*. Crookall (1969) specifically lists the Jockie's Syke specimens within the synonymy of the latter species and notes that, in Britain, *C. carinatus* is particularly common in the upper Bolsovian and Westphalian D. Kidston (1903b) also lists *Calamites undulatus*, now generally recognized as a preservational form of a number of different calamite species.

Kidston (1903b) also recorded *Annularia radiata* (Brongniart) Sternberg, which he regarded as the foliage borne by *Calamites ramosus* (syn. *C. carinatus*). However, *A. radiata*-type foliage rarely occurs this high stratigraphically, and in any case is a fairly artificial concept (Walton, 1936). In view of its association here with *Annularia stellata*, whose smaller leaves can often resemble *A. radiata*, the record has been deleted from the list.

The brief comments by Kidston (1903b) on the pecopteroid species appear to confirm that they belong to one of the *Cyathocarpus* species with small pinnules, such as *C. arborescens*. Unfortunately, however, he did not figure any specimens, either at that time or in his later monographic treatment of the British pecopteroid species (Kidston, 1924, 1925). In view of the complications surrounding this difficult group of species (Barthel, 1980), the identification of the Jockie's Syke material has been left open.

Kidston's (1903b) records of neuropteroid taxa, although unillustrated, are of easily recognized species which occur commonly in the Westphalian D of Britain.

Kidston's (1903b) record of *Alethopteris aquilina* (Brongniart) Goppert was transferred to *Alethopteris davreuxii* var. *friedelii* Bertrand by Crookall (1955). However, *A. friedelii* Bertrand is a later synonym of *Alethopteris grandinii* (Brongniart) Goppert (Wagner, 1968; Cleal, 1984b) and has nothing to do with *A. davreuxii* (Brongniart) Zeiller. Wagner (1968) instead transferred Kidston's specimen (which had been figured by Crookall, 1955, pl. 10 fig. 2) to *Alethopteris ambigua*. This is a rare species in Britain, which according to Wagner is otherwise only known from the late Westphalian D of the Bristol and Somerset Coalfield.

Crookall (1955) questioned Kidston's (1903b) record of *Alethopteris serlii* (Brongniart) Goppert, and placed the specimens in *Alethopteris grandinii*. Crookall's own interpretation of *A. grandinii* is clearly flawed, however, since the two species that he illustrated under that name (neither from Jockie's Syke) belong to *Alethopteris grandinioides* and *Alethopteris quadrata* Wagner (Wagner, 1968). From Crookall's comments on the Jockie's Syke specimens, it is likely that they in fact belong to *A. grandinioides*, a relatively common species in the Westphalian D of Britain.

Plant fossils of this age are rare in Britain. In southern Britain, there are few strata of this age, due to the effects of the Leonian folding phase. Only in South Wales are there reliable records of *L. micromiltoni* Zone plant fossils, from the Swansea Beds in the western part of the main coalfield, and these were all collected from underground workings (Cleal, 1978). In North Wales and the English Midlands, strata of this age are either missing or occur as the Etruria facies, in which plants are not normally preserved. In most of northern England and Scotland, there is little evidence of coeval strata.

In the rest of Europe, this zone is also very rare. Sedimentation appears to have ceased in the early Westphalian D throughout most of the paralic basin that stretches from northern France through Germany to Poland. In the Iberian Peninsula, the effects of the Leonian folding, mentioned above, again seem to have disrupted the non-marine stratigraphical record at this level. The only undoubted record of the zone comes from the intra-montane basin of Saar-Lorraine, and this is based on material collected from underground mine workings (Cleal, 1984b; Laveine, 1989). It is evident, therefore, that Jockie's Syke is one of the very few sites in Europe yielding plant fossils of this age.

## Conclusion

Jockie's Syke is one of the few sites in Europe, and the only one in Britain, to yield this particular assemblage of plant fossils from the Upper Coal Measures, rocks deposited as sediments about 300 million years before the present. Two species are found here (*Alethopteris ambigua* and *Cyathocarpus* aff. *arborescens*) which do not occur in older floras; their appearance represents the start of the gradual change from the typical Coal Measures vegetation represented by the British fossil floras, to the later (Stephanian) vegetation, more typically found in central and southern Europe. This vegetational change was probably triggered by topographic and possible climatic changes, which in turn were the result of the collision between two major continental plates (Gondwana and Laurussia). These large-scale environmental changes brought an end to the swamp-forest vegetation that had typified Coal Measures (Westphalian) times across northern Europe, and eventually culminated in the more arid conditions of the Permian, whose vegetation is the subject of the next chapter.

# Chapter 7

# *Permian*

# *Permian vegetation*

No other period in Earth's history was as traumatic for terrestrial vegetation as the Permian. It saw the extinction of much of the Late Palaeozoic terrestrial vegetation, which had been dominated by arborescent lycopsids and equisetopsids, and a number of 'primitive' gymnosperm groups (e.g. Trigonocarpales, Lagenostomaleans, Cordaitales, Dicranophyllales). During the succeeding Early Triassic, terrestrial vegetation, at least as revealed by the fossil record, was generally sparse and of poor diversity (Dobruskina, 1980). During the middle and late Triassic, however, there was the progressive introduction of a number of gymnosperm groups regarded as characteristically Mesozoic in aspect (Umkomasiaceae, Leptostrobaceae, Caytoniaceae, Bennettitaceae), of many of the modern families of ferns (e.g. Matoniaceae, Dipteridaceae, Polypodiaceae, Dicksoniaceae) and conifers (e.g. Podocarpaceae, Pinaceae), and even the first putative angiosperms (Cornet, 1989; Cleal, 1993). Unfortunately, this change, known as the Palaeophytic-Mesophytic transition (Figure 7.1), cannot be particularly well demonstrated in Britain, which at that time suffered arid conditions that supported only a sparse vegetation. Nevertheless, the broad pattern of the transition can be discerned and its stratigraphical position has potentially important consequences for understanding the underlying mechanism.

## PALAEOGEOGRAPHICAL SETTING

By the Early Permian, the Laurussia and Gondwana continental plates had fused to form part of the Pangaea 'super-continent' (Figure 7.2). Britain lay on the eastern margins of Pangaea and, although it had drifted north relative to its position in the Carboniferous, it was still within tropical latitudes (probably *c.* 20° north by the end of the Permian). The Lower Permian is mainly represented by red-beds, indicating arid climatic conditions, and generally not conducive to the preservation of plant fossils. The Upper Permian consists of carbonate beds deposited in the Zechstein and *Bakevellia* inland seas.

There is greater provincialism in Permian plant fossil distribution than in any other part of the Palaeozoic (Vakhrameev *et al.*, 1978; Meyen, 1987; Allen and Dineley, 1988; Cleal, 1991). Most currently accepted palaeophytogeographical models recognize five discrete palaeokingdoms for the period (Figure 7.2). The British assemblages belong to the Euramerian Palaeokingdom, which extends from eastern North America, through Europe to southern Kazakhstan. The Euramerian assemblages are generally rare and of limited diversity, dominated mainly by conifers and peltasperms. Nevertheless, there are a number of well-documented Lower Permian examples, such as in the Autunian of France (Doubinger, 1956) and Germany (Barthel, 1976; Kerp and Fichter, 1985). The Upper Permian is mainly represented by the Zechstein assemblages of Germany, France and Britain.

## STRATIGRAPHICAL BACKGROUND

Details of British Permian stratigraphy are discussed by Smith *et al.* (1974). The stages currently recognized by the IUGS Commission on Stratigraphy are based on the marine sequences in the Ural Mountains (Figure 7.6). However, they are difficult to use in the sequences found in much of Europe, and so a separate set of stages have been introduced for these strata. The approximate correlation between the European scheme and the marine stages is shown in Figure 7.6, based partly on data provided by Kozur (1984).

The Carboniferous–Permian boundary has still to be formally defined. In this volume, it is taken to correspond to the Stephanian–Autunian boundary, which appears to correlate broadly with the Gzhelian–Asselian boundary in the standard marine sequences (Doubinger and Bouroz, 1984).

## PERMIAN VEGETATION

The British Permian plant fossils represent the vegetation growing in the equatorial parts of the Pangaean 'super-continent' (Figure 7.2). Pteridophytic plants were on the whole rare in this vegetation. There were remnant populations of arborescent lycopsids and ferns in the Autunian, similar to those found in the Carboniferous, but they had largely disappeared by the Saxonian. There is palynological evidence that osmundacean ferns were present here in the Late Permian (Schweitzer, 1986), but macrofossils have yet to be found.

Equisetopsids similar to those found in the Carboniferous persisted into the Autunian, including the large calamostachyaleans (Kerp, 1984b) and the herbaceous bowmanitaleans (Kerp,

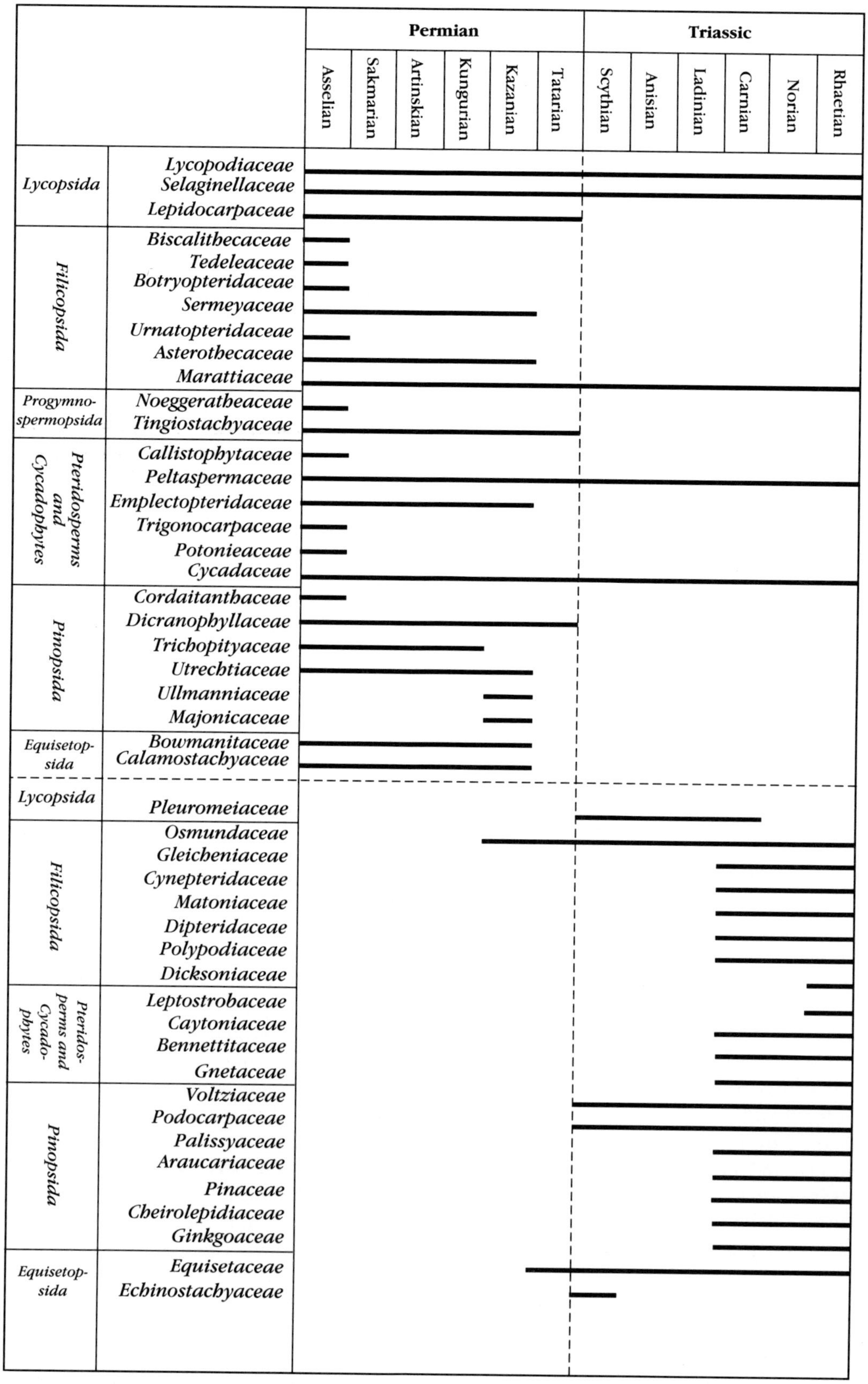

|  |  | Permian |  |  |  |  |  | Triassic |  |  |  |  |  |
|--|--|--|--|--|--|--|--|--|--|--|--|--|--|
|  |  | Asselian | Sakmarian | Artinskian | Kungurian | Kazanian | Tatarian | Scythian | Anisian | Ladinian | Carnian | Norian | Rhaetian |
| Lycopsida | Lycopodiaceae | ━ | ━ | ━ | ━ | ━ | ━ | ━ | ━ | ━ | ━ | ━ | ━ |
|  | Selaginellaceae | ━ | ━ | ━ | ━ | ━ | ━ | ━ | ━ | ━ | ━ | ━ | ━ |
|  | Lepidocarpaceae | ━ | ━ | ━ | ━ | ━ | ━ |  |  |  |  |  |  |
| Filicopsida | Biscalithecaceae | ━ |  |  |  |  |  |  |  |  |  |  |  |
|  | Tedeleaceae | ━ |  |  |  |  |  |  |  |  |  |  |  |
|  | Botryopteridaceae | ━ |  |  |  |  |  |  |  |  |  |  |  |
|  | Sermeyaceae | ━ | ━ | ━ | ━ | ━ |  |  |  |  |  |  |  |
|  | Urnatopteridaceae | ━ |  |  |  |  |  |  |  |  |  |  |  |
|  | Asterothecaceae | ━ | ━ | ━ | ━ | ━ |  |  |  |  |  |  |  |
|  | Marattiaceae | ━ | ━ | ━ | ━ | ━ | ━ | ━ | ━ | ━ | ━ | ━ | ━ |
| Progymno-spermopsida | Noeggeratheaceae | ━ |  |  |  |  |  |  |  |  |  |  |  |
|  | Tingiostachyaceae | ━ | ━ | ━ | ━ | ━ | ━ |  |  |  |  |  |  |
| Pteridosperms and Cycadophytes | Callistophytaceae | ━ |  |  |  |  |  |  |  |  |  |  |  |
|  | Peltaspermaceae | ━ | ━ | ━ | ━ | ━ | ━ | ━ | ━ | ━ | ━ | ━ | ━ |
|  | Emplectopteridaceae | ━ | ━ | ━ | ━ | ━ |  |  |  |  |  |  |  |
|  | Trigonocarpaceae | ━ |  |  |  |  |  |  |  |  |  |  |  |
|  | Potonieaceae | ━ |  |  |  |  |  |  |  |  |  |  |  |
|  | Cycadaceae | ━ | ━ | ━ | ━ | ━ | ━ | ━ | ━ | ━ | ━ | ━ | ━ |
| Pinopsida | Cordaitanthaceae | ━ |  |  |  |  |  |  |  |  |  |  |  |
|  | Dicranophyllaceae | ━ | ━ | ━ | ━ | ━ | ━ |  |  |  |  |  |  |
|  | Trichopityaceae | ━ | ━ | ━ | ━ |  |  |  |  |  |  |  |  |
|  | Utrechtiaceae | ━ | ━ |  |  |  |  |  |  |  |  |  |  |
|  | Ullmanniaceae |  |  |  |  | ━ | ━ |  |  |  |  |  |  |
|  | Majonicaceae |  |  |  |  | ━ | ━ |  |  |  |  |  |  |
| Equisetopsida | Bowmanitaceae | ━ | ━ | ━ | ━ | ━ | ━ |  |  |  |  |  |  |
|  | Calamostachyaceae | ━ | ━ | ━ | ━ | ━ | ━ |  |  |  |  |  |  |
| Lycopsida | Pleuromeiaceae |  |  |  |  |  |  | ━ | ━ | ━ |  |  |  |
| Filicopsida | Osmundaceae |  |  |  |  | ━ | ━ | ━ | ━ | ━ | ━ | ━ | ━ |
|  | Gleicheniaceae |  |  |  |  |  |  |  |  | ━ | ━ | ━ | ━ |
|  | Cynepteridaceae |  |  |  |  |  |  |  |  | ━ | ━ | ━ | ━ |
|  | Matoniaceae |  |  |  |  |  |  |  |  | ━ | ━ | ━ | ━ |
|  | Dipteridaceae |  |  |  |  |  |  |  |  | ━ | ━ | ━ | ━ |
|  | Polypodiaceae |  |  |  |  |  |  |  |  | ━ | ━ | ━ | ━ |
|  | Dicksoniaceae |  |  |  |  |  |  |  |  | ━ | ━ | ━ | ━ |
| Pteridosperms and Cycadophytes | Leptostrobaceae |  |  |  |  |  |  |  |  | ━ | ━ | ━ | ━ |
|  | Caytoniaceae |  |  |  |  |  |  |  |  | ━ | ━ | ━ | ━ |
|  | Bennettitaceae |  |  |  |  |  |  |  |  | ━ | ━ | ━ | ━ |
|  | Gnetaceae |  |  |  |  |  |  |  |  | ━ | ━ | ━ | ━ |
| Pinopsida | Voltziaceae |  |  |  |  |  |  | ━ | ━ | ━ | ━ | ━ | ━ |
|  | Podocarpaceae |  |  |  |  |  |  | ━ | ━ | ━ | ━ | ━ | ━ |
|  | Palissyaceae |  |  |  |  |  |  |  |  | ━ | ━ | ━ | ━ |
|  | Araucariaceae |  |  |  |  |  |  |  |  | ━ | ━ | ━ | ━ |
|  | Pinaceae |  |  |  |  |  |  |  |  | ━ | ━ | ━ | ━ |
|  | Cheirolepidiaceae |  |  |  |  |  |  |  |  | ━ | ━ | ━ | ━ |
|  | Ginkgoaceae |  |  |  |  |  |  |  |  | ━ | ━ | ━ | ━ |
| Equisetopsida | Equisetaceae |  |  |  |  |  |  | ━ | ━ | ━ | ━ | ━ | ━ |
|  | Echinostachyaceae |  |  |  |  |  |  | ━ | ━ |  |  |  |  |

**Figure 7.1** The distribution of the principal families of vascular plants in the Permian and Triassic. Based on data from Cleal (1993).

## Permian vegetation

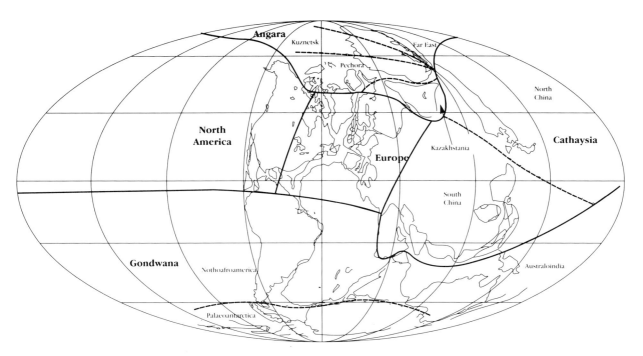

**Figure 7.2** The palaeogeography of the Permian, showing the distribution of the major floristic zones (phytochoria). Based on Scotese and McKerrow (1990) and Cleal and Thomas *in* Cleal (1991).

1984a). By the Late Permian, however, only small equisetopsids are found (*Neocalamites*). The latter were similar in general morphology to the recent *Equisetum*, but their affinities remain uncertain in the absence of fructifications.

Some characteristic pteridospermous groups of the Late Carboniferous tropical vegetation (Trigonocarpales, Callistophytales) also continued into the earliest Permian, but became extinct by the end of the Autunian. The dominant pteridosperms during the rest of the Permian Period were bushes or shrubs of the Peltaspermales. Although so far not reliably identified from Britain, remains of their fronds are common fossils in the Lower Permian of continental Europe, and are known as *Autunia* (*Callipteris auct.*), amongst other names (Kerp, 1986). In the Upper Permian of Britain, on the other hand, peltasperms are well represented, belonging to the form-genus *Peltaspermum* (Townrow, 1960; Poort and Kerp, 1990). They appear most closely related to the Carboniferous pteridosperms, the Callistophytales.

There is unequivocal evidence that cycads were present in Cathaysia (Gao and Thomas, 1989), but their presence in Pangaea is less certain. Kerp (1983) has described fertile leaves of what may be a primitive cycad-like plant (*Sobernheimia*), although Crane (1985) has queried their cycad credentials. There have also been reports of typical-looking cycad leaves, especially from the Upper Permian of Britain (Stoneley, 1958; Schweitzer, 1986), but there are neither cuticles nor fructifications preserved to confirm their identity.

Evidence of ginkgophytes in tropical Pangaea, including Britain, is also limited to rare fossils of foliage (*Sphenobaiera*). In this case, however, there is epidermal evidence to support their ginkgophyte affinities (Schweitzer, 1986). *Trichopitys*, an ovuliferous structure described from the Lower Permian of France as a primitive ginkgo (Florin, 1949), is now thought to be allied either with the peltasperms or dicranophylls (Meyen, 1987; Archangelsky and Cúneo, 1990).

By far the most abundant plants in the vegetation of tropical Pangaea were conifers (Figures 7.3 and 7.4). In the Early Permian they mainly belong to the Walchiaceae, which is the most primitive known conifer family, with a fossil record extending back into the Westphalian (Florin, 1938–1945; Clement-Westerhof, 1988). In the Late Permian, they were largely replaced by the Ullmanniaceae and Majonicaceae, whose ovuliferous cones have a simpler and thus 'more advanced' structure.

Outside Europe, Permian tropical vegetation is best represented in China, which seems to have

# Permian

**Figure 7.3** Reconstruction of Early Permian conifer shoot, *Walchia*. Based on Florin (1951).

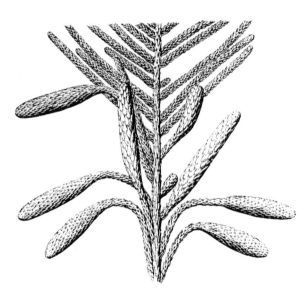

**Figure 7.4** Reconstruction of Early Permian conifer cones. Based on Florin (1951).

escaped the dramatic environmental changes that caused the extinction of the swamp-forests of Europe and North America. Arborescent lycopsids, equisetopsids, cordaites and marattialean ferns all persist here into the Late Permian ('Gu and Zhi', 1974; Li, 1980; Asama, 1984), and are

the real successors of the Carboniferous tropical forests. As noted above, unequivocal remains of cycads make their first appearance in the Permian of China. However, the most characteristic elements are of the gigantopterid complex (*Gigantopteris, Emplectopteris, Emplectopteridium*, etc.). Although almost certainly gymnospermous, the affinities of these plants are not certain. A possible link with the trigonocarpalean form-genus *Callipteridium* has been suggested, but the evidence remains far from conclusive. The plant fossils representing this tropical vegetation are referred to as the Cathaysia Palaeokingdom.

In parts of western North America (Figure 7.2), plant fossils in the Lower Permian appear to have a distinctly Cathaysian aspect, especially through the presence of leaves apparently belonging to the gigantopterid complex. This is referred to as the North America Palaeokingdom (Read and Mamay, 1964). It remains uncertain whether this represents a true Cathaysian-style vegetation, or merely reflects plants that underwent convergent evolution.

The Permian saw the first major development of vegetation in areas outside the palaeoequatorial belt. In the southern hemisphere, this is seen in the growth of forests dominated by arberialean trees, with leaves belonging to form-genera such as *Glossopteris* and *Gangamopteris*. In the southern polar regions (by this time there was no ice-cap), the vegetation consisted almost exclusively of these trees (Archangelsky, 1990). In lower latitudes (South Africa and South America), however, they are also associated with a range of other plant groups, including lycopsids, calamostachyalean equisetopsids and marattialean ferns and conifers (Archangelsky and Arrondo, 1969, 1975; Archangelsky *et al.*, 1981; Archangelsky, 1990; Anderson and Anderson, 1985; Clement-Westerhof, 1988).

In the northern hemisphere, temperate conditions produced a diverse range of vegetation in what was Angara (now Siberia). The dominant group consisted of plants related to the cordaitanthids of the Carboniferous tropical forests, and referred to the families Rufloriaceae and Vojnovskyaceae (Meyen, 1966, 1988). There were also endemic equisetopsids (Tchernoviaceae) and pteridosperms (Cardiolepidaceae), as well as a range of ferns, conifers and cycadophytes (Meyen, 1971, 1976, 1982, 1987). Much of the lowland parts of Angara were probably covered by forest, although some areas of more open vegetation also probably existed, including floras dominated by mosses (Fefilova, 1978).

## THE PALAEOPHYTIC–MESOPHYTIC TRANSITION

The Palaeophytic–Mesophytic floristic transition is traditionally placed at the stratigraphical level where the first fossils of conifers and other advanced gymnosperms occur (Gothan and Weyland, 1954; Frederiksen, 1972). It does not mark a major phylogenetic development; most of the typical Mesophytic gymnosperms probably existed in the Carboniferous, in the extra-basin habitats. Rather, it indicates an environmental change, which facilitated the migration of extra-basinal vegetation into the lowland habitats.

In most of the Eurameria Palaeokingdom, including Britain, the transition probably occurs in the Saxonian (?Sakmarian/Artinskian). Saxonian assemblages are very rare, but one example reported from Lodeve in France appears to be transitional between the Palaeophytic and Mesophytic (Doubinger and Krusemann, 1966). The transition appears to relate to the destruction of the lycopsid forests, which had dominated the landscape of tropical Pangaea during most of the Late Carboniferous. This had been caused by topographical uplift, following the collision of the Laurussia and Gondwana continental plates, which had drained the lowland swamps and made them unsuitable for the lycopsids (Cleal, 1991). They were partially replaced by stands of conifers, but there is no evidence that these were anywhere near as extensive as the earlier lycopsid forests.

Elsewhere in the world, the Palaeophytic–Mesophytic transition occurs at significantly higher stratigraphical levels, at or just above the Permian–Triassic boundary (Roy Chowdhury et al., 1975; Sadovnikov, 1981; Wang, 1989). It coincides with the start of a period of climatic aridity, which influenced much of the globe during the early Triassic. Despite the difference in timing, it is tempting to link the changes in tropical Pangaea with those elsewhere in the world. A significant reduction in the tropical biomass may have instigated a global climatic change through a 'greenhouse effect', which only later had a significant impact on the vegetation of other parts of the world (Cleal, 1991). If this model is correct, it has clear implications for understanding what might follow the further destruction of the present-day tropical forests.

## PERMIAN PLANT FOSSILS IN BRITAIN

There are relatively few Permian plant fossil sites in Britain (Vakhrameev et al., 1978). The few significant sites are shown in Figure 7.5. Only one Lower Permian site, in the Mauchline Volcanic Group of south-west Scotland is extant. The conifer-dominated assemblages from the Keele and Enville formations and their equivalents in the English Midlands were regarded as Early Permian (e.g. Wagner, 1983), but palynological evidence now suggests that they are late Westphalian D or Cantabrian (B. Besly, pers. comm., 1989). Rather more sites are known in the Upper Permian, particularly from the Marl Slate of northern England (Stoneley, 1958; Schweitzer, 1986). Even here, however, the plant fossils are not particularly common. Those Permian sites selected for their palaeobotanical interest to form part of the GCR network, are shown in Figure 7.6. The geographical positions of the two Upper Permian sites are shown in Figure 7.5.

## STAIRHILL

### Highlights

Stairhill has yielded the only known plant fossil assemblage from the Lower Permian of Britain, and is comparable to coeval assemblages from continental Europe.

### Introduction

Permian sandstones are exposed in the bed of the River Ayr near Stairhill, Strathclyde Region (NS 452243) (see Figure 7.7). Plant fossils were first reported by Mykura (1960, 1965), and preliminary identifications provided by Chaloner (in Mykura, 1965). Some of these identifications were subsequently challenged by Wagner (1966; Wagner in Smith et al., 1974), and a full systematic analysis of the assemblage was given by Wagner (1983).

### Description

*Stratigraphy*

The plant fossils occur in a one metre thick tuffaceous interval of mudstones and sandstones within the Mauchline Sandstones, and thus

# *Permian*

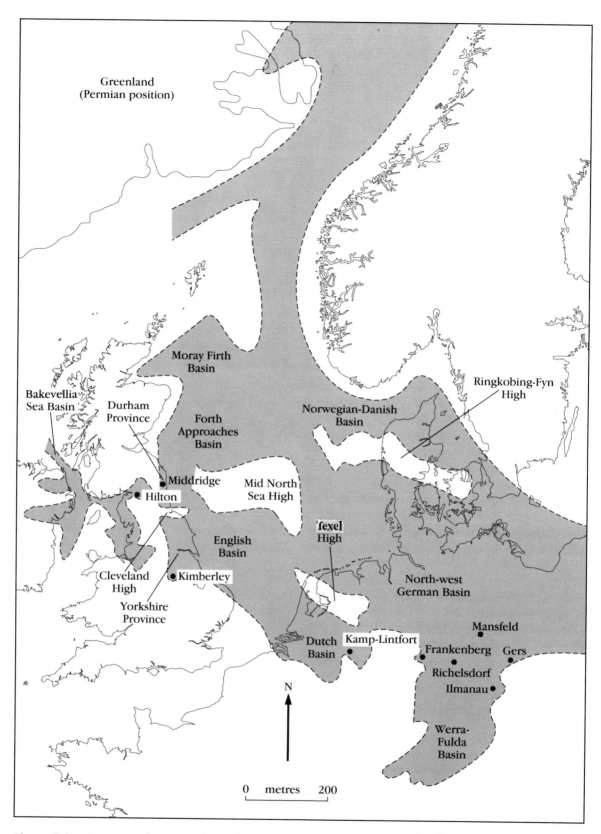

**Figure 7.5** Main areas of Late Permian sedimentation in north-west Europe, showing principal palaeobotanical sites including the two GCR sites (Kimberley and Middridge). Based on Schweitzer (1986, figure 1).

| International Chronostratigraphy || European non-marine 'stages' | GCR Palaeobotany sites |
|---|---|---|---|
| Series | Stages | | |
| Upper Permian | Tatarian | Zechstein | Middridge Quarry Kimberley Cutting |
| | Kazanian | | |
| Lower Permian | Kungurian | Saxonian | |
| | Artinskian | | |
| | Sakmarian | | |
| | Asselian | Autunian | Stairhill |

**Figure 7.6** Chronostratigraphical classification of the Permian and the positions of the GCR palaeobotany sites.

comprise part of the New Red Sandstone of the Mauchline Basin. Their chronostratigraphical position has been the matter of some debate, but the plant fossils suggest a probable early Autunian (Asselian) age (Wagner, 1983).

### Palaeobotany

The plant fossils are preserved as impressions. The following species are known:

Equisetopsida:
 *Annularia stellata* (Sternberg) Wood
 *A.* cf. *spicata* Gutbier
 *Asterophyllites equisetiformis* Brongniart

Filicopsida:
 *Lobatopteris geinitzii* (Gutbier) Wagner
 *Remia pinnatifida* (Gutbier) Knight
 *Pecopteris* cf. *monyii* Zeiller

Cycadopsida:
 *Odontopteris subcrenulata* var. *gallica* Doubinger and Remy

## Interpretation

A variety of equisetopsid foliage was reported from Stairhill by Wagner (1983), who assigned most of that identified to the wide-ranging species *Annularia stellata* and *Asterophyllites equisetiformis*. A single specimen with smaller and stiffer-looking leaves was assigned to *Annularia* cf. *spicata*. However, these identifications may have to be revised in the light of the work of Kerp (1984b), who described rather similar foliage from the Lower Permian of Germany as *Annularia carinata* (Gutbier) Schimper. In the absence of any evidence of fructifications, or even of the morphology of the stem that bore the leaves, the status of these Stairhill leaves is difficult to judge.

The dominant fern in the assemblage is *Lobatopteris geinitzii*. Wagner (1983) gave a detailed systematic discussion of this species, particularly as to its position relative to the very similar *Lobatopteris polypodioides* (Sternberg) Knight, from which it differs by having somewhat larger, squatter pinnules. The Stairhill specimens provide no evidence of fertile structures, but German material (Göppert, 1864) shows *Cyathocarpus*-like sporangia. Together with the predominance of pinnatifid pinnules in the frond, this supports Wagner's inclusion of this species within the form-genus *Lobatopteris*.

Wagner (1983) described some other fern fragments with pinnatifid pinnules similar to those of

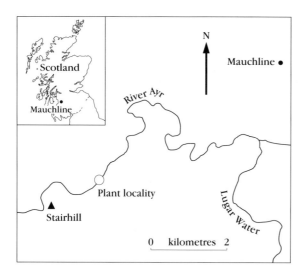

**Figure 7.7** Location map for Stairhill. Based on Wagner (1983).

*L. geinitzii*. However, these have a denser venation with a different forking pattern, and so Wagner identified them as *Pecopteris pinnatifida*. No fertile specimens have so far been found at Stairhill. They are clearly similar, however, to the frond fragments from Spain, which bear sporangia known as *Remia* (Knight, 1985). Knight proposed that the species should be transferred to *Remia*, a view which is followed here.

A single, near-terminal pinna fragment from Stairhill has been compared with *Pecopteris monyii* by Wagner (1983), although he also pointed out that there was a possible comparison with *Pecopteris densifolia* Göppert. The Stairhill specimen has rather tapered pinnules, and so Wagner made the main comparison with *P. monyii*. A definite identification awaits the discovery of more specimens.

Pteridosperm remains all appear to belong to the medullosan frond species *Odontopteris subcrenulata*. Wagner (1983) assigned this species to the form-genus *Mixoneura*. The distinction between *Odontopteris* and *Mixoneura* is far from clear, however, and needs to be confirmed by additional evidence, such as epidermal structures. The traditional view is therefore accepted here, and the species retained in *Odontopteris*. The Stairhill specimens of *O. subcrenulata* have more or less 'square' lateral pinnules with a short midvein, and large apical pinnules. Wagner therefore assigned them to *O. subcrenulata* var. *gallica*, following the classification of Doubinger and Remy (1958).

Stairhill has yielded the only known Early Permian plant fossils from Britain. Assemblages from the Keele Formation of the English Midlands have been thought to be of similar age (Howell, 1859; Vernon, 1912; Dix, 1935), but recent palynological evidence suggests that at least part of the Keele Formation is late Westphalian D or Cantabrian (B. Besly pers. comm., 1989).

The Stairhill assemblage is in many ways typical of the Eurameria Palaeokingdom (i.e. semi-arid palaeoequatorial) in the Lower Permian. In continental Europe, comparable assemblages have been reported from the uppermost Stephanian and Autunian of both France and Germany (reviewed by Doubinger, 1956; Barthel, 1976; Kerp and Fichter, 1985). In particular, the occurrence of *Lobatopteris geinitzii* and *Odontopteris subcrenulata* var. *gallica* supports this comparison. In some ways, the Stairhill assemblage may be regarded as a remnant of the vegetation that flourished on the river levée-banks in the Carboniferous tropical forests, although the lycopsid-dominated swamp vegetation itself had already disappeared (except in China - see Li, 1980). In contrast, many of the coeval assemblages from continental Europe are dominated by conifers (e.g. *Walchia*) and peltasperms (e.g. *Autunia*), suggesting markedly drier edaphic conditions (Bouroz and Doubinger, 1977).

There is also some comparison with assemblages from the basal Permian of eastern North America, as reviewed by Read and Mamay (1964), although many of these have not been fully documented. The American assemblages share species of *Annularia*, *Odontopteris* and various pecopteroids with Stairhill. However, unlike Stairhill, they often have abundant peltasperms ('callipterids') and conifers. Further west, in Texas, Arizona and New Mexico, fossils are of plants belonging to the North America Palaeokingdom and are quite different from anything seen in Europe, with the occurrence of leaves resembling the gigantopterids of China. Elsewhere, the Lower Permian fossil floras are totally different from what is seen at Stairhill, with the gigantopteroid assemblages of the Cathaysia Palaeokingdom, the *Cordaites/Rufloria* assemblages of the Angara Palaeokingdom and the arberialean assemblages of the Gondwana Palaeokingdom (Vakhrameev *et al.*, 1978; Meyen, 1987; Allen and Dineley, 1988; Cleal and Thomas *in* Cleal, 1991).

## Conclusion

Stairhill has yielded the only plant fossils of indisputably Early Permian age (about 295 million years old) in Britain. They can be compared with some of the floras from continental Europe, such as from southern France and Saarland, which are also dominated by horsetails, ferns and a group of seed plants known as the Trigonocarpales. However, most European floras of this age were growing in rather drier habitats, and as a consequence tended to be dominated by other types of seed plants, particularly conifers and the now extinct peltasperms. Plants at Stairhill represent the last, poor remnants of the luxurious tropical swamp-forests that were at their maximum development in the Late Carboniferous (see previous chapter).

# MIDDRIDGE QUARRY

## Highlights

Middridge Quarry has yielded the most diverse assemblage of plant fossils from the Upper Permian of Britain. It includes the best British examples of the conifer families Ullmanniaceae and Majonicaceae, and what may be the earliest British example of cycad foliage.

## Introduction

The Upper Permian exposures near Bishop Auckland, County Durham, have long been famous for their plant fossils (Sedgwick, 1829; King, 1850; Kirkby and Duff, 1872; Howse, 1890; Stoneley, 1958; Mills and Hull, 1976). Most of the specimens were collected during the nineteenth century from localities which are no longer available. In 1977, however, a new exposure was made at Middridge Quarry, between Bishop Auckland and Newton Aycliffe (NZ 249252), which allowed Late Permian fossils to be collected in this area for the first time in nearly 100 years (Bell *et al.*, 1979). A useful summary of the palaeobotany of this site is provided by Schweitzer (1986).

## Description

### Stratigraphy

The geology of this site is described by Bell *et al.* (1979). The 'Basal' Permian Breccia lies unconformably on Langsettian (Upper Carboniferous) sandstones and shales, and is overlain by about 2.5 metres of mainly calcareous siltstones known as the Marl Slate. These are in turn overlain by massive dolomitic limestones of the Lower Magnesian Limestone. The plant fossils occur in the Marl Slate, which here represents a shallow water deposit formed near the margin of the Zechstein Sea. The plant remains were presumably transported into the sea by rivers. The Marl Slate has been interpreted as a lateral equivalent of the Kupferschiefer Formation of Germany, and thus probably belongs to the Kazanian Stage.

### Palaeobotany

The plant fossils are preserved as adpressions, often with cuticles still present. The following species have been described to date:

Algae:
 *Algites virgatus* (Munster) Stoneley

Equisetopsida:
 *Neocalamites mansfeldicus* Weigelt

Peltaspermales:
 *Peltaspermum martinsii* (Germar) Poort and Kerp

Cycadopsida(?):
 *Pseudoctenis middridgensis* Stoneley

Pinopsida (?Ginkgoales):
 *Sphenobaiera digitata* (Brongniart) Florin

Pinopsida:
 *Pseudovoltzia liebeana* (Geinitz) Florin
 *Ullmannia frumentaria* (Brongniart) Göppert
 *U. bronnii* Göppert

Uncertain affinities:
 '*Neuropteris*' *buttoniana* King
 *Psygmophyllum cuneifolium* (Kutorga) Schimper
 *Lesleya eckhardtii* (Kurtze) Remy and Remy
 *Plagiozamites middridgensis* Schweitzer

## Interpretation

A single Middridge specimen was described briefly by Stoneley (1958) as *Algites virgatus*. It shows slender, forking axes originating from a basal patch of carbonized tissue (?holdfast). Two of the axes bear a terminal 'plume-like' structure, although no details were preserved. Similar, albeit smaller, bodies are present in the holotype specimen of the species figured by Munster (1842).

Stems and foliage of the equisetopsid *Neocalamites mansfeldicus* were first recorded from Middridge by Bell *et al.* (1979). Stoneley (1958) recorded examples of similar stems from other localities in north-east England as *Paracalamites kutorgai* (Geinitz) Zalessky, which she regarded as an earlier synonym of *N. mansfeldicus*. However, *P. kutorgai* probably belongs to the equisete family Tchernoviaceae, which is restricted to Angaran assemblages (Meyen, 1971) and would thus not be expected to occur in Britain. As the fructifications of *N. mansfeldicus* are unknown, there seems little justification at this stage for combining it in synonymy with this Angaran species.

Both King (1850) and Bell *et al.* (1979) record '*Neuropteris*' *buttoniana* from Middridge, but the specimens have not been illustrated. Stoneley (1958) referred to this taxon as *Mixoneura* sp., but it does not have the broadly attached pinnules seen in the type-species of this form-genus (*Odontopteris subcrenulata* (Rost) Zeiller). Equally, however, it is unlikely to belong to *Neuropteris* (as originally assigned by King, 1850) as interpreted by Cleal *et al.* (1991), since this is mainly restricted to Westphalian assemblages. Stoneley suggested a comparison with *Neurocallipteris neuropteroides* (Göppert) Cleal *et al.*, but that species has never previously been reported from higher than the Lower Permian. Establishing the taxonomic position of this Middridge species will have to wait until larger specimens become available to determine the frond architecture, and also preferably with cuticles preserved.

Although neither Stoneley (1958) nor Townrow (1960) record *Peltaspermum martinsii* from Middridge, it is listed (but not figured) by Bell *et al.* (1979), Schweitzer (1986) and Poort and Kerp (1990).

Another type of foliage was described by Stoneley (1958) as ?*Psygmophyllum cuneifolium*, a species belonging to the pteridosperm order Peltaspermales. Although showing some similarities to fragments of cycad frond, they have a more complex architecture, consisting of both sub-dichotomous and pinnate branching. Fragments of cuticles prepared from the Middridge specimens show sunken, cyclocytic stomata not arranged in distinct bands, and thus quite compatible with the epidermal structure usually associated with *Psygmophyllum* (Meyen, 1987). Whether the comparison with *P. cuneifolium*, a species more usually associated with Angaran assemblages, is valid will have to await the discovery of more complete material.

Stoneley (1958) based her description of *Pseudoctenis middridgensis* on a single fragment from Middridge, although Bell *et al.* (1979) report that they have since found additional material. Small fragments of adaxial cuticle prepared from the Stoneley specimen showed simple intercellular flanges (anticlinal walls), which preclude the assignment of the frond fragment to the Bennettitales, and Schweitzer (1986) argued that it probably belonged to the cycads. However, no evidence of abaxial cuticle was found. Consequently the stomatal structure, which is a key feature for recognizing *Pseudoctenis* (Harris, 1964), is unknown. If correctly assigned, it would be the only cycad frond found below the Mesozoic in Britain, although cycad remains have been reported from the Lower Permian of China (Gao and Thomas, 1989).

*Sphenobaiera digitata* is also listed, but not figured, from here by Stoneley (1958) and Bell *et al.* (1979). None of the Middridge specimens were found to yield cuticles, but Stoneley figures examples from Hilton Beck with sunken stomata surrounded by prominent papillae, of a type normally associated with this species. Stoneley (1958) and Schweitzer (1986) referred the species to the Ginkgoales, following Florin (1936), which would have made it the earliest example of the order known from Britain. In the absence of evidence of its fructifications, its taxonomic position must be regarded as provisional (the 'capsule' from Hilton Beck described by Stoneley is neither attached to *Sphenobaiera* foliage nor does it show any structure).

By far the most abundant plant fossils found at Middridge are conifer twigs, belonging to *Ullmannia* (Ullmanniaceae) and *Pseudovoltzia* (Majonicaceae). These two form-genera are best distinguished by their fertile structures, the epidermal structures being essentially indistinguishable. The naming of the Middridge conifers has relied on the leaf morphology, which must clearly raise some doubt as to the reliability of the identifications.

*Pseudovoltzia* shoots tend to be extremely heterophyllous, with needle-like leaves anything up to 20 mm long. Although not reported to date from Middridge, the characteristic lobed cone-scales of this group of conifers has been found at other Marl Slate localities, such as Kimberley Railway Cutting (see below). Stoneley (1958) identified all the Marl Slate specimens as *P. liebeana*, the best known species of the form-genus.

The leaves of *Ullmannia frumentaria* (Figure 7.8) are very similar to the above, except that they tend to be laxer and lie more parallel to the stem. Attached to one such shoot, Stoneley (1958) described what may be a female cone, similar in shape to *U. frumentaria* cones reported from Germany (Florin, 1938–1945), although no structure is preserved.

A specimen with smaller, broader leaves was described by Stoneley (1958) as *Ullmannia bronnii*, and further specimens as *Hiltonia rivulii* Stoneley. They were distinguished from *U. bronnii* by the leaves having a rounder apex, a decurrent base and no clearly marked vein. Also mentioned were differences in the thickness of the walls of the subsidiary cells surrounding the stomata. However, Schweitzer (1962) has shown that these differences are not sufficient to justify the separation of the two species, and that they should be regarded as synonymous.

A single problematic specimen was described by Stoneley (1958) as *Taeniopteris eckhardtii* Kurtze. Although fragmentary and without cuticles preserved, it shows a thick mid-vein and simple, straight lateral veins, very similar to the specimens from the Kupferschiefer of Mansfeld in Germany, the type area of this species (Schweitzer, 1968, fig. 4). Remy and Remy (1977) have argued that this type of nervation pattern is more characteristic of the form-genus *Lesleya*, to which they transferred Kurtze's species. The taxonomic position of this species is totally unclear. Schweitzer (1986) argued that it was most likely cycad foliage, but Meyen (1987) regarded at least some *Lesleya* leaves as belonging to the Dicranophyllales, a primitive order of conifer-like plants, that became extinct at the end of the Permian.

*Plagiozamites bellii*, for which Middridge is the type locality, has been interpreted as shoots with helically arranged leaves up to 90 mm long (Schweitzer, 1986). Although some details of the epidermal structure are known, including stomata, the affinities of the shoots are unclear beyond that they are almost certainly gymnospermous.

Plant fossils are known from a number of exposures of the Marl Slate and its lateral equivalents in northern England (evidence reviewed by Stoneley, 1958). Of these, Middridge Quarry has yielded the most diverse assemblage of abundant plant remains. The nearest comparison is with the Hilton Beck Plant Bed in Cumbria (Murchison and Harkness, 1864; Stoneley, 1958), but that locality has not yielded the *Pseudoctenis*, *Lesleya*, *Psygmophyllum*, *Plagiozamites* or '*Neuropteris*' species found at Middridge. The assemblage from Kimberley Railway Cutting in Nottinghamshire (see below) is also not as diverse as that from Middridge, although it has yielded rather better examples of the peltasperm *Peltaspermum*.

The Marl Slate assemblage belongs to the Atlantic Palaeoarea of Vakhrameev *et al.* (1978) (also called the Zechstein 'flora' by Meyen, 1987), which probably extended throughout present-day Europe, although the fossil evidence is relatively sparse. This is a subdivision of the Europe Palaeokingdom of Cleal and Thomas (*in* Cleal, 1991; see also Figure 7.2). The closest comparison outside of Britain is with the flora from the Kupferschiefer of the Lower Rhine, Saxony and Thuringia in Germany (Schweitzer, 1962, 1968, 1986), and Pec (formerly Funfkirchen) in Hungary (Heer, 1876). These continental localities yield

**Figure 7.8** *Ullmannia frumentaria* (Brongniart) Göppert. Conifer shoot; Natural History Museum, London, specimen V.35130. Marl Slate (Upper Permian), Middridge Quarry. × 0.5. (Photo: Photographic Studio, Natural History Museum, London.)

most of the species contained in the British assemblages. However, they also include a number of species not found in Britain, such as the conifers *Quadrocladus*, *Culmitzischia* and *Rhenania*.

The vegetation represented by the Atlantic Palaeoarea is thought to reflect the relatively sparse vegetation growing in semi-arid conditions around the hypersaline Zechstein Sea. It contrasts markedly with the situation further east in the palaeoequatorial belt, such as in China, where the vegetation was dominated by lush, lycopsid-dominated swamp-forests (Li, 1980). This difference in vegetation was probably the result of the collision between the Gondwana and Laurussia continental plates during the very late Carboniferous. The collision would have had its maximum effect in Europe and North America, and the resulting topographical changes there would have caused a marked lowering of the water-table, making conditions unsuitable for the growth of the swamp-forests. China escaped the effects of this orogenic event, however, thus allowing the lycopsid-dominated swamp-forests to persist there.

## Conclusion

Middridge Quarry is the best site in Britain for fossils of Late Permian plants, about 270 million years old. Similar fossil floras have been reported from Germany and Hungary, and are thought to represent the vegetation surrounding a large, inland sea that covered parts of northern and central Europe – the Zechstein Sea. Conifers are the dominant plants in the flora, mainly preserved as fragments of leafy shoot, but occasionally with reproductive cones. They belong to two families (Ullmanniaceae and Majonicaceae) that flourished in the Late Permian, but then became extinct at the end of the period. Far less abundant, but nevertheless of considerable evolutionary interest, are fragments that may represent early cycads and ginkgos that later, in the Mesozoic, became major components of the vegetation in Britain.

## KIMBERLEY RAILWAY CUTTING

## Highlights

Kimberley Railway Cutting is the best British locality for the Late Permian *Peltaspermum martinsii*, the only good example of the Peltaspermaceae known from this country.

## Introduction

There are a number of early records of plant remains from the now disused railway cutting (SK 500452) in the village of Kimberley, Nottinghamshire (Wilson, 1876, 1881; Smith, 1913; Carr, 1914) and Carruthers (*in* Wilson, 1881) compared them with the Permian floras of continental Europe. The first species list was not, however, published until Stoneley's (1958) monograph on the British Permian plant fossils. Most recently, Townrow (1960) described specimens from here in his study on the pteridosperm family Peltaspermaceae.

## Description

### Stratigraphy

About 10 metres of Upper Permian strata are exposed, lying unconformably on Westphalian (Upper Carboniferous) shales. A thin 'Basal' Permian Breccia is overlain by Marl Slate (the Lower Permian Marls of some authors), which in turn is overlain by Lower Magnesian Limestone. Plant fragments occur throughout the Marl Slate part of the sequence, but the most abundant and well-preserved material occurs in clay bands just below the junction with the Lower Magnesian Limestone. As at Middridge Quarry (see above), the Marl Slate here probably belongs to the Kazanian Stage, and was deposited in a shallow marine environment.

### Palaeobotany

The plant fossils are preserved here as adpressions, sometimes with cuticles preserved. The following species have been reported so far:

Peltaspermales:
    *Peltaspermum martinsii* (Germar) Poort and Kerp

Pinopsida:
    *Pseudovoltzia liebeana* (Geinitz) Florin
    *Ullmannia bronnii* Göppert
    *Samaropsis triangularis* (Geinitz) Seward

# Kimberley Railway Cutting

## Interpretation

*P. martinsii* is now taken as a 'natural, whole-plant' species (Poort and Kerp, 1990). This is based on the presumed connection between fronds previously known as *Lepidopteris martinsii* (Germar) Townrow and ovulate structures identified as *Strobilites bronnii* Solms-Laubach. So far, a connection between these organs has not been proved, but there is a consistent association between them, including at Kimberley Cutting, and a close similarity in their epidermal structure. The ovulate structures were initially regarded as fragments of conifer male-cone (Florin, 1938–1945). However, Poort and Kerp (1990) demonstrated that they have ovules attached to small, peltate discs, similar to those of *Peltaspermum ottonis* (Harris) Poort and Kerp from the Rhaetian of Greenland.

Fragments of *P. martinsii* foliage described by Stoneley (1958) and Townrow (1960) have

**Figure 7.10** *Pseudovoltzia liebeana* (Geinitz) Florin. Cone scale from a conifer; Natural History Museum, London, specimen V.35128. Marl Slate (Upper Permian), Kimberley Railway Cutting. × 3. (Photo: Photographic Studio, Natural History Museum, London.)

yielded details of the frond architecture, particularly the presence of intercalated pinnules on the main rachis. They have also yielded cuticles, showing that stomata occur on both abaxial and adaxial surfaces of the leaf. Stoneley argued that the British specimens differed from those from Germany described by Gothan and Nagalhard (1922) in not having a papilla on each of the epidermal cells, but, by staining the cuticles with safranin, Townrow was able to demonstrate that epidermal papillae are present in at least some of the British specimens. The taxonomic significance of these epidermal papillae is thus doubtful (Townrow suggested that their presence or absence might be controlled by pinnule shape), and the German and British specimens are almost certainly specifically identical.

In contrast to Middridge Quarry (see above) and other sites in Durham, the dominant conifer remains at Kimberley Railway Cutting are shoots of *Ullmannia bronnii* (synonym *Hiltonia rivulii* Stoneley, 1956). *U. frumentaria*, the dominant form at Middridge, has not so far been reported here.

Small, winged seeds were identified by Stoneley (1958) as *Samaropsis triangularis* (Figure 7.9).

**Figure 7.9** *Samaropsis triangularis* (Geinitz) Seward. Conifer seed; Natural History Museum, London, V.6209. Marl Slate (Upper Permian), Kimberley Railway Cutting. × 3. (Photo: Photographic Studio, Natural History Museum, London.)

Some authors have suggested that these might be the seeds of *Ullmannia frumentaria* (e.g. Weigelt, 1928), but Florin (1938-1945) reported rather differently shaped seeds attached to the dwarf shoot of that species. That *U. frumentaria* foliage has yet to be found at Kimberley is further evidence against this suggestion. It is instead more likely that they are the seeds of *U. bronnii*.

Stoneley (1958) reported examples of the lobed 'cone scales' of *Pseudovoltzia liebeana* (Figure 7.10), but she illustrated them by only a diagrammatic silhouette, revealing nothing of their fine detail (if any was preserved). This is unfortunate as the detailed structure of the bract and dwarf-shoot is a key feature for justifying the position of this species within the Majonicaceae, one of the most important families of conifer found in the palaeoequatorial Permian (Clement-Westerhof, 1987, 1988).

The assemblage from Kimberley Railway Cutting is meagre compared with that of Middridge Quarry (see above) and of the classic Kupferschiefer sites in Germany (Schweitzer, 1962, 1968). It is of significance, however, as the best British locality for *Peltaspermum martinsii*. This species is the only well-documented British representative of the Peltaspermaceae, one of the very few families of land plants that survived the Palaeophytic-Mesophytic floral transition, discussed in the introduction to the present chapter.

## Conclusion

Kimberley Railway Cutting is the best British locality for the group of seed plants with fern-like fronds, known as the Peltaspermaceae. The family is now extinct, but during the Permian and the succeeding Triassic Period (290-205 Ma) it was an important element of the equatorial vegetation. These British examples are about 270 million years old. It was one of the very few families of land plant to have survived the large-scale vegetational change that occurred between the Palaeozoic and Mesozoic eras, around 250 million years ago, as discussed earlier in this chapter.

# References

Allen, J.R.L. (1979) Old Red Sandstone in external basins, with particular reference to southern Britain. In *The Devonian System* (eds M.R. House, C.T. Scrutton and M.G. Bassett), *Special Papers in Palaeontology*, **23**, 65-80.

Allen, J.R.L. (1985) Marine to fresh water: the sedimentology of the interrupted transition (Ludlow-Siegenian) in the Anglo-Welsh region. *Philosophical Transactions of the Royal Society of London*, **B 309**, 85-104.

Allen, K.C. (1961) *Lepidostrobophyllum fimbriatum* (Kidston, 1883) from the Drybrook Sandstone (Lower Carboniferous). *Geological Magazine*, **98**, 225-9.

Allen, K.C. and Dineley, D.L. (1988) Mid-Devonian to mid-Permian floral and faunal regions and provinces. In *The Caledonian-Appalachian Orogeny* (eds A.L. Harris and D.J. Fettes), *Geological Society of London, Special Publication*, **38**, 531-48.

Allen, K.C. and Marshall, J.E.A. (1986) *Svalbardia* and the 'corduroy' plant from the Devonian of the Shetland Islands, Scotland. *Special Papers in Palaeontology*, **35**, 7-20.

Alvin, K.L. (1965) A new fertile lycopod from the Lower Carboniferous of Scotland. *Palaeontology*, **8**, 281-93.

Alvin, K.L. (1966) Two cristate megaspores from the Lower Carboniferous of Scotland. *Palaeontology*, **9**, 488-91.

Amerom, H.W.J. van (1975) Die eusphenopteridischen Pteridophyllen aus der Sammlung des Geologischen Bureaus in Heerlen, unter besonderer Berücksichtigung ihrer Stratigraphie bezüglich des südlimburger Kohlenreviers. *Mededelingen van's Rijks Geologischen Dienst*, Series C III-1-7, 1-208.

Ananiev, A.R. (1957) Novie iskopaemie rasteniya iz nizhnedevonskikh otlozhenii u s. Torgashino b yugo-vostochnoi chasti zapadnoi Sibiri. *Botanichnŷi Zhurnal SSSR*, **42**, 691-702. [In Russian].

Ananiev, A.R. and Stepanov, S.A. (1969) Pervaya nakhodka psilofitoboi flori v nizhnem devone Salairskogo kryazha (Zap. Sibir'). *Treatise of the Tomsk Order of the Worker's Red Banner*, **203**, 13-28. [In Russian].

Anderson, J.M. and Anderson, H.M. (1983) Vascular plants from the Devonian to Lower Cretaceous in southern Africa. *Bothalia*, **14**, 337-44.

Anderson, J.M. and Anderson, H.M. (1985) *Palaeoflora of southern Africa. Prodromus of South African megafloras Devonian to Lower Cretaceous*, A.A. Balkema, Amsterdam.

Andrew, G. (1925) Note on the occurrence of *Pachytheca* in the Buildwas Beds (Shropshire), *Memoirs of the Manchester Literary and Philosophical Society*, **69**, 57-60.

Andrews, H.N. (1940) A new cupule from the Lower Carboniferous of Scotland. *Bulletin of the Torrey Botanical Club*, **67**, 595-601.

Andrews, H.N. (1948) Some evolutionary trends in pteridosperms. *Botanical Gazette*, **110**, 13-31.

Andrews, H.N. (1961) *Studies in Paleobotany*, John Wiley, New York.

Andrews, H.N. (1963) Early seed plants. *Science*, **142**, 925-31.

Andrews, H.N. (1980) *The fossil hunters. In search of ancient plants*, Cornell University Press, Ithaca and London.

Andrews, H.N., Kasper, A.E., Forbes, W.H., Gensel, P.G. and Chaloner, W.G. (1977) Early Devonian flora of the Trout Valley Formation, Northern Maine. *Review of Palaeobotany and Palynology*, **23**, 255-85.

# References

Andrews, H.N. and Murdy, W.H. (1958) *Lepidophloios* and ontogeny in arborescent lycopods. *American Journal of Botany*, **45**, 552-60.

Anisimova, O.I. (1979) *Flora and phytostratigraphy of the Middle Carboniferous of North Caucasus,* Naukova Dumka, Kiev [In Russian].

Arber, E.A.N. (1904a) The Silurian plants. *Proceedings of the Geologists' Association*, **18**, 458-9.

Arber, E.A.N. (1904b) The fossil flora of the Culm Measures of north-west Devon, and the palaeobotanical evidence with regard to the age of the beds. *Philosophical Transactions of the Royal Society of London*, **197**, 291-325.

Arber, E.A.N. (1912) On the fossil flora of the Forest of Dean Coalfield (Gloucestershire), and the relationships of the coalfields of the west of England and South Wales. *Philosophical Transactions of the Royal Society of London*, **202**, 233-81.

Arber, E.A.N. (1914) On the fossil floras of the Wyre Forest, with special reference to the geology of the coalfield and its relationships to the neighbouring Coal Measure areas. *Philosophical Transactions of the Royal Society of London*, **204**, 363-445.

Arber, E.A.N. (1916) On the fossil floras of the Coal Measures of south Staffordshire. *Philosophical Transactions of the Royal Society of London*, **208**, 127-55

Arber, E.A.N. (1921) *Devonian floras. A study of the origin of Cormophyta.* Cambridge University Press, Cambridge.

Arber, E.A.N. and Goode, R.H. (1915) On some fossil plants from the Devonian rocks of north Devon. *Proceedings of the Cambridge Philosophical Society*, **18**, 89-104.

Arbey, F. (1973) Les milieux de sédimentation des series ordoviciennes terminales des Mouts d'Ougarta (Sahara algérien). *Compte Rendu Hebdomadaire des Séances de l'Académie des Sciences, Paris*, **273**, 1265-7.

Archangelsky, S. (1990) Plant distribution in Gondwana during the Late Palaeozoic. In *Antarctic paleobiology* (eds T.N. Taylor and E.L. Taylor), Springer, Berlin, pp. 102-117.

Archangelsky, S. and Arrondo, O.G. (1966) Elementos floristicos del Pérmico argentino. II. *Rhacopteris chubutiana* n. sp. de la Formación Nueva Lubecka, Provincia de Chubut. *Revista Museo de La Plata*, **NS 5**, 1-15.

Archangelsky, S. and Arrondo, O.G. (1969) The Permian taphofloras of Argentina with some considerations about the presence of 'northern' elements and their possible significance. In *Gondwana stratigraphy, IUGS Symposium, Buenos Aires (Mar del Plata), October 1967.* UNESCO, pp. 197-212.

Archangelsky, S. and Arrondo, O.G. (1975) Paleogeografiá y plantas fósiles en el Pérmico inferior Austrosudamericano. *Actas I Congreso Argentina de Paleontologie y Bioestratigrafie* (Tucumán, 1974), **1**, 479-96.

Archangelsky, S., Azcuy, C.L. and Wagner, R.H. (1981) Three dwarf lycophytes from the Carboniferous of Argentina. *Scripta Geologica*, **64**, 1-35.

Archangelsky, S. and Cúneo, R. (1990) *Polyspermum*, a new Permian gymnosperm from Argentina, with considerations about the Dicranophyllales. *Review of Palaeobotany and Palynology*, **63**, 117-35.

Armstrong, M. and Paterson, I.B. (1970) The Lower Old Red Sandstone of the Strathmore Region. *Report of the Institute of Geological Sciences*, **70/12**, 1-23.

Arnold, C.A. (1952) A specimen of *Prototaxites* from the Kettle Point black shale of Ontario. *Palaeontographica*, **B 93**, 45-56.

Artis, E.T. (1825) *Antediluvian phytology, illustrated by a collection of the fossil remains of plants peculiar to the coal formations of Great Britain,* London (reprinted 1838).

Asama, K. (1984) *Gigantopteris* flora in China and south-east Asia. *Geology and palaeontology of south-east Asia*, **24**, 311-23.

Baker, J.W. and Hughes, C.P. (1979) Summer (1973) field meeting in central Wales 31 August to 7 September 1973. *Proceedings of the Geologists' Association*, **90**, 65-79.

Ball, H.W. and Dineley, D.L. (1961) The Old Red Sandstone of Brown Clee Hill and the adjacent area. 1. Stratigraphy. *Bulletin of the British Museum of Natural History (Geology)*, **5**, 177-242.

Banks, H.P. (1960) Notes on Devonian lycopods. *Senckenbergiana*, **41**, 59-88.

Banks, H.P. (1968) The early history of land plants. In *Evolution and environment* (ed. E.T. Drake), Yale University Press, New Haven, pp. 73-107.

Banks, H.P. (1972) The stratigraphic occurrence of early land plants. *Palaeontology*, **15**, 365-77.

Banks, H.P. (1973) Occurrence of *Cooksonia*, the oldest vascular land plant macrofossil, in the upper Silurian of New York State. *Journal of*

# References

the *Indian Botanical Society Golden Jubilee Volume*, **50A**, 227-35.

Banks, H.P. (1975a) The oldest vascular land plants: a note of caution. *Review of Palaeobotany and Palynology*, **20**, 13-25.

Banks, H.P. (1975b) Reclassification of Psilophyta. *Taxon*, **24**, 401-13.

Banks, H.P. (1980) Floral assemblage zones in the Siluro-Devonian. In *Biostratigraphy of fossil plants: successional and paleoecological analysis* (eds D. Dilcher and T.N. Taylor), Dowden, Hutchinson and Ross, Stroudsburg, pp. 1-24.

Banks, H.P. and Davis, M.R. (1969) *Crenaticaulis*, a new genus of Devonian plants allied to *Zosterophyllum*, and its bearing on the classification of early land plants. *American Journal of Botany*, **56**, 436-49.

Barber, C.A. (1889) The structure of *Pachytheca*. *Annals of Botany*, **3**, 141-8.

Barber, C.A. (1891) The structure of *Pachytheca*. II. *Annals of Botany*, **5**, 145-62.

Barber, C.A. (1892) *Nematophycus storrei*, nov. sp. *Annals of Botany*, **6**, 329-38.

Barghoorn, F.S. and Darrah, W.C. (1938) *Horneophyton*, a necessary change of name for *Hornea*. *Harvard University Botanical Museum Leaflets*, **6**, 142-4.

Barker, W.R. and Whittle, W.L. (1944) The Coal Measure strata of Nostell, near Wakefield. *Proceedings of the Yorkshire Geological Society*, **25**, 175-89.

Barnard, P.D.W. (1959) On *Eosperma oxroadense* gen. et sp. nov.; a new Lower Carboniferous seed from East Lothian. *Annals of Botany*, **NS 23**, 285-96.

Barnard, P.D.W. (1960a) Studies on some Lower Carboniferous plants from east Lothian. Unpublished PhD Thesis, University of London.

Barnard, P.D.W. (1960b) *Calathospermum fimbriatum* sp. nov., a Lower Carboniferous pteridosperm cupule from Scotland. *Palaeontology*, **3**, 265-75.

Barnard, P.D.W. (1962) Revision of the genus *Amyelon* Williamson. *Palaeontology*, **5**, 213-24.

Barnard, P.D.W. and Long, A.G. (1973) On the structure of a petrified stem and some associated seeds from the Lower Carboniferous rocks of East Lothian, Scotland. *Transactions of the Royal Society of Edinburgh*, **69**, 91-108.

Barnard, P.D.W. and Long, A.G. (1975) *Triradioxylon* - a new genus of Lower Carboniferous petrified stems and petioles together with a review of the classification of early Pterophytina. *Transactions of the Royal Society of Edinburgh*, **69**, 231-49.

Barthel, M. (1976) Die Rotliegendflora Sachsens. *Abhandlungen der Staatlichen Museums für Mineralogie und Geologie zu Dresden*, **24**, 1-190.

Barthel, M. (1980) *Pecopteris*-Arten E.F. von Schlotheims aus Typuslokalitaten in der DDR. *Schriftenreihe für geologisches Wissenschaft*, **16**, 275-304.

Bassett, M.G. (1984) Lower Palaeozoic Wales - a review of studies in the past 25 years. *Proceedings of the Geologists' Association*, **95**, 291-311.

Bassett, M.G. and Edwards, D. (1973) Fossil plants from Wales. *Bulletin of the National Museum of Wales*, **13**, 2-27. (Also published in 1982 as a separate revised and expanded volume by the National Museum of Wales, Geological Series 2.)

Bateman, R.M. (1988) Palaeobotany and palaeoenvironments of Lower Carboniferous floras from two volcanigenic terrains in the Scottish Midland Valley. Unpublished PhD Thesis, University of London.

Bateman, R.M. (1991) Palaeobiological and phylogenetic implications of anatomically-preserved *Archaeocalamites* from the Dinantian of Oxroad Bay and Loch Humphrey Burn, southern Scotland. *Palaeontographica*, **B 223**, 1-59.

Bateman, R.M. (1992) Morphometric reconstruction, palaeobiology and phylogeny of *Oxroadia gracilis* Alvin emend. and *O. conferta* sp. nov., anatomically-preserved rhizomorphic lycopsids from the Dinantian of Oxroad Bay, SE Scotland. *Palaeontographica*, **B 228**, 29-103.

Bateman, R.M., DiMichele, W.A. and Willard, D.A. (1992) Experimental cladistic analysis of anatomically preserved arborescent lycopsids from the Carboniferous of Euramerica: an essay in paleobotanical phylogenetics. *Annals of the Missouri Botanical Garden*, **79**, 500-59.

Bateman, R.M. and DiMichele, W.A. (in press) Heterospory: the most iterative key innovation in the evolutionary history of the Plant Kingdom. *Biological Review*, **69**.

Bateman, R.M. and Rothwell, G.W. (1990) A reappraisal of the Dinantian floras at Oxroad Bay, East Lothian, Scotland. 1. Floristics and development of whole plant concepts. *Transactions of the Royal Society of Edinburgh (Earth Sciences)*, **81**, 127-59.

Bateman, R.M. and Scott, A.C. (1990) A reappraisal of the Dinantian floras at Oxroad Bay, East Lothian, Scotland, 2. Palaeoenvironments and

# References

palaeoecology. *Transactions of the Royal Society of Edinburgh (Earth Sciences)*, **81**, 161-94.

Batenburg, L.H. (1977) The *Sphenophyllum* species in the Carboniferous flora of Holz (Westphalian D, Saar Basin, Germany). *Review of Palaeobotany and Palynology*, **24**, 69-99.

Batenburg, L.H. (1981) Vegetative anatomy and ecology of *Sphenophyllum zwickaviense, S. emarginatum*, and other 'compression species' of *Sphenophyllum. Review of Palaeobotany and Palynology*, **32**, 275-313.

Beck, C.B. (1958) *Levicaulis arranensis*, gen. et sp. nov., a lycopsid axis from the Lower Carboniferous of Scotland. *Transactions of the Royal Society of Edinburgh*, **63**, 445-57.

Beck, C.B. (1960) Connection between *Archaeopteris* and *Callixylon. Science*, **131**, 1524-5.

Beck, C.B. (1962) Reconstructions of *Archaeopteris* and further consideration of its phylogenetic position. *American Journal of Botany*, **49**, 373-82.

Beck, C.B. (1964) Predominance of *Archaeopteris* in Upper Devonian flora of Western Catskills and adjacent Pennsylvania. *Botanical Gazette*, **125**, 126-8.

Beck, C.B. (1970) The appearance of gymnospermous structure. *Biological Review*, **45**, 379-400.

Beck, C.B. (1971) On the anatomy and morphology of lateral branch systems of *Archaeopteris. American Journal of Botany*, **58**, 758-84.

Beck, C.B. (1976) Current status of the Progymnospermopsida. *Review of Palaeobotany and Palynology*, **21**, 5-23.

Bell, J., Holden, J., Pettigrew, T.H. and Sedman, K.W. (1979) The Marl Slate and Basal Permian Breccia at Middridge, County Durham. *Proceedings of the Yorkshire Geological Society*, **42**, 439-60.

Bell, W.A. (1960) Mississippian Horton Group of type Windsor-Horton District, Nova Scotia. *Memoir of the Geological Survey, Canada*, **215**.

Benson, M. (1904) *Telangium scotti* a new species of *Telangium (Calymmatotheca)* showing structure. *Annals of Botany*, **18**, 161.

Benson, M. (1908) The sporangiophore, a unit of structure in the Pteridophyta. *New Phytologist* **1**, 143-9.

Benson, M. (1911) New observations on *Botryopteris antiqua* Kidston. *Annals of Botany*, **25**, 1045-57.

Benson, M. (1914) *Sphaerostoma ovale*, a Lower Carboniferous ovule from Pettycur, Scotland. *Transactions of the Royal Society of Edinburgh*, **50**, 1-15.

Benson, M. (1922) *Heterotheca grievii*, the microsporange of *Heterangium grievii. Botanical Gazette*, **74**, 121-42.

Benson, M. (1933) New evidence of the nature of the reproductive bodies and habit of the Lower Carboniferous pteridosperms. *Proceedings of the Linnean Society, London*, **1**, 38-40.

Benson, M. (1935a) The ovular apparatus of *Sphenopteridium affine* and *bifidum* and of *Diplopteridium (Sphenopteridium) teilianum* (Walton). *New Phytologist*, **34**, 232-44.

Benson, M. (1935b) The fructification *Calathiops bernhardti*, n. sp. *Annals of Botany*, **49**, 155-60.

Bertrand, P. (1907) Principaux caractères de la fronde du *Stauropteris oldhamia* Binney. *Compte Rendu Hebdomadaire des Séances de l'Académie des Sciences, Paris*, **145**, 147-9.

Bertrand, P. (1909) Études sur la fronde des Zygoptéridées. Thesis, University of Lille.

Bertrand, P. (1911a) Structure des stipes d'*Asterochlæna laxa. Mémoires de la Société Géologique du Nord*, **7**, 1-72.

Bertrand, P. (1911b) L'étude anatomique des fougères anciennes et les problèmes qu'elle soulève. *Progressus Rei Botanicae*, **4**, 258.

Bertrand, P. (1935) Contribution à l'étude des Cladoxylées de Saalfeld. *Palaeontographica*, **B 80**, 101-70.

Besly, B.M. and Turner, P. (1983) Origin of red beds in a moist tropical climate (Etruria Formation, Upper Carboniferous, UK). In *Residual deposits* (ed. R.C.L. Wilson), *Geological Society of London, Special Publication*, **11**, 131-47.

Bhutta, A.A. (1969) Studies on the flora of the Rhynie Chert. Unpublished PhD Thesis, University of Wales, Bangor.

Bhutta, A.A. (1972) Observations on the sporangia of *Horneophyton lignieri* (Kidston and Lang) Barghoorn and Darrah, 1938. *Pakistan Journal of Botany*, **4**, 27-34.

Bhutta, A.A. (1973a) On the spores (including germinating spores) of *Horneophyton (Hornea) lignieri* (Kidston and Lang) Barghoorn and Darrah, 1938. *Pakistan Journal of Botany*, **5**, 45-55.

Bhutta, A.A. (1973b) On the spores (including germinating spores) of *Rhynia major* Kidston and Lang. *Biologia*, **19**, 47-57.

Binney, E.W. (1871) Observations on the structure of fossil plants found in the Carboniferous strata. Part II. *Lepidostrobus* and some allied cones. *Palaeontographical Society (Monographs)* [for 1870], 33-62.

Black, G.P. (1988) Geological conservation: a review of past problems and future promise. *Special Papers in Palaeontology*, **40**, 105-11.

Boersma, M. (1972) The heterogeneity of the form-genus *Mariopteris* Zeiller. A comparative morphological study with special reference to the frond composition of west-European species. Thesis, Rijksuniversiteit te Utrecht.

Boullard, B. and Lemoigne, Y. (1971) Les champignons endophytes du *Rhynia gwynne-vaughanii* K. et L.: Étude morphologique et deductions sur leur biologie. *Botaniste*, **54**, 49-89.

Boulter, M.C. (1968) On a species of compressed lycopod sporophyll from the Upper Coal Measures of Somerset. *Palaeontology*, **11**, 445-57.

Boulter, M.C., Spicer, R.A. and Thomas, B.A. (1988) Patterns of plant extinction from some palaeobotanical evidence. In *Extinction and survival in the fossil record* (ed. G.P. Larwood), *Systematics Association Special Volume*, **34**, 1-36.

Boureau, E. (1953) Sur la présence d'une flore carbonifère dans l'Air (Sahara central). *Bulletin de la Société Géologique de France*, **6**(3), 293-8.

Boureau, E. and Doubinger, J. (1975) *Traite de paléobotanique IV (2), Pteridophylla (première partie)*, Masson et Cie, Paris.

Boureau, E., Jovet-ast, S., Høeg, O.A. and Chaloner, W.G. (1967) *Traité de paléobotanique. Tome II. Bryophyta, Psilophyta, Lycophyta.* Masson et Cie, Paris.

Boureau, E., Lejal-nicol, A. and Massa, D. (1978) A propos du Silurien et du Dévónien en Libye. Il faut reporter au Silurien la date d'apparition des plantes vasculaires. *Compte Rendu Hebdomadaire des Séances de l'Académie des Sciences, Paris*, **286D**, 1567-71.

Bouroz, A. and Doubinger, J. (1977) Report on the Stephanian-Autunian boundary and on the contents of upper Stephanian and Autunian in their stratotypes. In *Symposium on Carboniferous stratigraphy* (eds V.M. Holub and R.H. Wagner), Geological Survey, Prague, pp. 147-169.

Bouroz, A., Einor, O.L., Gordon, M., Meyen, S.V. and Wagner, R.H. (1978) Proposals for an international chronostratigraphic classification of the Carboniferous. *Compte Rendu 8e Congrès International de Stratigraphie et de Géologie Carbonifère* (Moscow, 1975), **1**, 36-69.

Bower, F.O. (1908) *The origin of a land flora. A theory based upon the facts of alteration*, Macmillan and Co, London.

Bower, F.O. (1920) The earliest known land flora. *Report of the Royal Institute of Great Britain for 1920*, 133-146.

Bowman, J.E. (1837) On the bone cave in Carboniferous limestone at Cefn in Denbighshire. *Report of the British Association for the Advancement of Science for 1836*, 88-90.

Brack, S.D. (1970) On a new structurally preserved arborescent lycopsid fructification from the Lower Pennsylvanian of North America. *American Journal of Botany*, **57**, 317-30.

Brack-Hanes, S.D. and Thomas, B.A. (1983) A re-examination of *Lepidostrobus* Brongniart. *Botanical Journal of the Linnean Society*, **86**, 125-33.

Bridge, J.S., Veen, P.M. van and Matten, L.C. (1980) Aspects of the sedimentology, palynology and palaeobotany of the Upper Devonian of southern Kerry Head, County Kerry, Ireland. *Geological Journal*, **15**, 143-70.

Brodie, P.B. (1869) On the occurrence of remains of *Eurypterus* and *Pterygotus* in the Upper Silurian rocks in Herefordshire. *Quarterly Journal of the Geological Society of London*, **25**, 235-7.

Brodie, P.B. (1871) On the 'Passage-Beds' in the neighbourhood of Woolhope, Herefordshire, and on the discovery of a new species of *Eurypterus*, and some new land-plants in them. *Quarterly Journal of the Geological Society of London*, **27**, 256-61.

Brousmiche, C. (1982) Sur la synonymie de *Crossotheca boulayi* Zeiller, 1886-88 et *Crossotheca bourozii* Danzé, 1956 avec l'espèce-type du genre: *Crossotheca crepinii* Zeiller, 1883. Une nouvelle interprétation de la fructification. *Geobios*, **15**, 679-703.

Brousmiche, C. (1983) Les fougères sphénoptéridiennes du Bassin Houiller Sarro-Lorraine. *Publication Société Géologique du Nord*, **10**, 1-480.

Brown, R. (1935) Note on the Glenarbuck beds with fossil plants. Unpublished manuscript in Botany Departmental Library, Glasgow University.

Bureau, E. (1913-1914) Bassin de la Basse Loire. II. Description des flores fossiles. *Études des Gîtes Minéraux de la France* (Atlas 1913, Text 1914).

Burgess, N.D. (1991) Silurian cryptospores and miospores from the type Llandovery area, south-west Wales. *Palaeontology*, **34**, 575-99.

# References

Burgess, N.D. and Edwards, D. (1988) A new Palaeozoic plant closely allied to *Prototaxites* Dawson. *Botanical Journal of the Linnean Society*, **97**, 189-203.

Burgess, N.D. and Richardson, J.B. (1991) Silurian cryptospores and miospores from the type Wenlock area, Shropshire, England. *Palaeontology*, **34**, 601-28.

Calder, M.G. (1934) Notes on the Kidston Collection of fossil plant slides. No. V. On the structure of two Lower Carboniferous lepidodendroid stems, one of the *Lepidophloios wunschianus* type and the other of the *Lepidodendron fuliginosum* type. No. VI. On the structure of two lepidodendroid stems from the Carboniferous flora of Berwickshire. *Transactions of the Royal Society of Edinburgh*, **58**, 113-24.

Calder, M.G. (1935) Further observations on the genus *Lyginorachis* Kidston. *Transactions of the Royal Society of Edinburgh*, **58**, 549-59.

Calder, M.G. (1938) On some undescribed species from the Lower Carboniferous flora of Berwickshire; together with a note on the genus *Stenomyelon* Kidston. *Transactions of the Royal Society of Edinburgh*, **59**, 309-31.

Campbell, D.H. (1895) *The structure and development of the mosses and ferns (Archegoniatæ)*, Macmillan and Co., London.

Campbell, R. (1913) The geology of south-eastern Kincardineshire. *Transactions of the Royal Society of Edinburgh*, **48**, 923-60.

Canright, J.E. (1959) Fossil plants of Indiana. *Indiana Geological Survey Report Program*, **14**, 1-45.

Carr, J.W. (1914) Permian. In *Report of an excursion to the Nottingham district* (eds J.W. Carr and H.H. Swinnerton), *Proceedings of the Geologists' Association*, **28**, 88-9.

Carruthers, R.G. (1927) The oil shales of the Lothians (3rd edn). *Memoir of the Geological Survey of Great Britain*.

Carruthers, W. (1869) On the structure of the stems of the arborescent Lycopodiaceæ of the Coal Measures. *Monthly Microscopical Journal*, **2**, 177.

Carruthers, W. (1870) On the nature of the scars in the stems of *Ulodendron*, *Bothrodendron* and *Megaphytum*. *Monthly Microscopical Journal*, **3**, 144-54.

Carruthers, W. (1873) On some lycopodiaceous plants from the Old Red Sandstone of the north of Scotland. *Journal of Botany, London*, **11**, 321-7.

Chaloner, W.G. (1953) On the megaspores of four species of *Lepidostrobus*. *Annals of Botany*, **17**, 264-73.

Chaloner, W.G. (1958) Isolated megaspore tetrads of *Stauropteris burntislandica*. *Annals of Botany*, **22**, 197-204.

Chaloner, W.G. (1960) The origin of vascular plants. *Science Progress*, **48**, 524-34.

Chaloner, W.G. (1968) The cone of *Cyclostigma kiltorkense* Haughton, from the Upper Devonian of Ireland. *Botanical Journal of the Linnean Society*, **61**, 25-36.

Chaloner, W.G. (1970) The rise of the first land plants. *Biological Reviews*, **45**, 353-77.

Chaloner, W.G. (1972) Devonian plants from Fair Isle, Scotland. *Review of Palaeobotany and Palynology*, **14**, 49-61.

Chaloner, W.G. (1986) Reassembling the whole plant fossil, and naming it. In *Systematic and taxonomic approaches in palaeobotany* (eds R.A. Spicer and B.A. Thomas), *Systematics Association Special Volume*, **31**, 79-89. Oxford University Press.

Chaloner, W.G. and Creber G.T. (1988) Fossil plants as indicators of late Palaeozoic plate positions. In *Gondwana and Tethys* (eds M.G. Audley-Charles and A. Hallam), *Geological Society of London, Special Publication*, **37**, 201-10.

Chaloner, W.G. and Creber G.T. (1990) Do fossil plants give a climatic signal? *Journal of the Geological Society of London*, **147**, 343-50.

Chaloner, W.G., Hill, A. and Rogerson, E.C.W. (1978) Early Devonian plant fossils from a southern England borehole. *Palaeontology*, **21**, 693-707.

Chaloner, W.G. and Lacey, W.S. (1973) The distribution of Late Palaeozoic floras. In *Organisms and continents through time* (ed. N.F. Hughes), *Special Papers in Palaeontology*, **12**, 271-89.

Chaloner, W.G. and Macdonald, P. (1980) *Plants invade the land*. HMSO, Edinburgh.

Chaloner, W.G. and Meyen, S.V. (1973) Carboniferous and Permian floras of the northern continents. In *Atlas of Palaeobiogeography* (ed. A. Hallam), Elsevier, Amsterdam, pp. 169-86.

Chaphekar, M. (1963) Some calamitean plants from the Lower Carboniferous of Scotland. *Palaeontology*, **6**, 408-29.

Chaphekar, M. (1965) On the genus *Pothocites* Paterson. *Palaeontology*, **8**, 107-12.

Chaphekar, M. and Alvin, K.L. (1972) On the fertile parts of the coenopterid fern *Metaclepsydropsis duplex* (Williamson). *Review of Palaeobotany and Palynology*, **14**, 63-76.

Chodat, R. (1912) Le *Bensonites fusiformis* D.H. Scott, glandes du *Stauropteris burntislandica* P. Bertrand. *Bulletin Société Botanique de Genève* (2e Sér.), **3**, 353-60.

Church, A.H. (1926) Reproductive mechanism in land flora. IV. Sporogonia. *Journal of Botany, London*, **64**, 99-103, 132-6, 172-8.

Cleal, C.J. (1978) Floral biostratigraphy of the upper Silesian Pennant Measures of South Wales. *Geological Journal*, **13**, 165-94.

Cleal, C.J. (1979) The Ravenhead Collection of fossil plants. *Amateur Geologist*, **9**, 12-23.

Cleal, C.J. (1984a) The recognition of the base of the Westphalian D Stage in Britain. *Geological Magazine*, **121**, 125-9.

Cleal, C.J. (1984b) The Westphalian D floral biostratigraphy of Saarland (Fed. Rep. Germany) and a comparison with that of South Wales. *Geological Journal*, **19**, 327-51.

Cleal, C.J. (1985) Kilmersdon Colliery coal tip. In *New sites for old. A student's guide to the geology of the east Mendips* (eds K.L. Duff, A.P. McKirdy and M.J. Harley), Nature Conservancy Council, Peterborough, pp. 61-4.

Cleal, C.J. (1986a) Plant macrofossils from the Edgehills Sandstone, Forest of Dean. *Bulletin of the British Museum of Natural History (Geology)*, **40**, 235-46.

Cleal, C.J. (1986b) Identifying plant fragments. In *Systematic and taxonomic approaches in palaeobotany* (eds R.A. Spicer and B.A. Thomas), *Systematics Association Special Volume*, **31**, Oxford University Press, 54-65.

Cleal, C.J. (1986c) Fossil plants of the Severn Coalfield and their biostratigraphical significance. *Geological Magazine*, **123**, 553-68.

Cleal, C.J. (1987a) This is the forest primaeval. *Nature, London*, **326**, 828.

Cleal, C.J. (1987b) Macrofloral biostratigraphy of the Newent Coalfield, Gloucestershire. *Geological Journal*, **22**, 207-17.

Cleal, C.J. (1988) British palaeobotanical sites. In *The use and conservation of palaeontological sites* (eds P.R. Crowther and W.A. Wimbledon), *Special Papers in Palaeontology*, **40**, 57-71.

Cleal, C.J. (1989) Evolution in hidden forests. *Nature, London*, **339**, 16.

Cleal, C.J. (1991) (ed.) *Plant fossils in geological investigation: the Palaeozoic*, Ellis Horwood, Chichester.

Cleal, C.J. (1993) Pteridophyta and Gymnospermophyta. In *The fossil record 2* (ed. M.J. Benton), Chapman and Hall, London, pp. 779-808.

Cleal, C.J. and Laveine, J.-P. (1988) The juvenile frond of the Middle Carboniferous pteridosperm *Paripteris* Gothan. *Geobios*, **21**, 245-50.

Cleal, C.J. and Shute, C.H. (1991) The Carboniferous pteridosperm frond *Neuropteris heterophylla* (Brongniart) Sternberg. *Bulletin of the British Museum of Natural History (Geology)*, **46**, 153-74.

Cleal, C.J. and Shute, C.H. (1992) Epidermal features of some Carboniferous neuropteroid fronds. *Review of Palaeobotany and Palynology*, **71**, 191-206.

Cleal, C.J., Shute, C.H. and Zodrow, E.L. (1991) A revised taxonomy for Palaeozoic neuropterid foliage. *Taxon*, **39**, 486-92.

Cleal, C.J. and Thomas, B.A. (1988) The Westphalian fossil floras from the Cattybrook Claypit, Avon (Great Britain). *Geobios*, **21**, 409-33.

Cleal, C.J. and Thomas, B.A. (1991) Lower Westphalian D fossil plants from the Nolton-Newgale Coalfield, Dyfed (Great Britain). *Geobios*, **25**, 315-22.

Cleal, C.J. and Zodrow, E.L. (1989) Epidermal structure of some medullosan *Neuropteris* foliage from the middle and upper Carboniferous of Canada and Germany. *Palaeontology*, **32**, 837-82.

Clement-Westerhof, J.A. (1987) Aspects of Permian palaeobotany and palynology. 7. The Majonicaceae, a new family of Late Permian conifers. *Review of Palaeobotany and Palynology*, **52**, 375-402.

Clement-Westerhof, J.A. (1988) Morphology and phylogeny of Paleozoic conifers. In *Origin and evolution of gymnosperms* (ed. C.B. Beck), Columbia University Press, New York, pp. 298-337.

Clouston, C. (1845) Parish of Sandwick. *The New Statistical Account of Scotland*, **15**, 41.

Cocks, L.R.M., Holland, C.H., Rickards, R.B. and Strachan, I. (1971) A correlation of Silurian rocks in the British Isles. *Geological Society of London Special Report*, **10**.

Collinson, M.E. (1986) Use of modern generic names for plant fossils. In *Systematic and taxonomic approaches in palaeobotany* (eds R.A. Spicer and B.A. Thomas), *Systematics Association Special Volume*, **31**, 91-104. Oxford University Press.

# References

Cornet, B. (1989) The reproductive morphology and biology of *Sanmiguella lewisii*, and its bearing on angiosperm evolution in the Late Triassic. *Evolution Trends in Plants*, **3**, 25-51.

Corsin, P. and Dubois, G. (1932) Caractères de la flore du Culm dinantien de Champenay, dans la haute vallée de la Bruche. *Compte Rendu Hebdomadaire des Séances de l'Académie des Sciences, Paris*, **194**, 1846-7.

Cox, A.H. and Heard, A. (1937) Week-end field meeting in the Cardiff district. *Proceedings of the Geologists' Association*, **48**, 52-60.

Crane, P.R. (1985) Phylogenetic analysis of seed plants and the origin of angiosperms. *Annals of the Missouri Botanical Garden*, **72**, 716-93.

Croft, W.N. and George, E.A. (1959) Blue-green algae from the Middle Devonian of Rhynie, Aberdeenshire. *Bulletin of the British Museum of Natural History (Geology)*, **3**, 341-53.

Croft, W.N. and Lang, W.H. (1942) The Lower Devonian flora of the Senni Beds of Monmouthshire and Breconshire. *Philosophical Transactions of the Royal Society of London*, **B 231**, 131-63.

Crookall, R. (1927) On a new species of *Sutcliffia*. *Summary of Progress of the Geological Survey of Great Britain* [for 1926], pp. 161-5.

Crookall, R. (1930) On some curious fossils from the Downtonian and Lower Old Red Sandstone of Scotland. *Proceedings of the Royal Society of Edinburgh*, **50**, 175-8.

Crookall, R. (1931a) The genus *Lyginorachis* Kidston. *Proceedings of the Royal Society of Edinburgh*, **51**, 27-34.

Crookall, R. (1931b) Petrified plants from the Pennant Rock. *Summary of Progress of the Geological Survey of Great Britain* [for 1930] (2), pp. 52-60.

Crookall, R. (1932) The stratigraphical distribution of British Lower Carboniferous plants. *Summary of Progress of the Geological Survey of Great Britain* [for 1931] (II), pp. 70-104.

Crookall, R. (1938) *The Kidston Collection of fossil plants, with an account of the life and work of Robert Kidston*, Geological Survey of Great Britain, London.

Crookall, R. (1939) Lycopodiaceous stems (?*Cyclostigma kiltorkense* Haughton) from Mitcheldean in the Forest of Dean. *Bulletin of the Geological Survey of Great Britain*, **2**, 72-7.

Crookall, R. (1955-1975) Fossil plants of the Carboniferous rocks of Great Britain [Second Section]. *Memoir of the Geological Survey of Great Britain, Palaeontology*, **4**. Pt.1 (1955), 1-84; Pt.2 (1959), 85-216; Pt.3 (1964), 217-354; Pt.4 (1966), 355-572; Pt.5 (1969), 573-792; Pt.6 (1970), 793-840; Pt.7 (1975), 841-1004.

Cross, A.T. and Hoskins, J.H. (1951) Palaeobotany of the Devonian-Mississippian black shales. *Journal of Paleontology*, **25**, 713-28.

Cummins, W.A. (1957) The Denbigh Grits; Wenlock greywackes in Wales. *Geological Magazine*, **94**, 433-51.

Daber, R. (1971) *Cooksonia* - one of the most ancient Psilophytes - widely distributed but rare. *Botanique*, **2**, 35-40.

Dalinval, A. (1960) Les *Pecopteris* du bassin houiller du Nord de la France. *Études Géologiques pour l'Atlas de Topographie Souterraine*, **1**(3), 1-222.

Danzé, J. (1955) Le genre *Crossotheca* Zeiller. *Compte Rendu Hebdomadaire des Séances de l'Académie des Sciences, Paris*, **241**, 1616-18.

Danzé, J. (1956) Contribution à l'étude des *Sphenopteris*: les fougères sphénoptéridiennes du bassin houiller du Nord de la France. *Études Géologiques pour l'Atlas de Topographie Souterraine*, **1**(2), 1-568.

Danzé, J. (1957) Découverte d'une échantillon de *Noeggerathia* dans le bassin houiller du Nord de la France. *Annales de la Société Géologiques Nord*, **77**, 197-212.

Danzé-Corsin, P. (1953) Contribution à l'étude des Marioptéridées. Les *Mariopteris* du Nord de la France. *Études Géologiques pour l'Atlas de Topographie Souterraine*, **1**(1), 1-269.

Danzé-Corsin, P. (1955) Contribution à l'étude des flores dévoniennes du Nord de la France. I. - Flore éodevonienne de Matringhem. *Annales de la Société Géologique du Nord*, **75**, 143-60.

Davies, A., McAdam, A.D. and Cameron, I.B. (1986) Geology of the Dunbar district. *Memoir of the Geological Survey of Great Britain*.

Davies, D. (1929) Correlation and palaeontology of the Coal Measures in east Glamorganshire. *Philosophical Transactions of the Royal Society of London*, **B 217**, 91-154.

Dawson, J.W. (1859) On fossil plants from the Devonian rocks of Canada. *Quarterly Journal of the Geological Society of London*, **15**, 477-88.

Dawson, J.W. (1870) The primitive vegetation of the earth. *Nature, London*, **2**, 85-8.

Dawson, J.W. (1871) *The fossil plants of the Devonian and Upper Silurian formations of Canada*, Geological Survey of Canada, Ottawa.

# References

Dawson, J.W. (1878) Notes on some Scottish Devonian plants. *Canadian Naturalist and Geologist*, **8**, 379-89.

Dawson, J.W. (1882) Notes on *Prototaxites* and *Pachytheca* discovered by Dr Hicks in the Denbighshire Grits of Corwen, N. Wales. *Quarterly Journal of the Geological Society of London*, **38**, 103-9.

Dawson, J.W. (1888) *The geological history of plants*. Kegan, Paul, Trench and Co, London.

Dawson, J.W. and Penhallow, D.P. (1891) *Parka decipiens* - Notes on specimens from the collections of James Reid, Esq., of Allan House, Blairgowrie, Scotland. *Transactions of the Royal Society of Canada*, **9**, 3-16.

Day, J.B.W. (1970) Geology of the country around Bewcastle. *Memoir of the Geological Survey of Great Britain*.

De la Beche, H.T. (1839) Report on the geology of Cornwall, Devon and west Somerset. *Memoir of the Geological Survey UK*.

Delevoryas, T. (1955) The Medullosae - structure and relationships. *Palaeontographica*, **B 97**, 114-67.

Delevoryas, T. (1962) *Morphology and evolution of fossil plants*, Holt, Rinehart and Winston, New York.

Dennis, R.L. (1974) Studies of Palaeozoic ferns: *Zygopteris* from the middle and upper Pennsylvanian of the United States. *Palaeontographica*, **B 148**, 95-136.

DiMichele, W.A. (1979) Arborescent lycopods of Pennsylvanian age coals: *Lepidophloios*. *Palaeontographica*, **B 171**, 57-77.

DiMichele, W.A. (1980) *Paralycopodites* Morey and Morey, from the Carboniferous of Euramerica - a reassessment of generic affinities and evolution of '*Lepidodendron*' *brevifolium* Williamson. *American Journal of Botany*, **67**, 1466-76.

DiMichele, W.A. and Demaris, P.J. (1987) Structure and dynamics of a Pennsylvanian-age *Lepidodendron* forest: colonizers of a disturbed swamp habitat in the Herrin (No. 6) Coal of Illinois. *Palaios*, **2**, 146-57.

DiMichele, W.A., Phillips, T.L. and Peppers, R.A. (1985) The influence of climate and depositional environment on the distribution and evolution of Pennsylvanian coal-swamp plants. In *Geological factors and the evolution of plants* (ed. B.H. Tiffney), Yale University Press, Harvard, pp. 223-56.

Dix, E. (1933) The succession of fossil plants in the Millstone Grit and the lower portion of the Coal Measures of the South Wales Coalfield (near Swansea) and a comparison with that of other areas. *Palaeontographica*, **B 78**, 158-202.

Dix, E. (1934) The sequence of floras in the Upper Carboniferous, with special reference to South Wales. *Transactions of the Royal Society of Edinburgh*, **57**, 789-821.

Dix, E. (1935) Note on the flora of the highest 'Coal Measures' of Warwickshire. *Geological Magazine*, **72**, 555-7.

Dixon, E.E.L. (1921) The geology of the South Wales Coalfield. Part XIII. The country around Pembroke and Tenby. *Memoir of the Geological Survey UK*.

Dobruskina, I.A. (1980) *Stratigraficheskoe polozhenie floronosnykh tolshch Triasa Evrazii*. Acadamy of Sciences USSR, Order of Red Banner of Labour Geological Institute, Moscow [In Russian].

Don, A.W.R. and Hickling, G. (1917) On *Parka decipiens*. *Quarterly Journal of the Geological Society of London*, **71**, 648-66.

Doubinger, J. (1956) Contribution à l'étude des flores autuno-stéphaniennes. *Mémoires de la Société Géologique de France*, **35**, 1-180.

Doubinger, J. and Alvarez Ramis, C. (1980) Nota sobre la flora de la Formación Ambó, Carbonifero inferior del Peru. *Acta II Congreso Argentinas de Paleontologie y Bioestratigraphie y I Congreso Latinoamericano Paleontalogie* (Buenos Aires, 1978), **4**, 89-101.

Doubinger, J. and Bouroz, A. (1984) Stéphanien-Autunien, Gzhélien-Assélien: zonations palynologiques et corrélations stratigraphiques. *Compte Rendu 9e Congrès International de Stratigraphie et de Géologie Carbonifère* (Washington and Urbana, 1979), **2**, 599-605.

Doubinger, J. and Krusemann, G.P. (1966) Sur la flore du Permien de la région de Lòdeve ('zone de transition' et Saxonien inférieur). *Bulletin de la Société Géologique de France*, **B7**, 541-8.

Doubinger, J. and Remy, W. (1958) Bemerkungen über *Odontopteris subcrenulata* Rost und *Odontopteris lingulata* Göppert. *Abhandlungen der deutschen Akademie der Wissenschaft Berlin, Klasse für Chemie Geologie und Biologie*, **5**, 7-14.

Douglas, J.G. and Lejal-Nicol, A. (1981) Sur les premières flores vasculaires terrestres datées du Silurien: Une comparaison entre la 'Flore à Baragwanathia' d'Australie et la 'Flore à

# References

Psilophytes et Lycophytes' d'Afrique du Nord. *Compte Rendu Hebdomadaire des Séances de l'Académie des Sciences, Paris*, **292**, 685-8.

Eames, A.J. (1936) *Morphology of vascular plants. Lower groups (Psilophytales to Filicales)*, McGraw Hill Book Co., New York.

Edwards, D. (1968) A new plant from the Lower Old Red Sandstone of South Wales. *Palaeontology*, **1**, 683-90.

Edwards, D. (1969a) Further observations on *Zosterophyllum llanoveranum* from the Lower Devonian of South Wales. *American Journal of Botany*, **56**, 201-10.

Edwards, D. (1969b) *Zosterophyllum* from the Lower Old Red Sandstone of South Wales. *New Phytologist*, **68**, 923-31.

Edwards, D. (1970a) Further observations on the Lower Devonian plant, *Gosslingia breconensis* Heard. *Philosophical Transactions of the Royal Society of London*, **B 258**, 225-43.

Edwards, D. (1970b) Fertile Rhyniophytina from the Lower Devonian of Britain. *Palaeontology*, **13**, 451-61.

Edwards, D. (1972) A *Zosterophyllum* fructification from the Lower Old Red Sandstone of Scotland. *Review of Palaeobotany and Palynology*, **14**, 77-83.

Edwards, D. (1973) Devonian floras. In *Atlas of Palaeobiogeography* (ed. A. Hallam), Elsevier, Amsterdam, pp. 105-15.

Edwards, D. (1975) Some observations on the fertile parts of *Zosterophyllum myretonianum* Penhallow from the Lower Old Red Sandstone of Scotland. *Transactions of the Royal Society of Edinburgh*, **69**, 251-65.

Edwards, D. (1976) On the systematic position of *Hicklingia edwardii* Kidston and Lang. *New Phytologist*, **76**, 173-81.

Edwards, D. (1977) A new non-calcified alga from the upper Silurian of mid-Wales. *Palaeontology*, **20**, 823-32.

Edwards, D. (1979a) A late Silurian flora from the Lower Old Red Sandstone of south-west Dyfed. *Palaeontology*, **22**, 23-52.

Edwards, D. (1979b) The early history of vascular plants based on late Silurian and early Devonian floras of the British Isles. In *The Caledonides of the British Isles - reviewed* (eds A.L. Harris, C.H. Holland and C.H. Leake), Geological Society of London, pp. 405-10.

Edwards, D. (1980a) Studies on Lower Devonian petrifactions from Britain. 1. Pyritised axes of *Hostinella* from the Brecon Beacons Quarry, Powys, South Wales. *Review of Palaeobotany and Palynology*, **29**, 189-200.

Edwards, D. (1980b) Early land floras. In *The terrestrial environment and the origin of land vertebrates* (ed. A.L. Panchen), *Systematics Association, Special Volume*, **15**, 55-85.

Edwards, D. (1981) Studies on Lower Devonian petrifactions from Britain. 2. *Sennicaulis*, a new form genus for sterile axes based on pyrite and limonite petrifactions from the Senni Beds. *Review of Palaeobotany and Palynology*, **32**, 207-26.

Edwards, D. (1982) Fragmentary non-vascular plant microfossils from the late Silurian of Wales. *Botanical Journal of the Linnean Society*, **84**, 223-56.

Edwards, D. (1984) Robert Kidston. The most professional palaeobotanist. A tribute on the 60th anniversary of his death. *Forth Naturalist and Historian*, **8**, 65-93.

Edwards, D. (1986) Preservation in early vascular plants. *Geology Today*, **2**, 176-81.

Edwards, D. (1990) Constraints on Silurian and Early Devonian phytogeographic analysis based on megafossils. In *Palaeozoic palaeogeography and biogeography* (eds W.S. McKerrow and C.R. Scotese), *Geological Society Memoir*, **12**, 233-42.

Edwards, D. and Banks, H.P. (1965) Branching in *Gosslingia breconensis*. *American Journal of Botany*, **52**, 636.

Edwards, D., Bassett, M.G. and Rogerson, E.C.W. (1979) The earliest vascular land plants: continuing the search for proof. *Lethaia*, **12**, 313-24.

Edwards, D. and Davies, E.C.W. (1976) Oldest recorded *in situ* tracheids. *Nature, London*, **263**, 494-5.

Edwards, D., Davies, K.L. and Axe, L.M. (1992) A vascular conducting strand in the early land plant *Cooksonia*. *Nature, London*, **357**, 683-5.

Edwards, D. and Edwards, D.S. (1986) A reconsideration of the Rhyniophytina Banks. In *Systematic and taxonomic approaches in palaeobotany* (eds R.A. Spicer and B.A. Thomas), *Systematics Association, Special Volume*, **31**, 199-220, Oxford University Press.

Edwards, D., Edwards, D.S. and Rayner, R. (1982) The cuticle of early vascular plants and its evolutionary significance. In *The Plant Cuticle* (eds D.F. Cutler, K.L. Alvin and C.E. Price), Academic Press, London, pp. 341-62.

Edwards, D. and Fanning, U. (1985) Evolution and environment in the late Silurian - early Devonian: the rise of the pteridophytes.

# References

*Philosophical Transactions of the Royal Society of London*, **B 309**, 147-65.

Edwards, D., Fanning, U. and Richardson, J.B. (1986) Stomata and sterome in early land plants. *Nature, London*, **323**, 438-40.

Edwards, D. and Feehan, J. (1980) Records of *Cooksonia*-type sporangia from late Wenlock strata in Ireland. *Nature, London*, **287**, 41-2.

Edwards, D., Feehan, J. and Smith, D.G. (1983) A late Wenlock flora from Co. Tipperary, Ireland. *Botanical Journal of the Linnean Society*, **86**, 19-36.

Edwards, D. and Kenrick, P. (1986) A new zosterophyll from the Lower Devonian of Wales. *Botanical Journal of the Linnean Society*, **92**, 269-83.

Edwards, D., Kenrick, P. and Carluccio, L.M. (1989) A reconsideration of cf. *Psilophyton princeps* (Croft and Lang, 1942), a zosterophyll widespread in the Lower Old Red Sandstone of South Wales. *Botanical Journal of the Linnean Society*, **100**, 293-318.

Edwards, D. and Richardson, J.B. (1974) Lower Devonian (Dittonian) plants from the Welsh Borderland. *Palaeontology*, **17**, 311-24.

Edwards, D. and Rogerson, E.C.W. (1979) New records of fertile Rhyniophytina from the late Silurian of Wales. *Geological Magazine*, **116**, 93-8.

Edwards, D. and Rose, V. (1984) Cuticles of *Nematothallus*: a further enigma. *Botanical Journal of the Linnean Society*, **88**, 35-54.

Edwards, D.S. (1973) Studies on the flora of the Rhynie Chert. Unpublished PhD Thesis, University of Wales, Cardiff.

Edwards, D.S. (1980) Evidence for the sporophytic status of the Lower Devonian plant *Rhynia gwynne-vaughanii* Kidston and Lang. *Review of Palaeobotany and Palynology*, **29**, 177-88.

Edwards, D.S. (1986) *Aglaophyton major*, a non-vascular land-plant from the Devonian Rhynie Chert. *Botanical Journal of the Linnean Society*, **93**, 173-204.

Edwards, D.S. and Lyon, A.G. (1983) Algae from the Rhynie Chert. *Botanical Journal of the Linnean Society*, **86**, 37-55.

Edwards, W.N. (1921) Note on *Parka decipiens*. *Annals and Magazine of Natural History*, **7**, 442-4.

Edwards, W.N. (1924) On the cuticular structure of the Devonian plant *Psilophyton*. *Botanical Journal of the Linnean Society*, **46**, 377-85.

Eggert, D.A. (1961) The ontogeny of Carboniferous arborescent Lycopsida. *Palaeontographica*, **B 108**, 43-92.

Eggert, D.A. (1974) The sporangium of *Horneophyton lignieri* (Rhyniophytina). *American Journal of Botany*, **61**, 405-13.

Eggert, D.A. and Taylor, T.N. (1966) Studies of Paleozoic ferns: on the genus *Tedelea* gen. nov. *Palaeontographica*, **B 118**, 52-73.

Eggert, D.A. and Taylor, T.N. (1971) *Telangiopsis* gen. nov., an Upper Mississippian pollen organ from Arkansas. *Botanical Gazette*, **132**, 30-7.

El-Saadawy, W.E.L.-S. (1966) Studies on the flora of the Rhynie Chert. Unpublished PhD Thesis, University Wales, Bangor.

El-Saadawy, W.E.L.-S. and Lacey, W.S. (1979a) Observations on *Nothia aphylla* Lyon ex. Høeg. *Review of Palaeobotany and Palynology*, **27**, 119-47.

El-Saadawy, W.E.L.-S. and Lacey, W.S. (1979b) The sporangia of *Horneophyton lignieri* (Kidston and Lang) Barghoorn and Darrah. *Review of Palaeobotany and Palynology*, **28**, 137-44.

Elles, G.L. (1900) The zonal classification of the Wenlock Shales of the Welsh Borderland. *Quarterly Journal of the Geological Society of London*, **56**, 370-414.

Elliott, G.F. (1971) A new fossil alga from the English Silurian. *Palaeontology*, **14**, 637-41.

Emberger, L. (1968) *Les plantes fossiles dans leurs rapports avec les végétaux vivants*, Masson, Paris.

Engel, B.A. (1989) S.C.C.S. Ballot results. *Newsletter on Carboniferous Stratigraphy*, **7**, 6-8.

Erben, H.K. (1964) Facies developments in the marine Devonian of the Old World. *Proceedings of the Ussher Society*, **1**, 92-118.

Erwin, D.H. (1990) End-Permian. In *Palaeobiology, a synthesis* (eds D.E.G. Briggs and P.R. Crowther), Blackwell, Oxford, pp. 187-94.

Etheridge, R. (1867) On the physical structure of west Somerset and north Devon, and on the palaeontological value of the Devonian fossils. *Quarterly Journal of the Geological Society of London*, **23**, 568-698.

Etheridge, R. (1874) On the remains of *Pterygotus* and other crustaceans from the Upper Silurian of the Pentland Hills. *Transactions of the Edinburgh Geological Society*, **2**, 314-6.

Evans, W.D. and Cox, A.H. (1956) An Old Red Sandstone-Carboniferous Limestone Series junction at Tongwynlais, north of Cardiff. *Geological Magazine*, **93**, 431-4.

Fairon-Demaret, M. (1977) A new lycophyte cone from the Upper Devonian of Belgium. *Palaeontographica*, **B 162**, 51-63.

# References

Fairon-Demaret, M. (1986a) Some uppermost Devonian megafloras: a stratigraphical review. *Annales de la Société Géologiques de Belgique*, **109**, 43-8.

Fairon-Demaret, M. (1986b) Les plantes emsiennes du Sart Tilman (Belgique), II. *Sartilmania jabachensis* (Kräusel et Weyland). *Review of Palaeobotany and Palynology*, **47**, 225-39.

Fairon-Demaret, M. and Scheckler, S.E. (1987) Typification and redescription of *Moresnetia zalesskyi* Stockmans, 1948, an early seed plant from the upper Famennian of Belgium. *Bulletin de l'Institut Royal des Sciences Naturelles de Belgique, Sciences de la Terre*, **57**, 183-99.

Fanning, U. (1987) Late Silurian - early Devonian plant assemblages in the Welsh Borderland. Unpublished PhD Thesis, University of Wales.

Fanning, U., Edwards, D. and Richardson, J.B. (1990) Further evidence for diversity in late Silurian land vegetation. *Journal of the Geological Society of London*, **147**, 725-8.

Fanning, U., Edwards, D. and Richardson, J.B. (1991) A new rhyniophytoid from the late Silurian of the Welsh Borderland. *Abhandlungen Neues Jahrbuch für Geologie und Paläontologie*, **183**, 37-47.

Fanning, U., Edwards, D. and Richardson, J.B. (1992) A diverse assemblage of early land plants from the Lower Devonian of the Welsh Borderland. *Botanical Journal of the Linnean Society*, **109**, 161-88.

Fanning, U., Richardson, J.B. and Edwards, D. (1988) Cryptic evolution in an early land plant. *Evolutionary trends in plants*, **2**, 13-24.

Fefilova, L.A. (1978) *Listostebel'nye mkhi permi evropeiskogo severa SSSR*. Komi Branch of the Geological Institute, Leningrad [In Russian].

Felix, C.J. (1954) Some American arborescent lycopod fructifications. *Annals of the Missouri Botanical Garden*, **41**, 351-94.

Filzer, P. (1948) Ein Beitrag zur ökologischen Anatomie von *Rhynia*. *Biologisches Zentralblatt*, **67**, 13-7.

Finlay, T.M. (1926) The Old Red Sandstone of Shetland. Part 1. South-eastern area. *Transactions of the Royal Society of Edinburgh*, **54**, 553-72.

Fleming, J. (1811) Mineralogical account of Papa Stour, one of the Zetland Islands. *Memoir of the Wernerian Society*, **1**, 162-75.

Fleming, J. (1831) On the occurrence of the scales of vertebrated animals in the Old Red Sandstone of Fifeshire. *Edinburgh Journal of Natural Geographical Science*, **3**, 81-6.

Florin, R. (1936) Die fossilen Ginkgophyten von Franz-Joseph-Land, nebst Erörterungen über vermeintliche Cordaitales mesozoischen Alters. *Palaeontographica*, **B 81**, 71-173.

Florin, R. (1938-1945) Die Koniferen des Oberkarbons und des unteren Perms. *Palaeontographica*, **B 85**, 1-729.

Florin, R. (1949) The morphology of *Trichopitys heteromorpha* Saporta, a seed-plant of Palaeozoic age, and the evolution of the female flowers in the Ginkgoinae. *Acta Horti Bergiana*, **15**, 80-109.

Florin, R. (1951) Evolution of cordaites and conifers. *Acta Horti Bergiana*, **15**, 285-388.

Ford, J. (1974) Palynology of the Upper Silurian and Lower Devonian of the Midland Valley of Scotland. Unpublished PhD Thesis, London University.

Fox, H. (1900) Geological notes. *Transactions of the Royal Geological Society of Cornwall*, **12**, 342-60.

Fox, H. (1901) Gunwalloe. *Transactions of the Royal Geological Society of Cornwall*, **12**, 434-7.

Fox, H. (1904) Geological notes. No. 2. Supplementary notes on the distribution of fossils on the north coast of Cornwall south of the Camel. *Transactions of the Royal Geological Society of Cornwall*, **12**, 753-9.

Frederiksen, N.O. (1972) The rise of the Mesophytic flora. *Geoscience and Man*, **4**, 17-28.

Frenguelli, J. (1944) Apuntes acerca del Paleozoico superior del noroeste argentino. *Revista Museo de La Plata* N.S. Geol. **2**, pp. 213-65.

Friend, P.F. and Williams, B.P.J. (1978) *A field guide to selected outcrop areas of the Devonian of Scotland, the Welsh Borderland and South Wales*, Palaeontologists' Association, London.

Fry, W.L. (1954) A study of the Carboniferous lycopod *Paurodendron*, gen. nov. *American Journal of Botany*, **41**, 415-28.

Galtier, J. (1964) Anatomie comparée et affinités de deux Zygoptéridacées du Carbonière inférieur. *Compte Rendu Hebdomadaire des Séances de l'Académie des Sciences, Paris*, **259**, 4764-7.

Galtier, J. (1967) Les sporanges de *Botryopteris antiqua* Kidston. *Compte Rendu Hebdomadaire des Séances de l'Académie des Sciences, Paris*, **264**, 897-900.

Galtier, J. (1968) Un nouveau type de fructification filicinéenne du Carbonifère inférieur. *Compte*

# References

*Rendu Hebdomadaire des Séances de l'Académie des Sciences, Paris*, **266**, 1004-7.

Galtier, J. (1969) Observations sur les structures foliaires et caulinaires de *Botryopteris antiqua* Kidston. *Compte Rendu Hebdomadaire des Séances de l'Académie des Sciences, Paris*, **268**, 3025-8.

Galtier, J. (1970) Recherches sur les végétaux à structure conservée du Carbonifère inférieur français. *Paléobiologie continentale*, **1**(4), 1-220.

Galtier, J. (1971) Sur les flores du Carbonifère inférieur d'Esnost et du Roannais. *Bulletin. Société d'Histoire Naturelle d'Autun*, **57**, 24-8.

Galtier, J. (1977) *Tristichia longii*, nouvelle ptéridospermale probable du Carbonifère de la Montagne Noire. *Compte Rendu Hebdomadaire des Séances de l'Académie des Sciences, Paris*, **284**, 2215-8.

Galtier, J. (1980) Données nouvelles sur la flore du Viséan d'Esnost prés d'Autun. *Bulletin. Société d'Histoire Naturelle l'Autun*, **95**, 27-33.

Galtier, J. (1981) Structures foliaires de fougères et ptéridospermales du Carbonifère inférieur et leur signification evolutive. *Palaeontographica*, **B 180**, 1-38.

Galtier, J. and Scott, A.C. (1985) Diversification of early ferns. *Proceedings of the Royal Society of Edinburgh*, **B 86**, 289-301.

Galtier, J. and Scott, A.C. (1986a) *Lyginorachis gordonii*, nouvelle Ptéridosperm probable du Carbonifère inférieur d'Ecosse. *Compte Rendu Hebdomadaire des Séances de l'Académie des Sciences, Paris*, **302**, 251-6.

Galtier, J. and Scott, A.C. (1986b) A partially permineralized *Lepidophloios* from the early Upper Carboniferous of Scotland. *Annals of Botany*, **58**, 617-26.

Gao, Z. and Thomas, B.A. (1989) A review of fossil cycad megasporophylls, with new evidence of *Crossozamia* Pomel and its associated leaves from the Lower Permian of Taiyuan, China. *Review of Palaeobotany and Palynology*, **60**, 205-23.

Garratt, M.J. (1979) New evidence for a Silurian (Ludlow) age for the earliest *Baragwanathia* flora. *Alcheringia*, **2**, 217-24.

Garratt, M.J. (1981) The earliest vascular land plants: comment on the age of the oldest *Baragwanathia* flora. *Lethaia*, **14**, 8.

Garratt, M.J., Tims, J.D., Rickards, R.B., Chambers, T.C. and Douglas, J.G. (1984) The appearance of *Baragwanathia* (Lycophytina) in the Silurian. *Botanical Journal of the Linnean Society*, **89**, 355-8.

Gastaldo, R.A. (1985) Upper Carboniferous paleoecological reconstructions: observations and reconsiderations. *Compte Rendu 10e Congrès International de Stratigraphie et de Géologie Carbonifère* (Madrid, 1983), **2**, 281-96.

Gastaldo, R.A. (1986) An explanation for lycopod configuration. 'Fossil Grove' Victoria Park, Glasgow. *Scottish Journal of Geology*, **22**, 77-83.

Gastaldo, R.A. (1987) Confirmation of Carboniferous clastic swamp communities. *Nature, London*, **326**, 869-71.

Gastaldo, R.A., Douglass, D.P. and McCarroll, S.M. (1987) Origin, characteristics, and provenance of plant macrodetritus in a Holocene crevasse splay, Mobile Delta, Alabama. *Palaios*, **2**, 229-40.

Gayer, R.A., Allen, K.C., Bassett, M.G. and Edwards, D. (1973) The structure of the Taff Gorge area, Glamorgan, and the stratigraphy of the Old Red Sandstone - Carboniferous Limestone transition. *Geological Journal*, **8**, 345-74.

Gensel, P.G. (1976) *Renalia hueberi*, a new plant from the Lower Devonian of Gaspé. *Review of Palaeobotany and Palynology*, **22**, 19-37.

Gensel, P.G. (1992) Phylogenetic relationships of the zosterophylls and lycopsids: evidence from morphology, paleoecology, and cladistic methods of inference. *Annals of the Missouri Botanical Gardens*, **79**, 450-73.

Gensel, P.G. and Andrews, H.N. (1984) *Plant life in the Devonian*, Praeger, New York.

Gensel, P.G., Andrews, H.N. and Forbes, W.H. (1975) A new species of *Sawdonia* with notes on the origin of microphylls and lateral sporangia. *Botanical Gazette*, **136**, 50-62.

George, T.N. (1970) *British Regional Geology: South Wales*, 3rd edn. Institute of Geological Sciences, HMSO, London.

George, T.N., Johnson, G.A.L., Mitchell, M., Prentice, J.E., Ramsbottom, W.H.C., Sevastopulo, G.D. and Wilson, R.B. (1976) A correlation of Dinantian rocks in the British Isles. *Geological Society of London, Special Report*, **7**, 1-87.

Gerrienne, P. (1988) Early Devonian plant remains from Marchin (north of Dinant Synclinorium, Belgium). 1. *Zosterophyllum deciduum* sp. nov. *Review of Palaeobotany and Palynology*, **55**, 317-35.

Gerrienne, P. (1990a) Les *Pachytheca* de la Gileppe et de Nonceveux (Dévonien) inférieur

# References

de Belgique). *Annals de la Société Géologique de Belgique*, **113**, 267-85.

Gerrienne, P. (1990b) Quelques spécimens de *Gosslingia breconensis* (Heard) Heard, 1927 du Dévonien inférieur de Wihéries (bord nord du Synclinorium de Dinant) Belgique. *Annals de la Société Géologique de Belgique*, **113**, 287-93.

Gerrienne, P. (1991) Les plantes fossiles du Dévonien inférieur de Marchin (bord nord du Synclinorium de Dinant, Belgique). II. *Forgesia curvata* gen, et sp. nov. *Compte Rendu Hebdomadaire des Séances de l'Académie des Sciences, Paris* Série II, **313**, 1213-9.

Gibson, G.A. (1877) The Old Red Sandstone of Shetland. A graduation thesis. Williams and Norgate, Edinburgh.

Gillespie, W.M. and Pfefferkorn, H.W. (1979) Distribution of commonly occurring plant megafossils in the proposed Pennsylvanian System stratotype. In *Proposed Pennsylvanian System stratotype, Virginia and West Virginia* (eds K.J. England, H.H. Arndt and T.W. Henry), *American Geological Institute Selected Guidebook Series*, **1**, 87-96.

Gillespie, W.M., Rothwell, G.W. and Scheckler, S.E. (1981) The earliest seeds. *Nature, London*, **293**, 462-4.

Goldring, R. (1970) The stratigraphy about the Devonian-Carboniferous boundary in the Barnstaple area of north Devon, England. *Compte Rendu 6e Congrès International de Stratigraphie et de Géologie Carbonifère* (Sheffield 1967), **2**, 807-16.

Goldring, R. (1971) Shallow water sedimentation as illustrated in the Upper Devonian Baggy Beds. *Memoirs of the Geological Society of London*, **5**, 1-80.

Good, C.W. (1981) A petrified fern sporangium from the British Carboniferous. *Palaeontology*, **24**, 483-92.

Goode, R.H. (1913) On the fossil flora of the Pembrokeshire portion of the South Wales coalfield. *Quarterly Journal of the Geological Society of London*, **69**, 252-76.

Göppert, H.R. (1864) Die fossil Flora der Permischen Formation. *Palaeontographica*, **12**, 1-316.

Gordon, W.T. (1908a) On the prothallus of *Lepidodendron veltheimianus*. *Transactions of the Botanical Society of Edinburgh*, **23**, 330-2.

Gordon, W.T. (1908b) On *Lepidophloios scottii* (a new species from the Carboniferous Sandstone Series at Pettycur, Fife), *Transactions of the Royal Society of Edinburgh*, **46**, 443-53.

Gordon, W.T. (1909) On the nature and occurrence of the plant-bearing rocks at Pettycur (Fife). *Transactions of the Edinburgh Geological Society*, **15**, 395-7.

Gordon, W.T. (1910a) On a new species of *Physostoma* from the Lower Carboniferous rocks of Pettycur (Fife). *Proceedings of the Cambridge Philosophical Society*, **15**, 395-7.

Gordon, W.T. (1910b) Note on the prothallus of *Lepidodendron veltheimianus*. *Annals of Botany*, **24**, 821-2.

Gordon, W.T. (1911a) On the structure and affinities of *Diplolabis roemeri* (Solms). *Transactions of the Royal Society of Edinburgh*, **47**, 711-36.

Gordon, W.T. (1911b) On the structure of and affinities of *Metaclepsydropsis duplex* (Williamson). *Transactions of the Royal Society of Edinburgh*, **48**, 163-90.

Gordon, W.T. (1912) On *Rhetinangium arberi*, a new genus of Cycadofilices from the Calciferous Sandstone Series. *Transactions of the Royal Society of Edinburgh*, **48**, 813-25.

Gordon, W.T. (1914) The country around Burntisland and Kirkaldy. *Proceedings of the Geologists' Association*, **25**, 34-9.

Gordon, W.T. (1935a) The genus *Pitys*, Witham, emend. *Transactions of the Royal Society of Edinburgh*, **58**, 279-311.

Gordon, W.T. (1935b) Plant life and the philosophy of geology. *Report of the British Association for the Advancement of Science* (Aberdeen, 1934), pp. 49-82.

Gordon, W.T. (1938) On *Tetrastichia bupatides* - a Carboniferous pteridosperm from East Lothian. *Transactions of the Royal Society of Edinburgh*, **59**, 351-70.

Gordon, W.T. (1941) On *Salpingostoma dasu*. A new Carboniferous seed from East Lothian. *Transactions of the Royal Society of Edinburgh*, **60**, 427-64.

Gorelova, S.G., Men'shikova, L.V. and Khalfin, L.L. (1973) Fitostratigrafiya i opredeliitel rastenii verkhnepaleozoiskikh uglenosnikh otlojenii Kuznetskogo Basseina. *Trudy S.N.I.I.G.G.I.M.Ca.*, **140**, 1-169 [In Russian].

Gosh, A.K. and Bose, A. (1952) Spores and tracheids from the Cambrian of Kashmir. *Nature, London*, **169**, 1056-7.

Gothan, W. (1913) Die oberschlesische Steinkohlenflora. 1. Teil, Farnen und farnähnlichen Gewachse. *Abhandlungen der Preussischen Geologischen Landesanstalt*, pp. 1-278.

# References

Gothan, W. (1937) Neuere Mitteilungen über die Kulmpflanzen von Rothwaltersdorf bei Neurode (Schles.). *Sitzungsberichte der Gesellschaft Naturforschende Freunde zu Berlin*, 1-7, 122-31.

Gothan, W. (1952) Der 'Florensprung' and die 'Erzgebirgische Phase' Kossmats. *Geologica*, **11**, 41-9.

Gothan, W. (1954) Pflanzengeographisches aus dem mitteleuropäischen Karbon. *Geologie*, **3**, 219-57.

Gothan, W., Leggewie, W. and Schonenfeld, W. (1959) Die Steinkohlenflora des westlichen paralischen Steinkohlenreviere Deutschlands. *Beihefte zum Geologischen Jahrbuch*, **36**, 1-90.

Gothan, W. and Nagalhard, K. (1922) Kupferschieferpflanzen aus dem niederrheinischen Zechstein. *Jahrbuch der Preussischen Geologischen Landesanstalt*, **57**, 507-13.

Gothan, W. and Weyland, H. (1954) *Lehrbuch der Paläobotanik*. Akademie, Berlin.

Gould, R.E. (1981) Palaeobotany is blooming: 1970-1979, a review. *Alcheringia*, **5**, 49-70.

Graham, R. (1935) An anatomical study of the leaves of the Carboniferous arborescent lycopods. *Annals of Botany*, **49**, 587-608.

Gray, J. (1985) The microfossil record of early land plants: advances in understanding of early terrestrialization, 1970-1984. *Philosophical Transactions of the Royal Society of London*, **B309**, 167-95.

Greguss, P. (1959) Abstract. *International Botanical Congress*, (Montreal, 1959), pp. 142.

Greuter, W., Burdet, H.M., Chaloner, W.G. *et al.* (1988) *International Code of Botanical Nomenclature*, Koeltz, Königstein.

Grierson, J.D. and Bonamo, P.M. (1979) *Leclercqia complexa*: the earliest ligulate lycopod (Middle Devonian). *American Journal of Botany*, **66**, 474-6.

'Gu and Zhi' (1974) *Palaeozoic plants from China*. Scientific Press, Beijing [In Chinese].

Guion, P.D. and Fielding, C.R. (1988) Westphalian A and B sedimentation in the Pennine Basin, UK. In *Sedimentation in a synorogenic basin complex. The Upper Carboniferous of Northwest Europe* (eds B.M. Besly and G. Kelling) Blackie, Glasgow, pp. 153-77.

Høeg, O.A. (1942) The Downtonian and Devonian flora of Spitsbergen. *Norges Svalbard-og Ishavs-undersøkalser, Skrifter*, **83**, 1-28.

Høeg, O.A., Bose, M.N. and Shukla, B.N. (1955) Some fossil plants from the Po Series of Spiti (NW Himalayas). *Palaeobotanist*, **4**, 10-13.

Hall, I.H.S. (1978) Loch Humphrey borehole. *Report of the Institute of Geological Sciences*, **79/12**, pp. 13.

Hall, T.M. (1867) On the relative distribution of fossils throughout the North Devon Series. *Quarterly Journal of the Geological Society of London*, **23**, 371-81.

Halle, T.G. (1916) Lower Devonian plants from Rörangen in Norway. *Kungliga Svenska Vetenskapsakademiens Handlingar*, **57**, 1-46.

Halstead, L.B. (1990) Cretaceous-Tertiary (terrestrial). In *Palaeobiology, a synthesis* (eds D.E.G. Briggs and P.R. Crowther), Blackwell, Oxford, pp. 203-7.

Harris, T.M. (1964) *The Yorkshire Jurassic flora. II Caytoniales, Cycadales and pteridosperms.* British Museum (Natural History), London.

Harris, W.H. (1884) Fossil plants in the Silurian formation near Cardiff. *Science Gossip*, **20**, 28-30.

Hartung, W. and Patteisky, K. (1960) Die Flora der Goniatiten-Zonen im Visé und Namur des ostsudetische Karbons. *Compte Rendu 4e Congrès International de Stratigraphie et de Géologie Carbonifère* (Heerlen, 1958), **1**, 247-62.

Harvey, R., Lyon, A.G. and Lewis, P.N. (1969) A fossil fungus from Rhynie Chert. *Transactions of the British Mycology Society*, **53**, 155-6.

Heard, A. (1926) *Psilophyton breconensis*. *Report of the British Association for the Advancement of Science* (Southampton, 1925), pp. 311-12.

Heard, A. (1927) On Old Red Sandstone plants showing structure from Brecon (South Wales). *Quarterly Journal of the Geological Society of London*, **83**, 195-207.

Heard, A. (1939) Further notes on Lower Devonian plants from South Wales. *Quarterly Journal of the Geological Society of London*, **95**, 223-9.

Heard, A. and Davies, R. (1924) The Old Red Sandstone of the Cardiff district. *Quarterly Journal of the Geological Society of London*, **80**, 489-519.

Heard, A. and Jones, J.F. (1931a) *Eohepatica dyfriensis*, a liverwort-like plant from the lower Downtonian of the Llandovery district. *Report of the British Association for the Advancement of Science* (Bristol, 1930), pp. 330-1.

Heard, A. and Jones, J.F. (1931b) A new plant (*Thallomia*) showing structure, from the Downtonian rocks of Llandovery, Carmarthenshire. *Quarterly Journal of the Geological Society of London*, **87**, 551-62.

# References

Heer, O. (1876) Ueber permischen Pflanzen von Fünfkirchen in Ungarn. *Mitteilungen aus dem Jarbuch der Königlich Ungarischen Geologischen Anstalt*, **5**, 3-18.

Hemsley, A.R. (1989) The ultrastructure of the spores of the Devonian plant *Parka decipiens*. *Annals of Botany*, **64**, 359-67.

Hemsley, A.R. (1990a) *Parka decipiens* and land plant spore evolution. *Historical Biology*, **4**, 39-50.

Hemsley, A.R. (1990b) The ultrastructure of the exine of the megaspores in two Palaeozoic seed-like structures. *Review of Palaeobotany and Palynology*, **63**, 137-52.

Hemsley, A.R., Galtier, J. and Clayton, G. (in press) Further studies on a late Tournaisian (Lower Carboniferous) flora from Loch Humphrey Burn, Scotland: spore taxonomy and ultrastructure. *Review of Palaeobotany and Palynology*.

Henderson, S.M.K. (1932) Notes on Lower Old Red Sandstone plants from Callander, Perthshire. *Transactions of the Royal Society of Edinburgh*, **57**, 277-85.

Hendriks, E.M.L. (1935) Rock succession and structure in south Cornwall: a revision. With notes on the central European facies and Variscan folding there present. *Quarterly Journal of the Geological Society of London*, **93**, 322-67.

Hendriks, E.M.L. (1966) Correlation of south and north Cornwall. *Proceedings of the Ussher Society*, **1**, 224-7.

Hendriks, E.M.L., House, M.R. and Rhodes, F.H.T. (1971) Evidence bearing on the stratigraphical succession in south Cornwall. *Proceedings of the Ussher Society*, **2**, 270-5.

Hickling, G. (1908) The Old Red Sandstone of Forfarshire, upper and lower. *Geological Magazine*, **5**, 396-408.

Hickling, G. (1912) On the geology and palaeontology of Forfarshire. *Proceedings of the Geologists' Association*, **23**, 302-11.

Hicks, H. (1869) Notes on a species of *Eophyton* (?) from the lower Arenig rocks of St Davids. *Geological Magazine*, **6**, 534-5.

Hicks, H. (1881) On the discovery of some remains of plants at the base of the Denbighshire Grits, near Corwen, North Wales. *Quarterly Journal of the Geological Society of London*, **37**, 482-96.

Hicks, H. (1882) Additional notes on the land plants from the Pen-y-glog Slate Quarry near Corwen, North Wales. *Quarterly Journal of the Geological Society of London*, **38**, 97-102.

Hind, W. (1907) Life-zones in the British Carboniferous rocks - Interim Report of Committee, with a note by A. Vaughan. *Report of the British Association for the Advancement of Science*, (York, 1906), pp. 302-13.

Hind, W. and Stobbs, J.T. (1906) The Carboniferous succession below the Coal-Measures in north Shropshire, Denbighshire and Flintshire. *Geological Magazine*, **43**, 385-400.

Hinxman, L.W. and Grant Wilson, J.S. (1902) The geology of Lower Strathspey. *Memoir of the Geological Survey UK*.

Hirmer, M. (1927) *Handbuch der Paläobotanik*, Berlin.

Hirmer, M. (1939) Die Pflanzen des Karbon und Perm und ihre stratigraphische Bedeutung. Teil 1: Einfuhrung und Unterkarbon-Flora des euramerischen Florenraumes. *Palaeontographica*, **B 84**, 45-102.

Hirmer, M. (1941) *Noeggerathia*, neuentdeckte verwandte Formen und ihre Stellung im System der Farne. *Biologia Generalis*, **15**, 134-71.

Hirmer, M. and Guthörl, P. (1940) Die Karbon-Flora des Saargebietes. 3. Filicales und Verwandte. 1. Noeggerathinae (Hirmer): *Rhacopteris* (Hirmer and Guthörl), *Palaeontographica Supplement*, **9**, 1-60.

Holden, H.S. (1960) The morphology and relationships of *Rachiopteris cylindrica*. *Bulletin of the British Museum of Natural History (Geology)*, **4**, 53-69.

Holden, H.S. (1962) The morphology of *Botryopteris antiqua*. *Bulletin of the British Museum of Natural History (Geology)*, **5**, 361-80.

Holland, C.H. and Bassett, M.G. (eds) (1989) *A global standard for the Silurian System*, National Museum of Wales, Cardiff (Geological Series, No. 9).

Holland, C.H., Lawson, J.D. and Walmsley, V.G. (1963) The Silurian rocks of the Ludlow district, Shropshire. *Bulletin of the British Museum of Natural History (Geology)*, **8**, 93-171.

Holmes, J.C. (1977) The Carboniferous fern *Psalixochlaena cylindrica* as found in Westphalian A coal balls from England. Part I. Structure and development of the cauline system. *Palaeontographica*, **B 164**, 33-75.

Holmes, J.C. (1981) The Carboniferous fern *Psalixochlaena cylindrica* as found in Westphalian A coal balls from England. Part II. The frond and fertile parts. *Palaeontographica*, **B 176**, 147-73.

# References

Holmes, J.C. (1989) Anomalous branching patterns in some fossil *Filicales*: implications in the evolution of the megaphyll and the lateral branch, habit and growth pattern. *Plant Systematics and Evolution*, **165**, 137-58.

Hooker, J.D. (1889) *Pachytheca*. *Annals of Botany*, **3**, 135-40.

Horne, J. (1917) The discovery of silicified peat beds in the Scottish Old Red Sandstone. *Scottish Geographical Magazine*, **33**, 385-92.

Horne, J. and Mackie, W. (1917) The plant-bearing cherts at Rhynie, Aberdeenshire. *Report of the British Association for the Advancement of Science* (Newcastle upon Tyne, 1916), pp. 206-16.

Horne, J. and Mackie, W. (1920a), Rhynie, Aberdeenshire. *Report of the British Association for the Advancement of Science* (Bournemouth, 1919), pp. 110-11.

Horne, J. and Mackie, W. (1920b) Old Red Sandstone rocks at Rhynie, Aberdeenshire. *Report of the British Association for the Advancement of Science* (Cardiff, 1920), pp. 261.

Hoskins, J.H. and Cross, A.T. (1951) Structure and classification of four plants from the New Albany Shale. *American Midland Naturalist*, **46**, 684-716.

Hoskins, J.H. and Cross, A.T. (1952) The petrifaction flora of the Devonian-Mississippian Black Shale. *Palaeobotanist*, **1**, 215-38.

House, M.R., Richardson, J.B., Chaloner, W.G., Allen, J.R.L., Holland, C.H. and Westoll, T.S. (1977) A correlation of Devonian rocks in the British Isles. *Geological Society of London, Special Report*, **7**, 1-110.

Howell, H.H. (1859) The geology of the Warwickshire Coalfield. *Memoir of the Geological Survey of Great Britain*.

Howse, R. (1890) Catalogue of the local fossils in the Museum of the Natural History Society. *Natural History Transactions of Northumberland*, **10**, 227-88.

Hueber, F.M. (1964) The psilophytes and their relationship to the origin of ferns. *Memoir of the Torrey Botanical Club*, **21**, 5-8.

Hueber, F.M. (1968) *Psilophyton*: the genus and the concept. In *Symposium on the Devonian System. Vol. 2.* (ed. D.H. Oswald), Alberta Society of Petroleum Geologists, Calgary, pp. 815-22.

Hueber, F.M. (1971) *Sawdonia ornata*: a new name for *Psilophyton princeps* var. *ornatum*. *Taxon*, **20**, 641-2.

Hueber, F.M. (1983) A new species of *Baragwanathia* from the Sextant Formation (Emsian) Northern Ontario, Canada. *Botanical Journal of the Linnean Society*, **86**, 57-79.

Hueber, F.M. (1992) Thoughts on the early lycopsids and zosterophylls. *Annals of the Missouri Botanical Gardens*, **79**, 474-99.

Hueber, F.M. and Banks, H.P. (1967) *Psilophyton princeps*: the search for organic connection. *Taxon*, **16**, 81-5.

Hughes, N.F. (1976) *Palaeobiology of Angiosperm Origins*. Cambridge University Press, Cambridge.

Ishchenko, T.A. (1965) Devonian flora of the Volynian-Podolian margin of the Russian platform. *Paleontologicheshii Sbornik*, **2**, 123-5 [In Russian]

Ishchenko, T.A. (1969) Kuksonievaya paleoflora v Skal'skom Gorizonte Podolii i ee stratigraficheskoe znachenie. *Geologicheshii Zhurnal*, **29**, 101-9. [In Russian].

Ishchenko, T.A. (1974) *Tirasophyton*, a new Early Devonian plant genus from Podolia. *Paleontologicheshii Zhurnal*, **1**, 112-6 [In Russian].

Ishchenko, T.A. (1975) *Pozdnesiluriiskaya flora Podolii*. Akad. Nauk. USSR, Institute of Geololoy, Kiev. [In Russian].

Ishchenko, T.A. and Shlyakov, R.N. (1979) Middle Devonian Liverworts (Marchantiidae) from Podolia. *Paleontologicheshii Zhurnal*, **13**, 369-80 [In Russian].

Jack, R.L. and Etheridge, R. (1877) On the discovery of plants in the Old Red Sandstone of the neighbourhood of Callander. *Quarterly Journal of the Geological Society of London*, **33**, 213-22.

Jacob, K., Jacob, C. and Shivastava, R.N. (1953) Spores and tracheids of vascular plants from the Vindhyan System, India: the advent of vascular plants. *Nature, London*, **172**, 166-7.

Jarzembowski, E.A. (1989) Writhlington Geological Nature Reserve. *Proceedings of the Geologists' Association*, **100**, 219-34.

Jenkins, T.B.H. (1962) The sequence and correlation of the Coal Measures of Pembrokeshire. *Quarterly Journal of the Geological Society of London*, **118**, 65-101.

Jennings, J.R. (1976) The morphology and relationships of *Rhodea*, *Telangium*, and *Heterangium*. *American Journal of Botany*, **63**, 1119-33.

Johnson, J.H. and Konishi, K. (1959) A review of Silurian (Gotlandian) algae. *Colorado School of Mines Quarterly*, **54**, 2-114.

Jongmans, W.J. (1940) Die Kohlenfelder von Gross Britannien. *Mededelingen Geologisch Bureau*

# References

*voor het Mijngebied te Heerlen* [for 1938/9], pp. 15-222.

Jongmans, W.J. (1952) Some problems on Carboniferous stratigraphy. *Compte Rendu 3e Congrès International de Stratigraphie et de Géologie Carbonifère* (Heerlen, 1951), **1**, 295-306.

Jongmans, W.J. (1956) Contribución al conocimiento de la flora carbonifera del SO de España. *Estudios Geologicos*, **12**, 19-58.

Jongmans, W.J., Gothan, W. and Darrah, W.C. (1937) Beitrage zur Kenntnis der Flora der Pocono-Schichten aus Pennsylvanien und Virginia. *Compte Rendu 2e Congrès International de Stratigraphie et de Géologie Carbonifère* (Heerlen, 1935), **1**, 423-44.

Jonker, F.P. (1976) The Carboniferous 'genera' *Ulodendron* and *Halonia* - an assessment. *Palaeontographica*, **B 157**, 97-111.

Jonker, F.P. (1979) *Prototaxites* in the Lower Devonian. *Palaeontographica*, **B 171**, 39-56.

Josten, K.-H. (1962) Die wichtigsten Pflanzen-Fossilien des Ruhrkarbons und ihre Bedeutung fur die Gliederung des Westfals. *Fortschritte in der Geologie von Rheinland und Westfalen*, **3**, 753-72.

Josten, K.-H. (1983) Die fossilen Floren im Namur des Ruhrkarbons. *Fortschritte in der Geologie von Rheinland und Westfalen*, **31**, 1-327.

Joy, K.W., Willis, A.J. and Lacey, W.S. (1956) A rapid cellulose peel technique in palaeobotany. *Annals of Botany*, **20**, 635-7.

Keeping, W. (1882) On some remains of plants, foraminifera and annelida. *Geological Magazine*, **19**, 485-91.

Keeping, W. (1883) On Silurian plants from central Wales. *Geological Magazine*, **20**, 192.

Kenrick, P. and Edwards, D. (1988a) The anatomy of Lower Devonian *Gosslingia breconensis* Heard based on pyritized axes, with some comments on the permineralization process. *Botanical Journal of the Linnean Society*, **97**, 95-123.

Kenrick, P. and Edwards, D. (1988b) A new zosterophyll from a recently discovered exposure of the Lower Devonian Senni Beds in Dyfed, Wales. *Botanical Journal of the Linnean Society*, **98**, 97-115.

Kerp, J.H.F. (1983) Aspects of Permian palaeobotany and palynology. I. *Sobernheimia jonkeri* nov. gen., nov. sp., a new fossil plant of cycadalean affinity from the Waderner Gruppe of Sobernheim. *Review of Palaeobotany and Palynology*, **38**, 173-83.

Kerp, J.H.F. (1984a) Aspects of Permian palaeobotany and palynology. III. A new reconstruction of *Lilpopia raciborskii* (Lilpop) Cornert et Schaarschmidt (Sphenopsida), *Review of Palaeobotany and Palynology*, **40**, 237-61.

Kerp, J.H.F. (1984b) Aspects of Permian palaeobotany and palynology. V. On the nature of *Asterophyllites dumasii* Zeiller, its correlation with *Calamites gigas* Brongniart and the problem concerning its sterile foliage. *Review of Palaeobotany and Palynology*, **41**, 301-17.

Kerp, J.H.F. (1986) On *Callipteris* Brongniart from the European Rotliegend basins. Thesis, University of Utrecht.

Kerp, J.H.F. and Fichter, J. (1985) Die Makrofloren des saarpfälzischen Rotliegend (? Ober-Karbon - Unter-Perm; SW-Deutschland). *Mainzer geowissenschaftliche Mitteilungen*, **14**, 159-286.

Kevan, P.G., Chaloner, W.G. and Savile, D.B.O. (1975) Interrelationships of early terrestrial arthropods and plants. *Palaeontology*, **18**, 391-417.

Kidston, R. (1882) On the affinities of the genus *Pothocites*, Paterson. *Annals and Magazine of Natural History (Series 5)*, **10**, 404-5.

Kidston, R. (1883a) Report on fossil plants collected by the Geological Survey of Scotland in Eskdale and Liddesdale. *Transactions of the Royal Society of Edinburgh*, **30**, 531-50.

Kidston, R. (1883b) On the affinities of the genus *Pothocites*, Paterson. With a description of a specimen from Glencartholm, Eskdale. *Annals and Magazine of Natural History (Series 5)*, **11**, 297-314.

Kidston, R. (1883c) On the affinities of the genus *Pothocites*, Paterson. With a description of a specimen from Glencartholm, Eskdale. *Transactions of the Botanical Society of Edinburgh*, **16**, 28-38.

Kidston, R. (1884) On a new species of *Lycopodites* Goldenberg (*L. stockii*) from the Calciferous Sandstone Series of Scotland. *Annals and Magazine of Natural History (Series 5)*, **14**, 111-7.

Kidston, R. (1885a) On the occurrence of *Lycopodites (Sigillaria) vanuxemi* Goeppert in Britain, with remarks on its affinities. *Botanical Journal of the Linnean Society*, **21**, 560-6.

Kidston, R. (1885b) On the relationship of *Ulodendron*, Lindley and Hutton, to *Lepidodendron*, Sternberg; *Bothrodendron*, Lindley and Hutton; *Sigillaria*, Brongniart; and *Rhytidodendron*, Boulay. *Annals and Magazine of Natural History (Series 5)* **16**, 123-39.

# References

Kidston, R. (1886) *Catalogue of the Palaeozoic plants in the Department of Geology and Palaeontology, British Museum (Natural History)*, British Museum (Natural History), London.

Kidston, R. (1888) Note on the nature of fossil trees found at Whiteninch. *Transactions of the Geological Society of Glasgow*, **8**, 235-6.

Kidston, R. (1889a) On some fossil plants from Teilia Quarry, Gwaenysgor, near Prestatyn, Flintshire. *Transactions of the Royal Society of Edinburgh*, **35**, 419-28.

Kidston, R. (1889b) Additional notes on some British Carboniferous lycopods. *Proceedings of the Royal Physical Society of Edinburgh*, **10**, 88-97.

Kidston, R. (1889c) Additional notes on some British Carboniferous lycopods. *Annals and Magazine of Natural History (Series 6)*, **4**, 60-7.

Kidston, R. (1889d) On the fossil plants in the Ravenhead Collection in the Free Public Library and Museum. *Transactions of the Royal Society of Edinburgh*, **35**, 391-417.

Kidston, R. (1892) Notes on some fossil plants from the Lancashire Coal Measures. *Transactions of the Manchester Geological Society*, **13**, 401-28.

Kidston, R. (1893) *Parka decipiens* - Notes on specimens from the collections of James Reid, Esq., of Allan House, Blairgowrie, Scotland. By Sir William Dawson, LL.D., F.R.S., and Professor Penhallow, B.Sc. 'Transactions of the Royal Society of Canada', Section IV., 1891, pp. 3-16, with a plate. *Annals of Scottish Natural History*, **2**, 252-4.

Kidston, R. (1894a) On the various divisions of British Carboniferous rocks as determined by their fossil floras. *Proceedings of the Royal Physical Society of Edinburgh*, **12**, 183-257.

Kidston, R. (1894b) Notes on some fossil plants from the Lancashire Coal Measures. *Transactions of the Manchester Geological Society*, **21**, 632-52.

Kidston, R. (1897) On *Cryptoxylon forfarense*, a new species of fossil plant from the Old Red Sandstone. *Proceedings of the Royal Physical Society of Edinburgh*, **13**, 102-11.

Kidston, R. (1901a) Notes on Carboniferous plants from Berwickshire. *Summary of Progress of the Geological Survey of Great Britain* [for 1900], pp. 174-5.

Kidston, R. (1901b) Carboniferous lycopods and sphenophylls. *Transactions of the Natural History Society of Glasgow*, **6**, 25-140.

Kidston, R. (1902a) Report on fossil plants from the Calciferous Sandstones of the Berwickshire border. *Summary of Progress of the Geological Survey of Great Britain* [for 1901], pp. 179-180.

Kidston, R. (1902b) The flora of the Carboniferous Period; Pt. II. *Proceedings of the Yorkshire Geological Society*, **14**, 344-99.

Kidston, R. (1903a) The fossil plants from the Canonbie Coal Field. *Summary of Progress of the Geological Survey of Great Britain* [for 1902], pp. 209-16.

Kidston, R. (1903b) The fossil plants of the Carboniferous rocks of Canonbie, Dumfriesshire, and of parts of Cumberland and Northumberland. *Transactions of the Royal Society of Edinburgh*, **40**, 741-833.

Kidston, R. (1907) Note on a new species of *Lepidodendron* from Pettycur (*Lepidodendron pettycurense*). *Proceedings of the Royal Society of Edinburgh*, **27**, 207-9.

Kidston, R. (1908) On a new species of *Dineuron* and of *Botryopteris* from Pettycur, Fife. *Transactions of the Royal Society of Edinburgh*, **46**, 361-4.

Kidston, R. (1922) Some fossil plants from the Old Red Sandstone at Rhynie. *Transactions of the Stirling Natural History and Archaeological Society*, **45**, 10-11.

Kidston, R. (1923-1925) Fossil plants of the Carboniferous rocks of Great Britain. *Memoir of the Geological Survey U.K., Palaeontology*, **2**, Pt. 1 (1923a), 1-110; Pt. 2 (1923b), 111-98; Pt. 3 (1923c), 199-274; Pt. 4 (1923d), 275-376; Pt. 5 (1924), 377-522; Pt. 6 (1925), 523-670.

Kidston, R. (1923e) On the vascular plants of the chert band of Rhynie, Aberdeenshire. *Transactions of the Edinburgh Geological Society*, **11**, 257-9.

Kidston, R., Gordon, W.T., Flett, J.S., Garwood, E.J., Horne, J. and Peach, B.N. (1917) Investigation of the Lower Carboniferous flora at Gullane. *Report of the British Association for the Advancement of Science* (Newcastle upon Tyne, 1916), p. 217.

Kidston, R. and Gwynne-Vaughan, D.T. (1912) On the Carboniferous flora of Berwickshire. I. *Stenomyleon tuedianum*. *Transactions of the Royal Society of Edinburgh*, **48**, 263-71.

Kidston, R. and Jongmans, W.J. (1917) *Calamites* of western Europe, part 1. In *Flora of the Carboniferous of the Netherlands and adjacent regions* (ed. W.J. Jongmans), Mededelingen, Rijksopsporing Delfstoffen, No. 7.

# References

Kidston, R. and Lang, W.H. (1917a) On *Rhynia gwynne-vaughanii*. *Report of the British Association for the Advancement of Science* (Newcastle upon Tyne, 1916), p. 493.

Kidston, R. and Lang, W.H. (1917b) On Old Red Sandstone plants showing structure, from the Rhynie Chert Bed, Aberdeenshire. Part I. *Rhynia gwynne-vaughanii*, Kidston and Lang. *Transactions of the Royal Society of Edinburgh*, **51**, 761-84.

Kidston, R. and Lang, W.H. (1920a) On Old Red Sandstone plants showing structure, from the Rhynie Chert Bed, Aberdeenshire. Part II. Additional notes on *Rhynia gwynne-vaughanii*, Kidston and Lang; with descriptions of *Rhynia major*, n. sp., and *Hornea lignieri*, n. g., n. sp. *Transactions of the Royal Society of Edinburgh*, **52**, 603-27.

Kidston, R. and Lang, W.H. (1920b) On Old Red Sandstone plants showing structure, from the Rhynie Chert Bed, Aberdeenshire. Part III. *Asteroxylon mackei*, Kidston and Lang. *Transactions of the Royal Society of Edinburgh*, **52**, 643-80.

Kidston, R. and Lang, W.H. (1921a) On Old Red Sandstone plants showing structure, from the Rhynie Chert Bed, Aberdeenshire. Part IV. Restorations of the vascular cryptogams, and discussion of their bearing on the general morphology of the Pteridophyta and the origin of the organisation of land-plants. *Transactions of the Royal Society of Edinburgh*, **52**, 831-54.

Kidston, R. and Lang, W.H. (1921b) On Old Red Sandstone plants showing structure, from the Rhynie Chert Bed, Aberdeenshire. Part V. The Thallophyta occurring in the peat-bed; the succession of the plants throughout a vertical section of the bed, and the conditions of accumulation and preservation of the deposit. *Transactions of the Royal Society of Edinburgh*, **52**, 855-902.

Kidston, R. and Lang, W.H. (1923a) Notes on fossil plants from the Old Red Sandstone of Scotland. I. *Hicklingia edwardi*, K. and L. *Transactions of the Royal Society of Edinburgh*, **53**, 405-7.

Kidston, R. and Lang, W.H. (1923b) On *Palaeopitys milleri*, M'Nab. *Transactions of the Royal Society of Edinburgh*, **53**, 409-17.

Kidston, R. and Lang, W.H. (1924) Notes on fossil plants from the Old Red Sandstone of Scotland. II. *Nematophyton forfarense*, Kidston sp. III. On two species of *Pachytheca* (*P. media* and *P. fasciculata*) based on the characters of the algal filaments. *Transactions of the Royal Society of Edinburgh*, **53**, 603-14.

King, W. (1850) A monograph of the Permian fossils of England. *Palaeontographical Society (Monographs)*.

King, W.W. (1925) Notes on the 'Old Red Sandstone' of Shropshire. *Proceedings of the Geologists' Association*, **36**, 385-9.

King, W.W. (1934) The Downtonian and Dittonian strata of Great Britain and north-western Europe, with a note on a new Cyathaspidian fish from the upper Downtonian rocks of Corvedale by A.S. Woodward. *Quarterly Journal of the Geological Society of London*, **90**, 526-70.

Kirkby, J.W. (1862) On the remains of fish and plants from the Upper Limestone of the Permian series of Durham. *Annals and Magazine of Natural History*, **3**, 267-9.

Kirkby, J.W. (1864) On some remains of fish and plants from the 'Upper Limestone' of the Permian series of Durham. *Quarterly Journal of the Geological Society of London*, **20**, 345-58.

Kirkby, J.W. (1867) On some remains of fish and plants from the 'Upper Limestone' of the Permian series of Durham. *Natural History Transactions of Northumberland, Durham and Newcastle upon Tyne*, **1**, 64-83.

Kirkby, J.W. and Duff, J. (1872) Notes on the geology of part of south Durham. *Natural History Transactions of Northumberland, Durham and Newcastle upon Tyne*, **4**, 151-98.

Klitzsh, E., Lejal-Nicol, A. and Massa, D. (1973) Le Siluro-Dévonien a psilophytes et lycophytes du bassin de Mourzouk (Libye). *Compte Rendu Hebdomadaire des Séances de l'Académie des Sciences, Paris*, **277**, 2465-7.

Knight, J.A. (1985) The stratigraphy of the Stephanian rocks of the Sabero Coalfield, León (NW Spain) and an investigation of the fossil flora. Part III Systematic palaeobotany; pecopterids. *Palaeontographica*, **B 197**, 1-80.

Koeniguer, J.C. (1975) Les *Prototaxites* (nematophytes) Ordoviciens et Dévoniens du Sahara central. *Actes Congrès National des Société Savantes*, (Bresançon 1974) Sciences **2**, 383-8.

Koslowski, R. and Greguss, P. (1959) Discovery of Ordovician land-plants. *Acta Palaeotologica Polonica*, **4**, 1.

Kozur, H. (1984) Carboniferous–Permian boundary in marine and continental sediments. *Compte Rendu 9e Congrès International de Stratigraphie et de Géologie Carbonifère* (Washington and Urbana, 1979), **2**, 577-86.

# References

Kräusel, R. and Weyland, H. (1926) Beiträge zur Kenntnis der Devonflora II. *Abhandlungen der Senckenbergischen Naturforschenden Gesellschaft*, **40**, 115-55.

Kräusel, R. and Weyland, H. (1929) Beiträge zur Kenntnis der Devonflora III. *Abhandlungen der Senckenbergischen Naturforschenden Gesellschaft*, **41**, 317-59.

Kräusel, R. and Weyland, H. (1930) Die Flora des deutschen Unterdevons. *Abhandlungen der Preussischen Geologischen Landesanstalt*, **131**, 1-92.

Kräusel, R. and Weyland, H. (1932) Pflanzenreste aus dem Devon II. *Senckenbergiana*, **14**, 185-90.

Kräusel, R. and Weyland, H. (1934) Algen im Deutschen Devon. *Palaeontographica*, **B 79**, 131-42.

Kräusel, R. and Weyland, H. (1935) Neue Pflanzenfunde im rheinischen Unterdevon. *Palaeontographica*, **B 83**, 172-95.

Kräusel, R. and Weyland, H. (1938) Neue Pflanzenfunde im Mitteldevon von Elberfeld. *Palaeontographica*, **83**, 172-195.

Kryshtofovich, A.N. (1953) Nakhodka plaunobrenogo pasteniya v kembrii vostochnoi Sibiri. *Doklady Akademii Nauk S.S.S.R.*, **91**, 1377-9 [In Russian].

Lacey, W.S. (1952a) Additions to the lower Carboniferous flora of North Wales. *Compte Rendu 3e Congrès International de Stratigraphie et de Géologie Carbonifère* (Heerlen, 1951), **2**, 375-7.

Lacey, W.S. (1952b) Correlation of the Lower Brown Limestone of North Wales with part of the Lower Carboniferous succession in Scotland and northern England. *International Geological Congress* (London, 1948), **10**, 18-25.

Lacey, W.S. (1952c) Additions to the Millstone Grit flora of Lancashire. *Compte Rendu 3e Congrès International de Stratigraphie et de Géologie Carbonifère* (Heerlen, 1951), **2**, 379-83.

Lacey, W.S. (1953) Scottish Lower Carboniferous plants: *Eristophyton waltoni* sp. nov. and *Endoxylon zonatum* (Kidston) Scott in Dumbartonshire. *Annals of Botany*, **17**, 579-96.

Lacey, W.S. (1962) Welsh Lower Carboniferous plants. 1. The flora of the Lower Brown Limestone in the Vale of Clwyd, North Wales. *Palaeontographica*, **B 111**, 126-60.

Lacey, W.S. (1963) Palaeobotanical techniques. In *Viewpoints in Biology*, **2**, Butterworths, London, pp. 202-43.

Lacey, W.S. (1969) Fossil bryophytes. *Biological Reviews*, **44**, 189-205.

Lacey, W.S., Joy, K.W. and Willis, A.J. (1957) Observations on the aphlebiae and megasporangia of *Stauropteris burntislandica* P. Bertrand. *Annals of Botany*, NS **21**, 621-5.

Lane, H.R., Bouckaert, J., Brenckle, P., Einor, O.L., Havlena, V., Higgins, A.C., Yang J.-Z., Manger, W.L., Nassichuk, W., Nemirovskaya, T., Owens, B., Ramsbottom, W.H.C., Reitlinger, E.A. and Weyant, M. (1985) Proposal for an international mid-Carboniferous boundary. *Compte Rendu 10e Congrès International de Stratigraphie et de Géologie Carbonifère* (Madrid, 1983), **4**, 323-39.

Lang, W.H. (1925) Contributions to the study of the Old Red Sandstone flora of Scotland. I. On plant-remains from the fish-beds of Cromarty. II. On a sporangium-bearing branch-system from the Stromness Beds. *Transactions of the Royal Society of Edinburgh*, **54**, 253-79.

Lang, W.H. (1926) Contributions to the study of the Old Red Sandstone flora of Scotland. III. On *Hostimella (Ptilophyton) thomsoni*, and its inclusion in a new genus, *Milleria*. IV. On a specimen of *Protolepidodendron* from the Middle Old Red Sandstone of Caithness. V. On the identification of the large 'Stems' in the Carmyllie Beds of the Lower Old Red Sandstone as *Nematophyton*. *Transactions of the Royal Society of Edinburgh*, **54**, 785-99.

Lang, W.H. (1927a) Contributions to the study of the Old Red Sandstone flora of Scotland. VI. On *Zosterophyllum myretonianum* Penh. and some other plant-remains from the Carmyllie Beds of the Lower Old Red Sandstone. VII. On a specimen of *Pseudosporochnus* from the Stromness Beds. *Transactions of the Royal Society of Edinburgh*, **55**, 443-55.

Lang, W.H. (1927b) The fossil plants of the Old Red Sandstone in Orkney. *Proceedings of the Orkney Natural History Society*, 1-8.

Lang, W.H. (1929) On fossil wood (*Dadoxylon hendricksi*, n. sp.) and other plant-remains from the Clay-Slates of south Cornwall. *Annals of Botany*, **43**, 663-81.

Lang, W.H. (1932) Contributions to the study of the Old Red Sandstone flora of Scotland. VIII. On *Arthrostigma*, *Psilophyton*, and some associated plant-remains from the Strathmore Beds of the Caledonian Lower Old Red Sandstone. *Transactions of the Royal Society of Edinburgh*, **57**, 491-521.

# References

Lang, W.H. (1937) On the plant remains from the Downtonian of England and Wales. *Philosophical Transactions of the Royal Society*, **B 227**, 245-91.

Lang, W.H. (1945) *Pachytheca* and some anomalous early plants (*Prototaxites, Nematothallus, Parka, Foerstia, Orvillea* n. gen.). *Journal of the Linnean Society*, **52**, 535-52.

Lang, W.H. (1952) Note on the interest of annular xylem in the Psilophytales. *Palaeobotanist*, **1**, 295-7.

Laveine, J.-P. (1967) Contribution à l'étude de la flore du terrain houillère. Les Neuroptéridées du Nord de la France. *Études Géologiques pour l'Atlas de Topographie Souterraine*, **1**(5), 1-344.

Laveine, J.-P. (1986) The size of the frond in the genus *Alethopteris* Sternberg (Pteridospermopsida, Carboniferous). *Geobios*, **19**, 49-56.

Laveine, J.-P. (1989) *Guide paléobotanique dans le terrain houiller Sarro-Lorraine.* Houillères du Bassin de Lorraine, Metz.

Lawson, J.A. and Lawson, J.D. (1976) *Geology explained around Glasgow and south-west Scotland, including Arran.* David and Charles, Newton Abbott.

Leary, R.L. and Thomas, B.A. (1989) *Lepidodendron aculeatum* with attached foliage: evidence of stem morphology and fossilization processes. *American Journal of Botany*, **76**, 283-8.

Leclercq, S. (1942) Quelques plantes fossiles recueillies dans le Dévonien inférieur des environs de Nonceveux (bordure orientale du bassin de Dinant). *Annales de la Société Géologique de Belgique*, **65**, 193-211.

Leclercq, S. (1970) Classe des Cladoxylopsida Pichi-Sermoli (1959). In *Traité de paléobotanique* (ed. E. Boureau), **4**, Masson, Paris.

Leclercq, S. and Banks, H.P. (1962) *Pseudosporochnus nodosus* sp. nov., a Middle Devonian plant with *Cladoxylon* affinities. *Palaeontographica*, **B 110**, 1-34.

Leclercq, S. and Bonamo, P.M. (1971) A study of the fructification of *Milleria (Protopteridium) thomsonii* Lang from the Middle Devonian of Belgium. *Palaeontographica*, **B 136**, 83-114.

Leitch, D., Owen, T.R. and Jones, D.G. (1958) The basal Coal Measures of the South Wales Coalfield from Llandybie to Brynmawr. *Quarterly Journal of the Geological Society of London*, **113**, 461-86.

Lele, K.M. and Walton, J. (1962a) Contributions to the knowledge of *Zosterophyllum myretonianum* Penhallow from the Lower Old Red Sandstone of Angus. *Transactions of the Royal Society of Edinburgh*, **64**, 469-75.

Lele, K.M. and Walton, J. (1962b) Fossil flora of the Drybrook Sandstone in the Forest of Dean, Gloucestershire. *Bulletin of the British Museum of Natural History (Geology)*, **7**, 135-52.

Lemoigne, Y. (1966) Sur un sporogone de bryale d'âge Dévonien. *Bulletin Mensuel de la Société Linnénne de Lyon*, **35**, 13-6.

Lemoigne, Y. (1968a) Observation d'archégones portés par des axes du type *Rhynia gwynne-vaughanii* Kidston et Lang. Existence de gamétophytes vascularisés du Dévonien. *Compte Rendu Hebdomadaire des Séances de l'Académie des Sciences, Paris*, **266**, 1655-7.

Lemoigne, Y. (1968b) Les genres *Rhynia* Kidston et Lang du dévonien et *Psilotum* Seward actuel appartiennent-ils au meme phylum? *Bulletin Société Botanique de France*, **115**, 425-44.

Lemoigne, Y. (1968c) Observation d'archégones portés par des axes vascularisés du type *Rhynia gwynne-vaughanii* Kidston et Lang. Existence de gametophytes vascularisés chez des Psilophytales du Dévonien. *Bulletin Mensuel de la Société Linnénne de Lyon*, **37**, 148-9.

Lemoigne, Y. (1969a) Organe assimilable à une anthéridie et stomates épidermiques, portés par des axes rampants du type *Rhynia gwynne-vaughanii* Kidston et Lang. *Compte Rendu Hebdomadaire des Séances de l'Académie des Sciences, Paris*, **269**, 1393-5.

Lemoigne, Y. (1969b) Contribution à la connaissance du gamétophyte *Rhynia gwynne-vaughanii* Kidston et Lang: problem des protubérances et processus de ramification. *Bulletin Mensuel de la Société Linnénne de Lyon*, **38**, 94-102.

Lemoigne, Y. (1970) Nouvelles diagnoses du genre *Rhynia* et de espèce *Rhynia gwynne-vaughanii*. *Bulletin. Société Botanique de France*, **117**, 307-20.

Lemoigne, Y. (1975) The present status of *Rhynia gwynne-vaughanii* (K. and L.) Y.L. *Palaeobotanist*, **22**, 39-46.

Lemoigne, Y. (1981) Confirmation de l'existence de gamétophytes vascularisés dans le Dévonien de Rhynie (Ecosse) et consideration sur leur nature. *Compte Rendu Hebdomadaire des Séances de l'Académie des Sciences, Paris*, **292**, 267-70.

# References

Lemoigne, Y. and Zdebska, D. (1980) Structures problematiques observées dans des axes provenant du chert Dévonien de Rhynie. *Acta Palaeobotanica*, **21**, 3-8.

Lhuyd, E. (1699) *Lithophylacii Britannici Ichnographia*, Oxford.

Li C. and Edwards, D. (1992) A new genus of early land plants with novel strobilar construction from the Lower Devonian Posongchong Formation, Yunnan Province, China. *Palaeontology*, **35**, 257-72.

Li X. (1980) The lepidophytic plants of the Cathaysia Flora in eastern Asia. *Scientia Sinica*, **23**, 634-41.

Li Z.-Z. and Cai C.-Y. (1978) A type section of Lower Devonian strata in south-west China with brief notes on the succession and correlations of its plant assemblages. *Acta Geologica Sinica*, **1**, 1-14 [In Chinese].

Lillie, D.G. (1910) On petrified plant remains from the Upper Coal Measures of Bristol. *Proceedings of the Cambridge Philosophical Society*, **15**, 411-2.

Lindley, J. and Hutton, W. (1831-1837) *The fossil flora of Great Britain*, J. Ridgway, London. Vol. 1 (1831), Vol. 2 (1833), Vol. 3 (1837).

Lindley, M. (1968) New plants from the past. *Nature, London*, **217**, 127.

Logan, K.J. and Thomas B.A. (1987) The distribution of lignin derivatives in fossil plants. *New Phytologist*, **105**, 157-73.

Long, A.G. (1959a) The fossil plants of Berwickshire: a review of past work. Part 1. *History of the Berwickshire Naturalists' Club*, **34**, 248-73.

Long, A.G. (1959b) On the structure of *Calymmatotheca kidstoni* Calder (emended) and *Genomosperma latens* gen. et sp. nov. from the Calciferous Sandstone Series of Berwickshire. *Transactions of the Royal Society of Edinburgh*, **64**, 29-44.

Long, A.G. (1960a) *Stamnostoma huttonense* gen. et sp. nov. - a pteridosperm seed and cupule from the Calciferous Sandstone Series of Berwickshire. *Transactions of the Royal Society of Edinburgh*, **64**, 201-15.

Long, A.G. (1960b) On the structure of *Samaropsis scotica* Calder (emended) and *Eurystoma angulare* gen. et sp. nov., petrified seeds from the Calciferous Sandstone Series of Berwickshire. *Transactions of the Royal Society of Edinburgh*, **64**, 261-80.

Long, A.G. (1961a) On the structure of *Deltasperma fouldenense* gen. et sp. nov., and *Camptosperma berniciense* gen. et sp. nov., petrified seeds from the Calciferous Sandstone Series of Berwickshire. *Transactions of the Royal Society of Edinburgh*, **64**, 281-95.

Long, A.G. (1961b) Some pteridosperm seeds from the Calciferous Sandstone Series of Berwickshire. *Transactions of the Royal Society of Edinburgh*, **64**, 401-19.

Long, A.G. (1962) *Tristichia ovensi* gen. et sp. nov., a protostelic Lower Carboniferous pteridosperm from Berwickshire and East Lothian, with an account of some associated seeds and cupules. *Transactions of the Royal Society of Edinburgh*, **64**, 477-88.

Long, A.G. (1963) Some specimens of *Lyginorachis papilio* Kidston associated with stems of *Pitys*. *Transactions of the Royal Society of Edinburgh*, **65**, 211-24.

Long, A.G. (1964a) Some specimens of *Stenomyelon* and *Kalymma* from the Calciferous Sandstone Series of Berwickshire. *Transactions of the Royal Society of Edinburgh*, **65**, 435-46.

Long, A.G. (1964b) On the structure of some petioles associated with *Rhetinangium* Gordon. *Transactions of the Royal Society of Edinburgh*, **66**, 1-7.

Long, A.G. (1964c) A petrified Lower Carboniferous *Lepidodendron* showing rooting organs identified with *Calamopsis* Solms-Laubach. *Transactions of the Royal Society of Edinburgh*, **66**, 35-48.

Long, A.G. (1965) On the cupule structure of *Eurystoma angulare*. *Transactions of the Royal Society of Edinburgh*, **66**, 111-28.

Long, A.G. (1966) Some Lower Carboniferous fructifications from Berwickshire, together with a theoretical account of the evolution of ovules, cupules, and carpels. *Transactions of the Royal Society of Edinburgh*, **66**, 345-75.

Long, A.G. (1967) Some specimens of *Protoclepsydropsis* and *Clepsydropsis* from the Calciferous Sandstone Series of Berwickshire. *Transactions of the Royal Society of Edinburgh*, **67**, 95-107.

Long, A.G. (1968a) Some specimens of *Mazocarpon*, *Achlamydocarpon* and *Cystosporites* from the Lower Carboniferous rocks of Berwickshire. *Transactions of the Royal Society of Edinburgh*, **67**, 359-72.

Long, A.G. (1968b) Some specimens of *Cladoxylon* from the Calciferous Sandstone

# References

Series of Berwickshire. *Transactions of the Royal Society of Edinburgh*, **68**, 45-61.

Long, A.G. (1969) *Eurystoma trigona* sp. nov., a pteridosperm ovule borne on a frond of *Alcicornopteris* Kidston. *Transactions of the Royal Society of Edinburgh*, **68**, 171-82.

Long, A.G. (1971) A new interpretation of *Lepidodendron calamopsoides* Long and *Oxroadia gracilis* Alvin. *Transactions of the Royal Society of Edinburgh*, **68**, 491-506.

Long, A.G. (1973) Presidential address delivered to the Berwickshire Naturalists' Club at Berwick, 4th October, 1972. *History of the Berwickshire Naturalists' Club*, **39**, 85-98.

Long, A.G. (1975) Further observations on some Lower Carboniferous seeds and cupules. *Transactions of the Royal Society of Edinburgh*, **69**, 267-93.

Long, A.G. (1976a) *Calathopteris heterophylla* gen. et spec. nov., a Lower Carboniferous pteridosperm bearing two types of petioles. *Transactions of the Royal Society of Edinburgh*, **69**, 329-36.

Long, A.G. (1976b) *Psalixochlaena berwickense* sp. nov., a Lower Carboniferous fern from Berwickshire. *Transactions of the Royal Society of Edinburgh*, **69**, 513-21.

Long, A.G. (1976c) Palaeobotanical reminiscences. *History of the Berwickshire Naturalists' Club*, **40**, 179-89.

Long, A.G. (1977a) Some Lower Carboniferous pteridosperm cupules bearing ovules and microsporangia. *Transactions of the Royal Society of Edinburgh*, **70**, 1-11.

Long, A.G. (1977b) Lower Carboniferous pteridosperm cupules and the origin of angiosperms. *Transactions of the Royal Society of Edinburgh*, **70**, 13-35.

Long, A.G. (1977c) Observations on Carboniferous seeds of *Mitrospermum*, *Conostoma* and *Lagenostoma*. *Transactions of the Royal Society of Edinburgh*, **70**, 37-61.

Long, A.G. (1979a) Observations on the Lower Carboniferous genus *Pitus* Witham. *Transactions of the Royal Society of Edinburgh*, **70**, 111-27.

Long, A.G. (1979b) The resemblance between the Lower Carboniferous cupules *Hydrasperma* cf. *tenuis* Long and *Sphenopteris bifida* Lindley and Hutton. *Transactions of the Royal Society of Edinburgh*, **70**, 129-37.

Long, A.G. (1984) *Oxroadopteris parvus* gen. et spec. nov.: a protostelic Lower Carboniferous pteridosperm from Oxroad Bay, East Lothian, Scotland. *Transactions of the Royal Society of Edinburgh*, **75**, 383-9.

Long, A.G. (1985) The Cupule-Carpel Theory. A defence. *Transactions of the Botanical Society of Edinburgh*, **44**, 281-5.

Long, A.G. (1986) Observations on the Lower Carboniferous lycopod *Oxroadia gracilis* Alvin. *Transactions of the Royal Society of Edinburgh*, **77**, 127-42.

Long, A.G. (1987) Observations on *Eristophyton* Zalessky, *Lyginorachis waltoni* Calder, and *Cladoxylon edromense* sp. nov. from the Lower Carboniferous Cementstone Group of Scotland. *Transactions of the Royal Society of Edinburgh*, **78**, 73-84.

Lowry, B., Lee, D. and Hébant, C. (1980) The origin of land plants: a new look at an old problem. *Taxon*, **29**, 183-97.

Lumsden, G.I., Tulloch, W., Howells, M.F. and Davies, A. (1967) The geology of the neighbourhood of Langholm. *Memoir of the Geological Survey of Great Britain*.

Lutz, J. (1933) Zur Kulmflora von Geigen bei Hof. *Palaeontographica*, **B 78**, 114-57.

Lyell, C. (1865) *Elements of geology (6th edn)*, John Murray, London.

Lyon, A.G. (1957) Germinating spores in the Rhynie Chert. *Nature, London*, **180**, 1219.

Lyon, A.G. (1962) On the fragmentary remains of an organism referable to the Nematophytales, from the Rhynie Chert, *Nematoplexus rhyniensis* gen. et sp. nov. *Transactions of the Royal Society of Edinburgh*, **65**, 79-87.

Lyon, A.G. (1964) The probable fertile region of *Asteroxylon mackei* K. and L. *Nature, London*, **203**, 1082-3.

Lyon, A.G. and Edwards, D. (1991) The first zosterophyll from the Lower Devonian Rhynie Chert, Aberdeenshire. *Transactions of the Royal Society of Edinburgh (Earth Sciences)*, **82**, 323-32.

Lyons, P.C. and Darrah, W.C. (1989) Earliest conifers of North America: upland and/or paleoclimatic indicators? *Palaios*, **4**, 480-6.

M'Nab, W.R. (1871) On the structure of a lignite from the Old Red Sandstone. *Transactions and Proceedings of the Botanical Society of Edinburgh*, **10**, 312-4.

MacGregor, M. and Walton, J. (1948) *The story of the fossil grove of Glasgow Public Parks and Botanical Gardens, Glasgow*. Glasgow D.C. Parks Department, Glasgow.

MacGregor, M. and Walton, J. (1972) *The story of the fossil grove of Glasgow Public Parks and*

# References

*Botanical Gardens, Glasgow.* (revised edition), Glasgow D.C. Parks Department, Glasgow.

Mackie, A. (1980) Sandstone quarrying in Angus – some thoughts on an old craft. *The Edinburgh Geologist*, **8**, 14-25.

Mackie, W. (1913) The rock series of Craigbeg and Ord Hill, Rhynie, Aberdeenshire. *Transactions of the Edinburgh Geological Society*, **10**, 205-36.

Malone, E.J. (1968) Devonian of the Anakie High Area, Queensland, Australia. In *International Symposium on the Devonian System* (ed. D.H. Oswald), Alberta Society of Petroleum Geologists, Calgary, pp. 93-97.

Mamay, S.H. (1957) *Biscalitheca*, a new genus of Pennsylvanian coenopterids based on its fructifications. *American Journal of Botany*, **44**, 229-39.

Mamay, S.H. and Bateman, R.M. (1991) *Archaeocalamites lazarii*, sp. nov.: the range of Archaeocalamitaceae extended from the lowermost Pennsylvanian to the mid-Lower Permian. *American Journal of Botany*, **78**, 489-96.

Mantell, G.A. (1852) On the supposed fossil eggs from the Devonian rocks of Forfarshire. *Quarterly Journal of the Geological Society of London*, **8**, 106-9.

Marshall, J.E.A. and Allen, K.C. (1982) Devonian miospore assemblages from Fair Isle, Shetland. *Palaeontology*, **25**, 277-312.

Marston, A. (1870) *A guide to the ferns and many of the rarer plants, growing around Ludlow. With a paper on the geology of the district.* Ludlow.

Martin, W., Gierl, A. and Saedler, H. (1989) Molecular evidence for pre-Cretaceous angiosperm origins. *Nature, London,* **339**, 46-8.

Matten, L.C. (1981) *Svalbardia banksii* sp. nov. from the Upper Devonian (Frasnian) of New York State. *American Journal of Botany*, **68**, 1383-91.

Matten, L.C., Lacey, W.S and Edwards, D. (1975) Discovery of one of the oldest gymnosperm floras containing cupulate seeds. *Phytologia*, **32**, 299-303.

Matten, L.C., Fine, T.I., Tanner, W.R. and Lacey, W.S. (1984a) The megagametophyte of *Hydrasperma tenuis* Long from the Upper Devonian of Ireland. *American Journal of Botany*, **71**, 1461-4.

Matten, L.C. and Lacey, W.S. (1981) Cupule organization in early seed plants. In *Geobotany II* (ed. R.C. Romans), Plenum, New York, pp. 221-34.

Matten, L.C., Lacey, W.S. and Lucas, R.C. (1980) Studies on the cupulate seed genus *Hydrasperma* Long from Berwickshire and East Lothian in Scotland and County Kerry in Ireland. *Botanical Journal of the Linnean Society*, **81**, 249-73.

Matten, L.C., Lacey, W.S., May, B.I. and Lucas, R.C. (1980) A megafossil flora from the uppermost Devonian near Ballyheigue, Co. Kerry, Ireland. *Review of Palaeobotany and Palynology*, **29**, 241-51.

Matten, L.C. and Schweitzer, H.-J. (1982) On the correct name for *Protopteridium (Rellimia) thomsonii* (fossil). *Taxon*, **31**, 322-6.

Matten, L.C., Tanner, W.R. and Lacey, W.S. (1984b) Additions to the silicified Upper Devonian/Lower Carboniferous flora from Ballyheigue, Ireland. *Review of Palaeobotany and Palynology*, **43**, 303-20.

McAdam, A.D. and Clarkson, E.N.K. (eds) (1986) *Lothian geology: an excursion guide.* Scottish Academic Press, Edinburgh.

McAdam, A.D. and Tulloch, W. (1985) Geology of the Haddington district. *Memoir of the Geological Survey UK.*

Mclean, A.C. (1973) Excursion 1: Fossil Grove. In *Excursion guide to the geology of the Glasgow District* (ed. B.J. Bluck), Geological Society of Glasgow, Glasgow.

Merker, H. (1958) Zum fehlenden Gliede der Rhynienflora. *Botaniska Notiser*, **111**, 608-18.

Merker, H. (1959) Analyse der Rhynien-Basis und Nachweis des Gametophyten. *Botaniska Notiser*, **112**, 441-52.

Merker, H. (1961) Entwurf zur Lebenskreis-Rekonstruktion der Psilophytales nebst phylogenetischen Ausblick. *Botaniska Notiser*, **114**, 88-102.

Meyen, S.V. (1966) *Kordaitovie verkhnego paleozoya severnoi evrazii.* Akademia Nauk SSSR, Moscow [In Russian].

Meyen, S.V. (1971) *Phyllotheca*-like plants from the Upper Palaeozoic flora of Angaraland. *Palaeontographica*, **B 133**, 1-33.

Meyen, S.V. (1972) Are there ligula and parichnos in Angaran Carboniferous lepidophytes? *Review of Palaeobotany and Palynology*, **14**, 149-57.

Meyen, S.V. (1976) Carboniferous and Permian lepidophytes of Angaraland. *Palaeontographica*, **B 157**, 112-57.

Meyen, S.V. (1978) Systematics, phylogeny and ecology of propteridophytes. *Trudy Moskovskogo Obshchestva Ispytatelei Prirody*, **83**, 72-84 [In Russian].

# References

Meyen, S.V. (1982) The Carboniferous and Permian floras of Angaraland. (A synthesis), *Biological Memoir*, **7**, 1-109.

Meyen, S.V. (1984) Basic features of gymnosperm systematics and phylogeny as shown by the fossil record. *Botanical Review*, **50**, 1-111.

Meyen, S.V. (1987) *Fundamentals of Palaeobotany*, Chapman & Hall, London.

Meyen, S.V. (1988) Gymnosperms of the Angara flora. In *Origin and evolution of gymnosperms* (ed. C.B. Beck), Columbia University Press, New York, pp. 338-81.

Meyer-Berthaud, B. (1986) *Melissiotheca*: a new pteridosperm pollen organ from the Lower Carboniferous of Scotland. *Botanical Journal of the Linnean Society*, **93**, 277-90.

Meyer-Berthaud, B. and Galtier, J. (1986a) Studies on a Lower Carboniferous flora from Kingswood, near Pettycur, Scotland. II. *Phacelotheca*, a new synangiate fructification of pteridosperm affinities. *Review of Palaeobotany and Palynology*, **48**, 181-98.

Meyer-Berthaud, B. and Galtier, J. (1986b) Une nouvelle fructification du Carbonifère inférieur d'Ecosse: *Burnitheca*, Filicinée ou Ptéridospermale. *Compte Rendu Hebdomadaire des Séances de l'Académie des Sciences, Paris*, **303**, 1263-8.

Mickle, J.E. (1984) Taxonomy of specimens of the Pennsylvanian-age marattialean fern *Psaronius* from Ohio and Illinois. *Illinois State Museum of Science, Paper*, **19**, 1-64.

Mickle, J.E. and Rothwell, G.W. (1979) *Bensoniotheca*, a new name for *Heterotheca* Benson. *Taxon*, **28**, 591.

Millay, M. and Taylor, T.N. (1979) Paleozoic seed fern pollen organs. *Botanical Review*, **45**, 301-75.

Miller, H. (1841) *The Old Red Sandstone; or, new walks in an old field*, A. and C. Black, Edinburgh.

Miller, H. (1849) *Footprints of the Creator or the Asterolepis of Stromness*, A. and C. Black, Edinburgh.

Miller, H. (1855) On the less-known fossil floras of Scotland. *Report of the British Association for the Advancement of Science*, (Edinburgh, 1855), pp. 83-5.

Miller, H. (1857) *The testimony of the rocks or geology in its bearing on the two theologies, natural and revealed*, Nimo, Edinburgh.

Mills, D.A.C. and Hull, J.H. (1976) Geology of the country around Barnard Castle. *Memoir of the Geological Survey UK*.

Moore, L.R. (1941) The presence of the Namurian in the Bristol District. *Geological Magazine*, **78**, 279-92.

Moore, L.R. and Trueman, A.E. (1942) The Bristol and Somerset coalfields with particular reference to their future development. *Proceedings of the South Wales Institute of Engineers*, **57**, 180-247, 303-5.

Morgan, J. (1959) The morphology and anatomy of American species of the genus *Psaronius*. *Illinois Biological Monographs*, **27**, 1-108.

Mortimer, M.G. (1967) Some Lower Devonian microfloras from southern Britain. *Review of Palaeobotany and Palynology*, **1**, 95-109.

Morton, D.J. (1976) Lower Old Red Sandstone sedimentation in the north-west Midland Valley and north Argyll areas of Scotland. Unpublished PhD Thesis, University of Glasgow.

Morton, G.H. (1871) On the Mountain Limestone of Flintshire and part of Denbighshire. *Report of the British Association for the Advancement of Science*, for 1870, 82.

Morton, G.H. (1886) The Carboniferous Limestone and Cefn-y-Fedw Sandstone of Flintshire. *Proceedings of Liverpool Geological Association*, **5**, 169-97.

Morton, G.H. (1898) The Carboniferous Limestone of the country around Llandudno (North Wales). *Quarterly Journal of the Geological Society of London*, **54**, 382-400.

Münster, G. (1842) Ueber die Fucoiden des Kupferschiefers. *Beiträge zur Petrefacten-Kunde, Bayreuth*, **5**, 100-2.

Murchison, R.I. (1839) *The Silurian System*. John Murray, London.

Murchison, R.I. (1859) On the succession of the older rocks in the northernmost counties of Scotland; with some observations on the Orkney and Shetland islands. *Quarterly Journal of the Geological Society of London*, **15**, 353-418.

Murchison, R.I. and Harkness, R. (1864) On the Permian rocks of the north-west of England and their extension into Scotland. *Quarterly Journal of the Geological Society of London*, **20**, 144-65.

Mykura, W. (1960) The replacement of coal by limestone and the reddening of Coal Measures in the Ayrshire coalfield. *Bulletin of the Geological Survey of Great Britain*, **16**, 69-109.

Mykura, W. (1965) The age of the lower part of the New Red Sandstone of south-west Scotland. *Scottish Journal of Geology*, **1**, 9-18.

Mykura, W. (1972) The Old Red Sandstone sediments of Fair Isle, Shetland Islands. *Bulletin of*

# References

the *Geological Survey of Great Britain*, **41**, 1-31.

Mykura, W. and Phemister, J. (1976) The geology of western Shetland. *Memoir of the Geological Survey UK*.

Nathorst, A.G. (1883a) On the so-called 'plant fossils' from the Silurian of central Wales. *Geological Magazine*, **20**, 33-4.

Nathorst, A.G. (1883b) On the so-called plant fossils from central Wales. *Geological Magazine*, **20**, 286-7.

Nathorst, A.G. (1915) Zur Devonflora des westlichen Norwegens. *Bergens Museums Arborog Afhandlinger og Arsberetning*, **9**, 1-34.

Neaverson, E. (1930) The Carboniferous rocks around Prestatyn, Dyserth and Newmarket (Flintshire), Historical review. *Proceedings of the Liverpool Geological Society*, **15**, 179-212.

Neaverson, E. (1945) The Carboniferous rocks between Abergele and Denbigh. *Proceedings of the Liverpool Geological Society*, **19**, 52-68.

Neuber, E. (1979) *Parka decipiens* Fleming: Grunalga oder Lebermoos? *Neues Jahrbuch für Geologie und Paläontologie, Monatsheftes* [for 1979], 681-9.

Nicholson, H.A. (1869) On the occurrence of plants in the Skiddaw Slates. *Geological Magazine*, **6**, 494-8.

Nicol, W. (1834) Observations on the structure of recent and fossil conifers. *Edinburgh New Philosophical Journal*, **16**, 137-58.

Niklas, K.J. (1976a) Morphological and ontogenetic reconstruction of *Parka decipiens* Fleming and *Pachytheca* Hooker from the Lower Old Red Sandstone, Scotland. *Transactions of the Royal Society of Edinburgh*, **69**, 483-99.

Niklas, K.J. (1976b) Chemical examinations of some non-vascular Palaeozoic plants. *Brittonia*, **28**, 113-37.

Niklas, K.J. (1976c) The chemotaxonomy of *Parka decipiens* from the Lower Old Red Sandstone, Scotland, UK. *Review of Palaeobotany and Palynology*, **21**, 202-7.

Niklas, K.J. (1981) Airflow patterns around some early seed plant ovules and cupules: implications concerning efficiency in wind pollination. *American Journal of Botany*, **68**, 635-50.

Niklas, K.J. and Banks, H.P. (1990) A reevaluation of the Zosterophyllophytina with comments on the origin of lycopods. *American Journal of Botany*, **77**, 274-83.

Niklas, K.J., Tiffney, B.H. and Knoll, A.H. (1980) Apparent changes in the diversity of fossil plants. A preliminary assessment. In *Evolutionary biology*, Vol. 12 (eds M.K. Hecht, W.C. Steere and B. Wallace), Plenum Press, New York, pp. 1-89.

North, F.J. (1931) *Coal, and the coalfields in Wales* (2nd edn), National Museum of Wales, Cardiff.

Novik, E.O. (1952) *Carboniferous flora of the European part of the USSR*. Akad. Nauk SSSR, Palaeontology [In Russian].

Oberste-Brink, K. (1913) Beitrage zur Kenntnis der Farne und farnähnliche Gewächse des Kulms von Europa. *Jahrbuch der Preussischen Geologischen Landesanstalt*, **35**, 63-153.

Obrhel, J. (1959) Ein Landpflanzfund im mittelböhmischen Ordovizium. *Geologie*, **8**, 535-41.

Obrhel, J. (1962) Die Flora der Pridoli-Schichten (Budnany-Stufe) des mittelböhmischen Silurs. *Geologie*, **11**, 83-97.

Obrhel, J. (1968) Die Silur- und Devonflora des Barrandiums. *Paläontologisches Abhandlungen (B)*, **4**, 635-793.

Owen, T.R., Rhodes, F.H.T., Jones, D.G. and Kelling, G. (1965) Summer (1964) field meeting in South Wales. *Proceedings of the Geologists' Association*, **76**, 463-95.

Pal, A.K. and Chaloner, W.G. (1979) A Lower Carboniferous *Lepidodendropsis* flora in Kashmir. *Nature, London*, **281**, 295-7.

Pant, D.D. (1962) The gametophyte of the Psilophytales. *Proceedings of the Summer School on Botany, Darjeeling 1960*, pp. 276-301.

Pant, D.D. and Walton, J. (1961) *Lycostachys protostelicus* gen. et sp. nov. and some associated megaspores from the Lower Carboniferous of Scotland. *Palaeontographica*, **B 108**, 1-10.

Paproth, E. (1980) The Devonian-Carboniferous boundary. *Lethaia*, **13**, 287.

Paterson, R. (1841) Description of *Pothocites grantonii*, a new fossil vegetable. *Transactions of the Botanical Society of Edinburgh*, **1**, 45.

Patteisky, K. (1929) *Die Geologie und Fossilfuhrung der mährisch-schlesischen Dachschiefer und Grauwacken Formation*. Naturwissenschaftliche Verein, Troppau.

Patteisky, K. (1957) Die phylogenetische Entwicklung der Arten von *Lyginopteris* und ihre Bedeutung fur die Stratigraphie. *Mitteilungen des Westfalischen Bergwerkschaftsklasse, Bochum*, **12**, 59-83.

Peach, B.N. and Horne, J. (1903) The Canonbie Coalfield: its geological structure and relations to the Carboniferous rocks of the north of

# References

England and central Scotland. *Transactions of the Royal Society of Edinburgh*, **40**, 835-77.

Peach, B.N. and Horne, J. (1910) The geology of the neighbourhood of Edinburgh (2nd edn), *Memoir of the Geological Survey UK*.

Peach, C.W. (1877) Notes on the fossil plants found in the Old Red Sandstone of Shetland, Orkney, Caithness, Sutherland, and Forfarshire. *Transactions of the Edinburgh Geological Society*, **3**, 148-52.

Penhallow, D.P. (1892) Additional notes on Devonian plants of Scotland. *Canadian Record of Science*, **5**, 1-13.

Petrosyan, N.M. (1968) Stratigraphic importance of the Devonian flora of the USSR. In *International Symposium on the Devonian System* (ed. D.H. Oswald), Alberta Society of Petroleum Geologists, Calgary, pp. 579-586.

Pettitt, J.M. (1969) Pteridophytic features in some Lower Carboniferous seed megaspores. *Botanical Journal of the Linnean Society*, **62**, 233-9.

Pettitt, J.M. and Beck, C.B. (1968) *Archaeosperma arnoldii* - a cupulate seed from the Upper Devonian of North America. *Contributions from the Museum of Paleontology. University of Michigan*, **22**, 139-54.

Pfefferkorn, H.W. (1976) Pennsylvanian tree fern compressions *Caulopteris*, *Megaphyton* and *Artisophyton* gen. nov. in Illinois. *Illinois State Geological Survey Circular*, **492**, 11-31.

Phillips, J. (1848) The Malvern Hills, compared with the Palaeozoic districts of Abberley, Woolhope, May Hill, Tortworth and Usk. *Memoir of the Geological Survey UK*, **2**, 1-330.

Phillips, J. and Salter, J.W. (1848) Palaeontological appendix to Professor John Phillips' memoir on the Malvern Hills compared with the Palaeozoic districts of Abberley, etc. *Memoir of the Geological Survey UK*, **2**, 331-86.

Phillips, T.L. (1979) Reproduction of heterosporous arborescent lycopods in the Mississippian-Pennsylvanian of Euramerica. *Review of Palaeobotany and Palynology*, **27**, 239-89.

Phillips, T.L. (1980) Stratigraphic and geographic occurrences of permineralized coal-swamp plants - Upper Carboniferous of North America and Europe. In *Biostratigraphy of fossil plants* (eds D.L. Dilcher and T.N. Taylor), Dowden, Hutchinson and Ross, Stroudsburg, pp. 25-92.

Phillips, T.L., Andrews, H.N. and Gensel, P.G. (1972) Two heterosporous species of *Archaeopteris* from the Upper Devonian of West Virginia. *Palaeontographica*, **B 139**, 47-71.

Phillips, T.L. and DiMichele, W.A. (1992) Comparative ecology and life-history biology of arborescent lycopsids in Late Carboniferous swamps of Euramerica. *Annals of the Missouri Botanical Garden*, **79**, 560-88.

Pirozynski, K.A. (1981) Interactions between fungi and plants through the ages. *Canadian Journal of Botany*, **59**, 1824-7.

Pirozynski, K.A. and Malloch, D.W. (1975) The origin of land plants: a matter of mycotrophism. *Biosystems*, **6**, 153-64.

Poort, R.J. and Kerp, J.H.F. (1990) Aspects of Permian palaeobotany and palynology. XI. On the recognition of true peltasperms in the Upper Permian of western and central Europe and a reclassification of species formerly included in *Peltaspermum* Harris. *Review of Palaeobotany and Palynology*, **63**, 197-225.

Potter, J.F. and Price, J.H. (1965) Comparative sections through rocks of Ludlovian - Downtonian age in the Llandovery and Llandeilo districts. *Proceedings of the Geologists' Association*, **76**, 379-402.

Pratt, L.M., Phillips, T.L. and Dennison, J.M. (1978) Evidence of non-vascular land plants from the early Silurian (Llandoverian) of Virginia, USA. *Review of Palaeobotany and Palynology*, **25**, 121-49.

Ramsbottom, W.H.C., Calver, M.A., Eagar, R.M.C., Hodson, F., Holliday, D.W., Stubblefield, C.J. and Wilson, R.B. (1978) A correlation of Silesian rocks in the British Isles. *Geological Society of London, Special Report*, **10**, 1-81.

Ramsbottom, W.H.C., Saunders, W.B. and Owens, B. (eds) (1982) *Biostratigraphic data for a mid-Carboniferous boundary*. IUGS Subcommission on Carboniferous Stratigraphy, Leeds.

Raup, D.M. and Sepkoski, J.J. (1986) Periodic extinctions of families and genera. *Science*, **231**, 833-6.

Raymond, A. (1985) Floral diversity, phytogeography and climatic amelioration during the early Carboniferous (Dinantian). *Paleobiology*, **11**, 293-309.

Raymond, A. (1987) Paleogeographic distribution of Early Devonian plant traits. *Palaios*, **2**, 113-32.

Raymond, A., Parker, W.C. and Barrett, S.F. (1985) Early Devonian phytogeography. In *Geological factors and the evolution of plants* (ed. H. Tiffney), Yale University Press, pp. 129-67.

# References

Raymond, A. and Parrish, J.T. (1985) Phytogeography and paleoclimate of the Early Carboniferous. In *Geological factors and the evolution of plants* (ed. B.H. Tiffney), Yale University Press, pp. 169-222.

Rayner, D.H. (1963) The Achanarras Limestone of the Middle Old Red Sandstone, Caithness, Scotland. *Proceedings of the Yorkshire Geological Society*, **34**, 117-38.

Rayner, D.H. (1982) Studies on Old Red Sandstone floras from Scotland. Unpublished PhD Thesis, University of Wales.

Rayner, D.H. (1983) New observations on *Sawdonia ornata* from Scotland. *Transactions of the Royal Society of Edinburgh*, **74**, 79-93.

Rayner, D.H. (1984) New finds of *Drepanophycus spinaeformis* Göppert from the Lower Devonian of Scotland. *Transactions of the Royal Society of Edinburgh*, **75**, 353-63.

Read, C.B. (1937) The flora of the New Albany Shale Part 2. The Calamopityeae and their relationships. *US Geological Survey, Professional Paper*, **186-E**, 81-91.

Read, C.B. and Campbell, G. (1939) Preliminary account of the New Albany Shale flora. *American Midland Naturalist*, **21**, 435-53.

Read, C.B. and Mamay, S.H. (1964) Upper Paleozoic floral zones and floral provinces of the United States. *US Geological Survey, Professional Paper*, **454-K**, 1-35.

Reid, J. (1895) The vegetable origin of *Parka decipiens*. *Transactions and Proceedings of the Perthshire Society of Natural Science*, **2**, 123-6.

Reid, J., Graham, W. and Macnair, P. (1898) *Parka decipiens*, its origin, affinities, and distribution. *Transactions of the Geological Society of Glasgow*, **11**, 105-21.

Reid, J. and Macnair, P. (1896) On the genera *Lycopodites* and *Psilophyton* of the Old Red Sandstone Formation of Scotland. *Transactions of the Geological Society of Glasgow*, **2**, 323-30.

Reid, J. and Macnair, P. (1899) On the genera *Psilophyton, Lycopodites, Zosterophyllum* and *Parka decipiens* of the Old Red Sandstone of Scotland. Their affinities and distribution. *Transactions of the Edinburgh Geological Society*, **7**, 368-80.

Remy, W. (1978) Der Dehiszenzmechanismus der Spornagien von *Rhynia*. *Argumenta Palaeobotanica*, **5**, 23-30.

Remy, W. (1980a) Der Generationswechsel der archegoniaten Pflanzen im Übergangsfeld von aquatischer zu terrestrischer Lebenweise. *Argumenta Palaeobotanica*, **6**, 139-55.

Remy, W. (1980b) Wechselwirkung von Vegetation und Boden im Palaophyticum. In *Festschrift für Gerhard Keller*. Wenner, Osnabruck, pp. 43-79.

Remy, W. (1991) Forebears of the terrestrial plants. *German Research*, **3/82**, 12-13.

Remy, W. and Hass, H. (1991a) Ergänzende Beobachtungen an *Lyonophyton rhyniensis*. *Argumenta Palaeobotanica*, **8**, 1-27.

Remy, W. and Hass, H. (1991b) *Langiophyton mackei* nov. gen., nov. spec., ein Gametophyt mit Archegoniophoren aus dem Chert von Rhynie (Unterdevon, Schottland). *Argumenta Palaeobotanica*, **8**, 69-117.

Remy, W. and Hass, H. (1991c) *Kidstonophyton discoides* nov. gen., nov. spec., ein Gametophyt aus dem Chert von Rhynie (Unterdevon, Schottland). *Argumenta Palaeobotanica*, **8**, 29-45.

Remy, W. and Remy, R. (1977) *Die Floren des Erdaltertums. Einfuhrung in Morphologie, Anatomie, Geobotanik und Biostratigraphie der Pflanzen des Paläophytikums*. Glückauf, Essen.

Remy, W. and Remy, R. (1980a) Devonian gametophytes with anatomically preserved gametangia. *Science*, **208**, 295-6.

Remy, W. and Remy, R. (1980b) *Lyonophyton rhyniensis* nov. gen. et nov. spec., ein Gametophyt aus dem Chert von Rhynie (Unterdevon, Schottland). *Argumenta Palaeobotanica*, **6**, 37-72.

Remy, W., Remy, R., Hass, H., Schultka, S.T. and Franzmeyer, F. (1980) *Sciadophyton* Steinmann - ein Gametophyt aus dem Siegen. *Argumenta Palaeobotanica*, **6**, 73-94.

Remy, W., Schultka, S.T., Hass, H. and Franzmeyer, F. (1980) *Sciadophyton* - Bestände im Siegen des Rheinischen Schiefergebirges als Beleg für festländische Bedingungen. *Argumenta Palaeobotanica*, **6**, 95-114.

Remy, W. and Spassov, C. (1959) Der paläbotanische Nachweiss von Oberdevon in Bulgarien. *Monatsberichte der Deutschen Akademie der Wissenschaften zu Berlin*, **1**, 384-6.

Renault, B. (1896) Bassin Houiller et Permien d'Autun et d'Epinac, 4. Flore fossile. *Études des Gîtes Minéraux de la France*, 1-578.

Renault, B. and Zeiller, R. (1888) Flore fossile de Commentry. *Bulletin Société Industrie Mineraux*, **41**, 1-746.

# References

Retallack, G.J. (1980) Late Carboniferous to Middle Triassic megafossil floras from the Sydney Basin. In *A guide to the Sydney Basin* (eds C. Herbert and R.J. Helby), *Geological Survey of New South Wales Bulletin*, **26**, 384-430.

Retallack, G.J. (1986) The fossil record of soils. In *Palaeosols: their recognition and interpretation* (ed. V.P. Wright), Blackwell Scientific Publications, Oxford, pp. 1-57.

Retallack, G.J. and Dilcher, D.L. (1988) Reconstructions of selected seed ferns. *Annals of the Missouri Botanical Garden*, **75**, 1010-57.

Rex, G.M. and Scott, A.C. (1987) The sedimentology, palaeoecology and preservation of the Lower Carboniferous plant deposits at Pettycur, Fife, Scotland. *Geological Magazine*, **124**, 43-66.

Richardson, J.B. (1964) Middle Old Red Sandstone spore assemblages from the Orcadian Basin, north-east Scotland. *Palaeontology*, **7**, 559-605.

Richardson, J.B. (1967) Some British Lower Devonian spore assemblages and their stratigraphic significance. *Review of Palaeobotany and Palynology*, **1**, 111-29.

Richardson, J.B. and Lister, T.R. (1969) Upper Silurian and Lower Devonian spore assemblages from the Welsh Borderland and South Wales. *Palaeontology*, **12**, 201-52.

Richardson, L. (1907) An outline of the geology of Herefordshire. *Woolhope Naturalists' Field Club* [for 1905], 1-68.

Ritchie, A. (1963) Palaeontological studies on Scottish Silurian fish beds. Unpublished PhD Thesis, University of Edinburgh.

Robertson, T. (1932) The geology of the South Wales Coalfield. Part V. The country around Merthyr Tydfil. 2nd Edn. *Memoir of the Geological Survey UK*.

Rogers, I. (1926) On the discovery of fossil fishes and plants in the Devonian rocks of north Devon. *Report of the Transactions of the Devonshire Association for the Advancement of Science*, **58**, 223-34.

Rolfe, W.D.I. (1969) Phyllocarida. In *Treatise on Invertebrate Paleontology R* (ed. R.C. Moore), Geological Society of America, Boulder, Colorado and University of Kansas Press, Lawrence, Kansas, pp. 296-331.

Roselt, G. (1962) Über die ältesten Landpflanzen und eine mögliche Landpflanzen aus dem Ludlow Sachsens. *Geologie*, **11**, 320-33.

Rosie, G. (1981) *Hugh Miller. Outrage and order. A biography and selected writings*, Mainstream Publishing Co, Edinburgh.

Rothwell, G.W. (1975) The Callistophytaceae (Pteridospermopsida): 1. Vegetative structures. *Palaeontographica*, **B 151**, 171-96.

Rothwell, G.W. (1981) The Callistophytales (Pteridospermopsida): reproductively sophisticated Palaeozoic gymnosperms. *Review of Palaeobotany and Palynology*, **32**, 103-21.

Rothwell, G.W. (1986) Classifying the earliest gymnosperms. In *Systematic and taxonomic approaches in palaeobotany* (eds R.A. Spicer and B.A. Thomas), *Systematics Association Special Volume 31*, Oxford University Press, pp. 137-61.

Rothwell, G.W. (1988) Cordaitales. In *Origin and evolution of gymnosperms* (ed. C.B. Beck), Columbia University Press, New York, pp. 273-97.

Rothwell, G.W. and Erwin, D.M. (1987) Origin of seed plants: an aneurophyte/seed-fern link elaborated. *American Journal of Botany*, **74**, 970-3.

Rothwell, G.W. and Scheckler, S.E. (1988) Biology of ancestral gymnosperms. In *Origin and evolution of gymnosperms* (ed. C.B. Beck), Columbia University Press, New York, pp. 85-134.

Rothwell, G.W., Scheckler, S.E. and Gillespie, W.H. (1989) *Elkinsia* gen. nov., a Late Devonian gymnosperm with cupulate ovules. *Botanical Gazette*, **150**, 170-89.

Rothwell, G.W. and Scott, A.C. (1992a) *Stamnostoma oliveri*, a gymnosperm with systems of ovulate cupules from the Lower Carboniferous (Dinantian) floras at Oxroad Bay, East Lothian, Scotland. *Review of Palaeobotany and Palynology*, **72**, 273-84.

Rothwell, G.W. and Scott, A.C. (1992b) Confronting the *Buteoxylon/Triradioxylon/Tristichia/Calathopteris/Oxroadopteris* exasperplexium (abstract). *American Journal of Botany*, **79**, 104.

Rothwell, G.W. and Warner, S. (1984) *Cordaixylon dumusum* n. sp. (Cordaitales), I. Vegetative structures. *Botanical Gazette*, **145**, 275-91.

Rothwell, G.W. and Wight, D.C. (1989) *Pullaritheca longii* gen. nov. and *Kerryia mattenii* gen. et sp. nov., Lower Carboniferous cupules with ovules of the *Hydrasperma tenuis*-type. *American Journal of Botany*, **60**, 295-309.

# References

Rowe, N.P. (1986) The fossil flora of the Drybrook Sandstone (Lower Carboniferous) from the Forest of Dean, Gloucestershire. Unpublished PhD Thesis, University of Bristol.

Rowe, N.P. (1988a) A herbaceous lycophyte from the Lower Carboniferous Drybrook Sandstone of the Forest of Dean, Gloucestershire. *Palaeontology*, **31**, 69-83.

Rowe, N.P. (1988b) New observations on the Lower Carboniferous pteridosperm *Diplopteridium* Walton and an associated synangiate organ. *Botanical Journal of the Linnean Society*, **97**, 125-58.

Rowe, N.P. (1988c) Two species of the lycophyte genus *Eskdalia* Kidston from the Drybrook Sandstone (Visean) of Great Britain. *Palaeontographica*, **B 208**, 81-103.

Rowe, N.P. (1992) The gymnosperm *Archaeopteridium tschermackii* and an associated glandular fructification from the upper Visean Drybrook Sandstone of Great Britain. *Palaeontology*, **35**, 875-900.

Rowley, D.B., Raymond, A., Parrish, J.T., Lottes, A.L., Scotese, C.R. and Ziegler, A.M. (1985) Carboniferous paleogeographic, phytogeographic and paleoclimatic reconstructions. *International Journal of Coal Geology*, **5**, 7-42.

Roy Chowdhury, M.K., Sastry, M.V.A., Shah, S.C., Gopal Singh and Ghosh, S.C. (1975) Triassic floral succession in the Gondwana of Peninsular India. In *Gondwana geology* (ed. K.S.W. Campbell), Australia National University Press, Melbourne, pp. 149-58.

Sadovnikov, G.N. (1981) Regional stratigraphic subdivisions of the Upper Permian and Lower Triassic of the Siberian platform and adjoining regions. *Sovetskaya Geologiya*, **6**, 74-84 [In Russian].

Salter, J.W. (1858) On some remains of terrestrial plants in the Old Red Sandstone of Caithness. *Quarterly Journal of the Geological Society of London*, **14**, 72-8.

Satterthwaite, D.F. and Schopf, J.W. (1972) Structurally preserved phloem zone tissue in *Rhynia*. *American Journal of Botany*, **59**, 373-6.

Scheckler, S.E. (1974) Systematic characters in Devonian ferns. *Annals of the Missouri Botanical Garden*, **61**, 462-73.

Scheckler, S.E. (1978) Ontogeny of progymnosperms. II. Shoots of Upper Devonian Archaeopteridales. *Canadian Journal of Botany*, **56**, 3136-70.

Scheihing, M.H. and Pfefferkorn, H.W. (1984) The taphonomy of land plants in the Orinoco delta: a model for the incorporation of plant parts in clastic sediments of late Carboniferous age in Euramerica. *Review of Palaeobotany and Palynology*, **41**, 205-40.

Schimper, W.P. (1869-1874) *Traité de paléontologie végétale*, Paris.

Schlanker, C.M. and Leisman, G.A. (1969) The herbaceous Carboniferous lycopod *Selaginella fraiponti* comb. nov. *Botanical Gazette*, **130**, 35-41.

Schopf, J.M. (1941) Contribution to Pennsylvanian paleobotany; *Mazocarpon oedipternum* sp. nov., and sigillarian relationships. *Illinois Geological Survey, Report of Investigations*, **75**, 1-53.

Schopf, J.M. (1975) Modes of fossil preservation. *Review of Palaeobotany and Palynology*, **20**, 27-35.

Schopf, J.M., Mencher, E., Boucot, A.J. and Andrews, H.N. (1966) Erect plants in the early Silurian of Maine. *US Geological Survey, Professional Paper*, **550-D**, 69-75.

Schweitzer, H.-J. (1962) Die Makroflora des niederrheinischen Zechsteins. *Fortschritte in der Geologie von Rheinland und Westfalen*, **6**, 1-46.

Schweitzer, H.-J. (1967) Die Oberdevon-Flora der Bäreninsel. 1. *Pseudobornia ursina* Nath. *Palaeontographica*, **B 120**, 116-37.

Schweitzer, H.-J. (1968) Die Flora des Oberen Perms in Mitteleuropa. *Naturwissenschaftliche Runddschau Stuttgart*, **21**, 93-102.

Schweitzer, H.-J. (1969) Die Oberdevon-Flora der Bäreninsel. 2. Lycopodiinae. *Palaeontographica*, **B 126**, 101-37.

Schweitzer, H.-J. (1983a) Der Generationswechsel der Psilophyten. *Bericht der Deutschen Botanischen Gesellschaft*, **96**, 483-96.

Schweitzer, H.-J. (1983b) Die Unterdevonflora des Rheinlandes. I. Teil. *Palaeontographica*, **B 189**, 1-138.

Schweitzer, H.-J. (1986) The land flora of the English and German Zechstein sequences. In *The English Zechstein and related topics* (eds G.M. Harwood and D.B. Smith), *Geological Society Special Publication*, **22**, 31-54.

Scotese, C.R. and McKerrow, W.S. (1990) Revised world maps and introduction. In *Palaeozoic palaeogeography and biogeography* (eds W.S. McKerrow and C.R. Scotese), *Geological Society of London Memoir*, **12**, 1-21.

Scott, A.C. (1977) A review of the ecology of upper Carboniferous plant assemblages, with

# References

new data from Strathclyde. *Palaeontology*, **20**, 447-73.

Scott, A.C. (1978) Sedimentological and ecological control of Westphalian B plant assemblages from West Yorkshire. *Proceedings of the Yorkshire Geological Society*, **41**, 461-508.

Scott, A.C. (1979) The ecology of Coal Measure floras from northern Britain. *Proceedings of the Geologists' Association*, **90**, 97-116.

Scott, A.C. (1985) Distribution of Lower Carboniferous floras in northern Britain. *Compte Rendu 9e Congrès International de Stratigraphie et de Géologie Carbonifère* (Washington and Urbana, 1979), **5**, 77-82.

Scott, A.C., Edwards, D. and Rolfe, W.D.I. (1976) Fossiliferous Lower Old Red Sandstone near Cardross, Dumbartonshire. *Proceedings of the Geological Society of Glasgow*, **117**, 4-5.

Scott, A.C. and Galtier, J. (1985) Distribution and ecology of early ferns. *Proceedings of the Royal Society of Edinburgh*, **86B**, 141-9.

Scott, A.C. and Galtier, J. (1988) A new Lower Carboniferous flora from East Lothian, Scotland. *Proceedings of the Geologists' Association*, **99**, 141-51.

Scott, A.C., Galtier, J. and Clayton, G. (1984) Distribution of anatomically-preserved floras in the Lower Carboniferous in western Europe. *Transactions of the Royal Society of Edinburgh, Earth Sciences*, **75**, 311-40.

Scott, A.C., Galtier, J. and Clayton, G. (1985) A new late Tournaisian (Lower Carboniferous) flora from the Kilpatrick Hills, Scotland. *Review of Palaeobotany and Palynology*, **44**, 81-99.

Scott, A.C. and Meyer-Berthaud, B. (1985) Plants from the Dinantian of Foulden, Berwickshire, Scotland. *Transactions of the Royal Society of Edinburgh, Earth Sciences*, **76**, 13-20.

Scott, A.C., Meyer-Berthaud, B., Galtier, J., Rex, G.M., Brindley, S.A. and Clayton, G. (1986) Studies on a new Lower Carboniferous flora from Kingswood, near Pettycur, Scotland. 1. Preliminary report. *Review of Palaeobotany and Palynology*, **48**, 161-80.

Scott, A.C. and Rex, G.M. (1987) The accumulation and preservation of Dinantian plants from Scotland and its Borders. In *European Dinantian environments* (eds J. Miller, A.E. Adams and V.P. Wright), John Wiley, London, pp. 329-44.

Scott, D.H. (1897) On the structure and affinities of fossil plants from the Palaeozoic rocks. Part. 1. On *Cheirostrobus*, a new type of fossil cone from the Lower Carboniferous strata (Calciferous Sandstone Series). *Philosophical Transactions of the Royal Society of London*, **189**, 1-34.

Scott, D.H. (1899) On the primary wood of certain Araucarioxylons. *Annals of Botany*, **13**, 615-9.

Scott, D.H. (1900) Note on the occurrence of a seed-like fructification in certain Palaeozoic lycophytes. *Proceedings of the Royal Society of London*, **67**, 306-9.

Scott, D.H. (1901) On the structure and affinities of fossil plants from the Palaeozoic rocks - IV. The seed-like fructification of *Lepidocarpon*. *Philosophical Transactions of the Royal Society of London*, **B 194**, 291-333.

Scott, D.H. (1902) On the primary structure of certain Palaeozoic stems with the *Dadoxylon* type of wood. *Transactions of the Royal Society of Edinburgh*, **40**, 331-65.

Scott, D.H. (1908-1909) *Studies in fossil botany* (2nd edn), A. and C. Black, London (2 vols).

Scott, D.H. (1911) Presidential address. *Proceedings of the Linnean Society of London*, **123**, 17-29.

Scott, D.H. (1918) Notes on *Calamopitys* Unger. *Botanical Journal of the Linnean Society*, **344**, 205-32.

Scott, D.H. (1920-1923) *Studies in fossil botany* (3rd edn), A. and C. Black, London (2 vols).

Scott, D.H. (1924a) *Extinct plants and problems of evolution.* Macmillan and Co, London.

Scott, D.H. (1924b) Fossil plants of the *Calamopitys* type from the Carboniferous rocks of Scotland. *Transactions of the Royal Society of Edinburgh*, **53**, 569-96.

Scott, D.H. (1926) The most remarkable plant I ever saw. *Conquest*, **7**, 101-3.

Scott, D.H. (1928) Notes on Palaeozoic palaeobotany 1907-1927. *Recueil des travaux botaniques neerlandes*, **25**, 346-85.

Scott, R. (1908) On *Bensonites fusiformis* sp. nov. a fossil associated with *Stauropteris burntislandica* P. Bertrand, and on the sporangia of the latter. *Annals of Botany*, **22**, 683-7.

Scrutton, C.T. (1978) *A field guide to selected outcrop areas of the Devonian of south-west England,* Palaeontological Association, London.

Sebby, W.S. and Matten, L.C. (1969) A reconstruction of the frond of *Kalymma*. *Transactions of the Illinois Academy of Science*, **62**, 356-61.

Sedgwick, A. (1829) On the geological relations and the internal structure of the Magnesian Limestone, and the lower portions of the New Red Sandstone series in their range through

# References

Nottinghamshire, Derbyshire, Yorkshire, and Durham, to the southern extremity of Northumberland. *Transactions of the Geological Society of London (Series 2)*, **3**, 37-124.

Sedgwick, A. (1848) On the organic remains found in the Skiddaw Slate, with some remarks on the classification of the older rocks of Cumberland and Westmoreland. *Quarterly Journal of the Geological Society of London*, **4**, 216-25.

Senkevich, M.A. (1963) Novie nakhodki flori verknego ordovika Kazakhstana. *Akademie Nauk S.S.S.R.* (Serie Geologiya), **5**, 67-81 [In Russian]. (Also, *International Geological Review*, **71**, 476-85 - in English).

Senkevich, M.A. (1975) New Devonian psilophytes from Kazakhstan. *Esheg Vses Palaeontol Obschestva*, **21**, 288-98 [In Russian].

Seward, A.C. (1895) Notes on *Pachytheca*. *Proceedings of the Cambridge Philosophical Society*, **8**, 278.

Seward, A.C. (1917) *Fossil plants. Vol. 3*, Cambridge University Press, Cambridge.

Seward, A.C. (1931) *Plant life through the ages*, Cambridge University Press, Cambridge.

Seward, A.C. and Hill, A.W. (1900) On the structure and affinities of a lepidodendroid stem from the Calciferous Sandstone of Dalmeny, Scotland, possibly identical with *Lepidophloios harcourtii* Witham. *Transactions of the Royal Society of Edinburgh*, **39**, 907-31.

Shadle, G.L. and Stidd, B.M. (1975) The frond of *Heterangium*. *American Journal of Botany*, **62**, 67-75.

Shute, C.H. and Cleal, C.J. (1989) The holotype of the Carboniferous marattialean fern *Lobatopteris miltoni* (Artis). *Bulletin of the British Museum of Natural History (Geology)*, **45**, 71-6.

Shute, C.H. and Edwards, D. (1989) A new rhyniopsid with novel sporangium organization from the Lower Devonian of South Wales. *Botanical Journal of the Linnean Society*, **100**, 111-37.

Simpson, S. (1959) *Lexique stratigraphique international. Europe 3, 2 vi, England, Wales and Scotland*. Centre Nationale de la Recherche Scientifique, Paris.

Skog, J.E. and Gensel, P.G. (1980) A fertile species of *Triphyllopteris* from the Early Carboniferous (Mississippian) of south-western Virginia. *American Journal of Botany*, **67**, 440-51.

Smith, A.H.V. (1962) The palaeoecology of Carboniferous peats based on the miospores and petrography of bituminous coals. *Proceedings of the Yorkshire Geological Society*, **33**, 423-74.

Smith, B. (1913) The geology of the Nottingham district. *Proceedings of the Geologists' Association*, **24**, 205-40.

Smith, B. and George, T.N. (1961) *North Wales, British Regional Geology*: 3rd edn. Institute of Geological Sciences, HMSO, London.

Smith, D.B., Brunstron, R.G.W., Manning, P.I., Simpson, S. and Shotton, F.W. (1974) Permian. A correlation of Permian rocks in the British Isles. *Geological Society of London, Special Report*, **5**, 1-45.

Smith, D.L. (1959) *Geminitheca scotica* gen. et sp. nov.: a pteridosperm from the Lower Carboniferous of Dunbartonshire. *Annals of Botany*, **23**, 477-91.

Smith, D.L. (1960) The Lower Carboniferous flora of the Kilpatrick Hills. Unpublished PhD Thesis, University of Glasgow.

Smith, D.L. (1962a) The spores of *Alcicornopteris hallei* Walton. *Annals of Botany*, **26**, 533-50.

Smith, D.L. (1962b) Three fructifications from the Scottish Lower Carboniferous. *Palaeontology*, **5**, 225-37.

Smith, D.L. (1962c) The stems of three species of lepidodendrid from the Scottish Lower Carboniferous. *Annals of Botany*, **26**, 533-50.

Smith, D.L. (1964a) The evolution of the ovule. *Biological Reviews*, **39**, 137-59.

Smith, D.L. (1964b) Two Scottish Lower Carboniferous floras. *Transactions of the Botanical Society of Edinburgh*, **39**, 460-6.

Smith, G.M. (1955) *Cryptogamic botany* (2nd edn), McGraw-Hill Book Co., New York.

Smith, J.A. (1862) Notes on fossils from the Old Red Sandstone of the south of Scotland. *Proceedings of the Royal Physical Society of Edinburgh*, **2**, 36-7.

Smith, S. and Reynolds, S.H. (1929) The Carboniferous section at Cattybrook, near Bristol. *Quarterly Journal of the Geological Society of London*, **85**, 1-8.

Snigirevskaya, N.S. (1972) Studies of coal balls of the Donets Basin. *Review of Palaeobotany and Palynology*, **14**, 197-204.

Solms-Laubach, H. de (1892) Über die in den Kalksteinen des Kulm von Glätzisch-Falkenberg in Schlesien erhaltenen strukturbietenden Pflanzenreste. *Botanisches Zentralblatt*, **50**, 49-113.

Solms-Laubach, H. de (1893) Über die in den Kalksteinen des Kulm von Glätzisch-Falkenberg in Schlesien erhaltenen

# References

strukturbietenden Pflanzenreste. *Botanisches Zentralblaft*, **51**, 197.

Solms-Laubach, H. de (1896) Über die seinerzeit von Unger beschriebenen strukturbietenden Pflanzenreste des Unterkulms von Saalfeld in Thuringen VII: *Sphenophyllum*. *Abhandlunger der Königlishen Preussischen Geologischen Landesanstalt*, **N.F. 23**, 1-100.

Solms-Laubach, H. de (1910) Über die in den Kalksteinen des Kulm von Glätzisch-Falkenberg in Schlesien erhaltenen strukturbietenden Pflanzenreste. IV. *Volkelia refracta, Stenoxylon ludwigii*. *Zeitschrift für Botanik*, **2**, 529-54.

Somerville, I.D., Strank, A.R.E. and Welsh, A. (1989) Chadian faunas and flora from Dyserth: depositional environments and palaeogeographical setting of Viséan strata in northeast Wales. *Geological Journal*, **24**, 49-66.

Sorby, H.C. (1858) On the microscopical structure of crystals, indicating the origin of minerals and rocks. *Quarterly Journal of the Geological Society of London*, **14**, 453-500.

Sorby, H.C. (1875) On the remains of a fossil forest in the Coal-measures at Wadsley, near Sheffield. *Quarterly Journal of the Geological Society of London*, **31**, 458-60.

Speck, T. and Vogellehner, D. (1988a) Biophysikalische Untersuchungen zur Mechanostabilität verschiedener Stelentypen und zur Art des Festigungssystems früher Gefässlandpflanzen. *Palaeontographica*, **B 210**, 91-126.

Speck, T. and Vogellehner, D. (1988b) Biophysical examinations of the bending stability of various stele types and the upright axes of early 'vascular' land plants. *Botanica Acta*, **101**, 262-8.

Spicer, R.A. (1980) The importance of depositional sorting to the biostratigraphy of plant megafossils. In *Biostratigraphy of Fossil Plants: succession and paleoecological Analyses* (eds D.L. Dilcher and T.N. Taylor), Dowden, Hutchinson and Ross, Stroudsburg, PA, USA, pp. 171-83.

Spicer, R.A. and Hill, C.R. (1979) Principal components and correspondence analyses of quantitative data from a Jurassic plant bed. *Review of Palaeobotany and Palynology*, **28**, 273-99.

Spicer, R.A. and Thomas, B.A. (1986) Systematic and taxonomic approaches in palaeobotany. *Systematics Association Special Volume 31*, Oxford University Press.

Spicer, R.A. and Thomas, B.A. (1987) A Mississippian Alaska-Siberia connection: evidence from plant megafossils. In *Alaskan North Slope Geology, Volume 1* (eds I. Tailleur and P. Weimer), Pacific Section, Society of Economic Paleontologists and Mineralogists, Bakersfield CA, and Alaska Geological Society, Anchorage, pp. 355-8.

Spinner, E. (1984) Further studies on the megaspore genus *Setispora* Butterworth and Spinner 1967. *Pollen et Spores*, **24**, 301-13.

Sprengel, A. (1828) *Commentatio de Psarolithus ligni fossilis genre*.

Squirrell, H.C. and Tucker, E.V. (1960) The geology of the Woolhope Inlier (Herefordshire). *Quarterly Journal of the Geological Society of London*, **116**, 139-85.

Squirrell, H.C. and Tucker, E.V. (1967) Itinerary II - North Woolhope. In *The Silurian inliers of the south-eastern Welsh Borderland* (eds M.L.K. Curtis, J.D. Lawson, H.C. Squirrell, E.V. Tucker and V.G. Walmsley), *Geologists' Association Field Guides*, **5**, 1-32.

Stamp, L.D. (1923) The base of the Devonian, with special reference to the Welsh Borderland. *Geological Magazine*, **60**, 276-82, 331-6, 367-72, 385-410.

Stepanov, S.A. (1975) *Fitostratigrafiya opornikh pazrezov Devona okrain Kuzbassa*. Siberian Scientific Research Institute [In Russian].

Stewart, W.N. (1960) More about the origin of vascular plants. *Plant Science Bulletin*, **6**, 1-4.

Stewart, W.N. (1981) The Progymnospermopsida: the construction of a concept. *Canadian Journal of Botany*, **59**, 1539-42.

Stewart, W.N. (1983) *Palaeobotany and the evolution of plants*. Cambridge University Press, Cambridge.

Stewart, W.N. and Rothwell, G.W. (1992) *Paleobotany and the evolution of plants*. Cambridge University Press, Cambridge.

Stockmans, F. (1940) Végétaux Eodévoniens de la Belgique. *Mémoires du Musée Royal d'Histoire Naturelle de Belgique*, **93**, 1-90.

Stockmans, F. (1948) Végétaux du Dévonien superieur de la Belgique. *Mémoires du Musée Royal d'Histoire Naturelle de Belgique*, **110**, 1-85.

Stockmans, F. and Willière, Y. (1952-1953) Végétaux Namuriens de la Belgique. *Publications. Association pour l'Étude de la Paléontologie et de la Stratigraphie Houillères*, **13**, 1-382.

Stoneley, H.M.M. (1956) *Hiltonia*, a new plant genus from the Upper Permian of England. *Annals and Magazine of Natural History*, (Series 12), **9**, 713-20.

# References

Stoneley, H.M.M. (1958) The Upper Permian flora of England. *Bulletin of the British Museum of Natural History (Geology)*, **3**, 295-337.

Stopa, S.Z. (1962) Subdivision stratigraphique du Houiller en Pologne. *Compte Rendu 4e Congrès International de Stratigraphie et de Géologie Carbonifère* (Heerlen, 1958), **3**, 683-96.

Storch, D. (1966) Die Arten der Gattung *Sphenophyllum* Brongniart im Zwickau-Lugau-Oelsnitzer Steinkohlenrevier: Ein Beitrag zur Revision der Gattung. *Paläontologische Abhandlungen*, **B 2**, 195-326.

Storrie, J. (1892) On the occurrence of *Pachytheca* at Tymawr Quarry, Rumney. *Report of the British Association for the Advancement of Science*, (Cardiff, 1891), pp. 652.

Strahan, A. (1885) Geology of the coasts adjoining Rhyl, Abergele and Colwyn. *Memoir of the Geological Survey UK*.

Strahan, A. and Cantrill, T.C. (1912) The geology of the South Wales Coalfield, Part III. The country around Cardiff (2nd edn), *Memoir of the Geological Survey UK*.

Straw, S.H. (1926) Some notes on the genus *Actinophyllum*. *Memoir and Proceedings of the Manchester Literary and Philosophical Society*, **70**, 133-9.

Straw, S.H. (1930) The Siluro-Devonian boundary in south-central Wales. *Journal of the Manchester Geological Association*, **11**, 79-102.

Straw, S.H. (1937) The higher Ludlovian rocks of the Builth district. *Quarterly Journal of the Geological Society of London*, **93**, 406-56.

Straw, S.H. (1953) The Silurian succession at Cwm Graig Ddu (Breconshire). *Liverpool and Manchester Geological Journal*, **1/2**, 208-19.

Streel, M., Higgs, K., Loboziak, S., Riegel, W. and Steemans, P. (1987) Spore stratigraphy and correlation with faunas and floras in the type marine Devonian of the Ardenne-Rhenisch regions. *Review of Palaeobotany and Palynology*, **50**, 211-19.

Stubblefield, S. and Banks, H.P. (1978) The cuticle of *Drepanophycus spinaeformis*, a long ranging Devonian lycopod from New York and eastern Canada. *American Journal of Botany*, **65**, 110-18.

Stur, D. (1875) *Die Culm-Flora der mährisch-schlesisch Dachschiefers*, Geological Survey, Vienna.

Sullivan, H.J. (1964) Miospores from the Drybrook Sandstone and associated measures in the Forest of Dean, Gloucestershire. *Palaeontology* 7, 351-92.

Sullivan, H.J. and Hibbert, A.F. (1964) *Tetrapterites visensis* - a new spore bearing structure from the Lower Carboniferous. *Palaeontology*, 7, 64-71.

Surange, K.R. (1952a) The morphology of *Stauropteris burntislandica* P. Bertrand and its megasporangium *Bensonites fusiformis* R. Scott. *Philosophical Transactions of the Royal Society of London*, **B 237**, 75-91.

Surange, K.R. (1952b) The morphology of *Botryopteris antiqua* with some observations on *Botryopteris ramosa*. *Palaeobotanist*, **1**, 420-34.

Tasch, P. (1957) Flora and fauna of the Rhynie Chert: a paleoecological re-evaluation of published evidence. *Bulletin of the University of Wichita*, **32**, 3-24.

Taylor, T.N. (1981) *Paleobotany. An introduction to fossil plant biology*, McGraw-Hill Book Co., New York.

Taylor, T.N. and Millay, M.A. (1981) Morphological variability of Pennsylvanian lyginopterid seed ferns. *Review of Palaeobotany and Palynology*, **25**, 151-62.

Taylor, T.N., Remy, W. and Hass, H. (1992a) Parasitism in a 400-million-year-old green alga. *Nature, London*, **357**, 493-4.

Taylor, T.N., Remy, W. and Hass, H. (1992b) Fungi from the Lower Devonian Rhynie Chert: Chytridiomycetes. *American Journal of Botany*, 79, 1233-41.

Taylor, T.N. and Taylor, E.L. (1992) *The biology and evolution of fossil plants*, Prentice Hall, New Jersey.

Thiselton-Dyer, W.T. (1891) Note on Mr Barber's paper on *Pachytheca*. *Annals of Botany*, **5**, 223-5.

Thomas, B.A. (1967a) The cuticle of two species of *Bothrodendron* (Lycopsida; Lepidodendrales). *Journal of Natural History*, **1**, 53-60.

Thomas, B.A. (1967b) *Ulodendron* Lindley and Hutton and its cuticle. *Annals of Botany*, **NS 31**, 775-82.

Thomas, B.A. (1968) A revision of the Carboniferous lycopod genus *Eskdalia*. *Palaeontology*, 11, 439-44.

Thomas, B.A. (1970) Epidermal studies in the interpretation of *Lepidodendron* species. *Palaeontology*, 13, 145-73.

Thomas, B.A. (1972) A probable moss from the Lower Carboniferous of the Forest of Dean, Gloucestershire. *Annals of Botany*, **36**, 155-61.

# References

Thomas, B.A. (1974) The lepidodendroid stomata. *Palaeontology*, **17**, 525-39.

Thomas, B.A. (1977) Epidermal studies in the interpretation of *Lepidophloios* species. *Palaeontology*, **20**, 273-93.

Thomas, B.A. (1978a) Carboniferous Lepidodendraceae and Lepidocarpaceae. *Botanical Reviews*, **44**, 321-64.

Thomas, B.A. (1978b) New British records of Carboniferous lycopod cones. *Geological Journal*, **13**, 11-14.

Thomas, B.A. (1980) Notes on *Bothrodendron depereti* Vaffier and a *Sigillariostrobus* with *Tuberculatisporites brevispiculus* megaspores. *Argumenta Palaeobotanica*, **6**, 157-64.

Thomas, B.A. (1981a) *Evolution of plants and flowers*. Eurobooks.

Thomas, B.A. (1981b) Structural adaptations shown by the Lepidocarpaceae. *Review of Palaeobotany and Palynology*, **32**, 377-88.

Thomas, B.A. (1986) The biochemistry and analysis of fossil plants and its use in taxonomy and systematics. In *Systematic and taxonomic approaches in palaeobotany* (eds R.A. Spicer and B.A. Thomas), *Systematics Association Special Volume*, **31**. Oxford University Press, pp. 39-51.

Thomas, B.A. (1990) Rules of nomenclature: disarticulated plant fossils. In *Palaeobiology. A synthesis* (eds D.E.G. Briggs and P.R. Crowther), Blackwell Scientific Publications, Oxford, pp. 421-2.

Thomas, B.A. (1992) Paleozoic herbaceous lycopsids and the beginnings of extant *Lycopodium* sens. lat. and *Selaginella* sens. lat. *Annals of the Missouri Botanical Garden*, **79**, 623-31.

Thomas, B.A. and Brack-Hanes, S.D. (1984) A new approach to family groupings in the lepipophytes. *Taxon*, **33**, 247-55.

Thomas, B.A. and Cleal, C.J. (1993) Middle Westphalian plant fossils from the West-Cumberland Coalfield, UK. *Geological Journal*, **28**, 101-23.

Thomas, B.A. and Cleal, C.J. (1994) Plant fossils from the Writhlington Geological Nature Reserve. *Proceedings of the Geologists' Association*, **105**, 15-32.

Thomas, B.A. and Crampton, D.M. (1971) A fertile *Zeilleria avoldensis* from the British Upper Carboniferous. *Review of Palaeobotany and Palynology*, **11**, 283-95.

Thomas, B.A. and Meyen, S.V. (1984) A reappraisal of the Lower Carboniferous lepidophyte *Eskdalia* Kidston. *Palaeontology*, **27**, 707-18.

Thomas, B.A. and Purdy, H.M. (1982) Additional fossil plants from the Drybrook Sandstone, Forest of Dean, Gloucestershire. *Bulletin of the British Museum of Natural History (Geology)*, **36**, 131-42.

Thomas, B.A. and Spicer, R.A. (1987) *The evolution and palaeobiology of land plants*, Croom Helm, London.

Thorpe, F.T. (1972) *A history of Middlewood Psychiatric Hospital*, Middlewood Hospital Printing and Bookbinding Department, Sheffield.

Tims, J.D. and Chambers, T.C. (1984) Rhyniophytina and Trimerophytina from the early land floras of Victoria, Australia. *Palaeontology*, **27**, 265-79.

Townrow, J.A. (1960) The Peltaspermaceae, a pteridosperm family of Permian and Triassic age. *Palaeontology*, **3**, 333-61.

Trewin, N.H. (1985) Mass mortalities of Devonian fish - the Achanarras Fish Bed, Caithness. *Geology Today*, **1**, 45-9.

Trewin, N.H. (1989) The Rhynie hot-spring deposit. *Earth science conservation*, **26**, 10-2.

Trewin, N.H. and Rice, C.M. (1992) Stratigraphy and sedimentology of the Devonian Rhynie chert locality. *Scottish Journal of Geology*, **28**, 37-47.

Trivett, M.L. and Rothwell, G.W. (1985) Morphology, systematics, and paleoecology of Paleozoic fossil plants: *Mesoxylon priapi*, sp. nov. (Cordaitales). *Systematic Botany*, **10**, 205-23.

Tufnell, H. (1853) Notice of the discovery of fossil plants in the Shetland Islands. *Quarterly Journal of the Geological Society of London*, **9**, 49-51.

Tyrrell, G.W. (1928) The geology of Arran. *Memoirs of the Geological Survey of Great Britain*.

Utting, J. and Neves, R. (1970) Palynology of the Lower Limestone Shale Group (Basal Carboniferous Limestone series) and Portishead Beds (Upper Old Red Sandstone) of the Avon Gorge, Bristol, England. In *Colloque sur la Stratigraphie du Carbonifère* (eds M. Streel and R.H. Wagner), Liége University Congrès et Colloques, **55**, 411-27.

Vakhrameev, V.A., Dobruskina, I.A. and Meyen, S.V. (1978) *Paläozoische und mesozoische Floren Eurasiens und die Phytogeographie dieser Zeit*. G. Fischer, Jena.

Vernon, R.D. (1912) On the geology and palaeontology of the Warwickshire Coalfield.

*Quarterly Journal of the Geological Society of London*, **68**, 587-638.

Wagner, R.H. (1961) Some Alethopterideae from the South Limburg Coalfield. *Mededelingen van de Geologisch Stichting*, **14**, 5-13.

Wagner, R.H. (1966) On the presence of probable upper Stephanian beds in Ayrshire, Scotland. *Scottish Journal of Geology*, **2**, 122-3.

Wagner, R.H. (1968) Upper Westphalian and Stephanian species of *Alethopteris* from Europe, Asia Minor, and North America. *Mededelingen van's Rijks Geologischen Dienst*, **C 3-1-6**, 1-188.

Wagner, R.H. (1974) The chronostratigraphic units of the Upper Carboniferous in Europe. *Bulletin de la Société Belge de Géologie, Paléontologie et d'Hydrologie*, **83**, 235-53.

Wagner, R.H. (1977) Comments on the Upper Westphalian and Stephanian floras of Czechoslovakia, with particular reference to their stratigraphic age. In *Symposium on Carboniferous stratigraphy* (eds V.M. Holub and R.H. Wagner), Geological Survey, Prague, pp. 441-57.

Wagner, R.H. (1983) A lower Rotliegend flora from Ayrshire. *Scottish Journal of Geology*, **19**, 135-55.

Wagner, R.H. (1984) Megafloral zones of the Carboniferous. *Compte Rendu 9e Congrès International de Stratigraphie et de Géologie Carbonifère* (Washington and Urbana, 1979), **2**, 109-34.

Wagner, R.H. and Spinner, E. (1972) The stratigraphic implication of the Westphalian D macro- and microflora of the Forest of Dean Coalfield (Gloucestershire). *International Geological Congress* (Montreal, 1972), **7**, 428-31.

Wagner, R.H., Villegas, F.J. and Fonolla, F. (1969) Description of the Lower Cantabrian stratotype near Tejerina (León, NW Spain). *Compte Rendu 6e Congrès International de Stratigraphie et de Géologie Carbonifère* (Sheffield, 1967), **1**, 115-38

Wagner, R.H., Winkler Prins, C.F. and Granados, L.F. (eds) (1985) *The Carboniferous of the world. II Australia, Indian Subcontinent, South Africa, South America and north Africa*. Instituto Geológico y Minero de España, Madrid (IUGS Publication No. 20).

Walton, J. (James) (1932) The Black Bed at Bradley Park, Huddersfield. *The Naturalist* [for 1932], 304-8.

Walton, J. (John) (1926) Contributions to the knowledge of Lower Carboniferous plants. *Philosophical Transactions of the Royal Society of London*, **B 215**, 201-24.

Walton, J. (John) (1928) A preliminary account of the Lower Carboniferous flora of North Wales and its relation to the floras of some other parts of Europe. *Compte Rendu Congrès pour l'Advancement des Études de Stratigraphie Carbonifère* (Heerlen, 1927), pp. 743-7.

Walton, J. (John) (1931) Contributions to the knowledge of Lower Carboniferous plants - (cont.) III - On the fossil-flora of the Black Limestones in Teilia Quarry, Gwaenysgor, near Prestatyn, Flintshire, with special reference to *Diplopteridium teilianum* Kidston sp. (gen. nov.) and some other fern-like fronds. *Philosophical Transactions of the Royal Society of London*, **B 219**, 347-79.

Walton, J. (John) (1935) Scottish Lower Carboniferous plants: the fossil hollow trees of Arran and their branches (*Lepidophloios wunschianus* Carruthers). *Transactions of the Royal Society of Edinburgh*, **58**, 313-37.

Walton, J. (John) (1936) On the factors which influence the external form of fossil plants; with the description of some species of the Palaeozoic equisetalean genus *Annularia* Sternberg. *Philosophical Transactions of the Royal Society of London*, **B 226**, 219-37.

Walton, J. (John) (1940) *Introduction to the study of fossil plants*, A. and C. Black, London.

Walton, J. (John) (1941) On *Cardiopteridium*, a genus of fossil plants of Lower Carboniferous age, with special reference to Scottish specimens. *Proceedings of the Royal Society of Edinburgh*, **B 61**, 59-66.

Walton, J. (John) (1949a) *Calathospermum scoticum* - an ovuliferous fructification of Lower Carboniferous age from Dumbartonshire. *Transactions of the Royal Society of Edinburgh*, **61**, 719-28.

Walton, J. (John) (1949b) On a Lower Carboniferous Equisetinae from the Clyde area. *Transactions of the Royal Society of Edinburgh*, **61**, 729-43.

Walton, J. (John) (1949c) A petrified example of *Alcicornopteris* (*A. hallei* sp. nov.) from the Lower Carboniferous of Dumbartonshire. *Annals of Botany*, **13**, 445-52.

Walton, J. (John) (1954) The evolution of the ovule in the pteridosperms. *Report of the British Association for the Advancement of Science* (Edinburgh, 1953), pp. 223-30.

# References

Walton, J. (John) (1957) On *Protopitys* (Göppert): with a description of a fertile specimen *Protopitys scotica* sp. nov. from the Calciferous Sandstone Series of Dumbartonshire. *Transactions of the Royal Society of Edinburgh*, **63**, 333-40.

Walton, J. (John) (1959) Palaeobotany in Great Britain. *Vistas in Botany*, **1**, 230-44.

Walton, J. (John) (1964a) On the morphology of *Zosterophyllum* and some other Early Devonian plants. *Phytomorphology*, **14**, 155-60.

Walton, J. (John) (1964b) The pteridosperms. *Transactions of the Botanical Society of Edinburgh*, **39**, 449-59.

Walton, J. (John) (1969) On the structure of a silicified stem of *Protopitys* and roots associated with it from the Carboniferous (Mississippian) of Yorkshire, England. *American Journal of Botany*, **56**, 808-13.

Walton, J. (John) and Long, A.G. (1964) *Excursion S7. Scottish Palaeozoic*. Excursion Guide, 10th International Botanical Congress, Edinburgh.

Walton, J. (John), Weir, J. and Leitch, D. (1938) A Summary of Scottish Carboniferous stratigraphy and palaeontology. *Compte Rendu 2e Congrès International de Stratigraphie et de Géologie Carbonifère* (Heerlen, 1935), **3**, 1343-56.

Wang, Z.-Q. (1989) Permian gigantic palaeobotanical events in north China. *Acta Palaeont. Sinica*, **28**, 314-43 [In Chinese].

Warden, A.J. (1881) *Angus or Forfarshire, the land and people, descriptive and historical. Vol. 2*. Alexander and Co., Dundee.

Weigelt, J. (1928) Die Pflanzenreste des mitteldeutschen Kupferschiefers und ihre Einschaltung ins Sediment. *Fortschritte der Geologie und Paläontologie*, **6**, 395-592.

Welch, F.B.A. and Trotter, F.M. (1961) Geology of the country around Monmouth and Chepstow. *Memoir of the Geological Survey UK*.

Westoll, T.S. (1951) The vertebrate-bearing strata of Scotland. *International Geological Congress (1948)*, **11**, 5-21.

Wight, D.C. (1987) Stelar morphology of early seed plants. *American Journal of Botany*, **74**, 693.

Williams, B.P.J., Allen, J.R.L. and Marshall, J.E.A. (1982) Old Red Sandstone facies of the Pembrokeshire peninsula south of the Ritec Fault. In *Geological excursions in Dyfed, south-west Wales* (ed. M.G. Bassett), National Museum of Wales, Cardiff, 151-74.

Williams, D. (1838) On some fossil wood and plants recently discovered by the author, low down in the Grauwacke of Devon. *Report of the British Association for the Advancement of Science*, (Liverpool, 1837), pp. 94-5.

Williamson, W.C. (1872) On the organisation of the fossil plants of the Coal Measures. Part III. Lycopodiaceæ (continued). *Philosophical Transactions of the Royal Society of London*, **162**, 283-318.

Williamson, W.C. (1873) On the organisation of the fossil plants of the Coal Measures. Part IV. *Dictyoxylon*, *Lyginodendron*, and *Heterangium*. *Philosophical Transactions of the Royal Society of London*, **163**, 377-407.

Williamson, W.C. (1874a) On the organisation of the fossil plants of the Coal Measures. Part V. *Asterophyllites*. *Philosophical Transactions of the Royal Society of London*, **164**, 41-81.

Williamson, W.C. (1874b) On the organisation of the fossil plants of the Coal Measures. Part VI. Ferns. *Philosophical Transactions of the Royal Society of London*, **164**, 675-703.

Williamson, W.C. (1877) On the organisation of the fossil plants of the Coal Measures. Part VIII. Ferns (continued) and gymnospermous stems and seeds. *Philosophical Transactions of the Royal Society of London*, **167**, 213-70.

Williamson, W.C. (1880) On the organisation of the fossil plants of the Coal Measures. Part X. *Philosophical Transactions of the Royal Society of London*, **171**, 493-539.

Williamson, W.C. (1883) On the organisation of the fossil plants of the Coal Measures. Part XII. *Philosophical Transactions of the Royal Society of London*, **174**, 459-75.

Williamson, W.C. (1887) A monograph on the morphology and histology of *Stigmaria ficoides*. *Palaeontographical Society (Monographs)*, **40**, 1-62.

Williamson, W.C. (1893) On the organisation of the fossil plants of the Coal Measures. Part XIX. *Philosophical Transactions of the Royal Society of London*, **184**, 1-38.

Williamson, W.C. (1895) On the light thrown upon the growth and development of the Carboniferous arborescent lepidodendra by a study of the details of their organization. *Proceedings of the Manchester Literary and Philosophical Society*, **39**, 31-65.

Williamson, W.C. (1896) *Reminiscences of a Yorkshire naturalist*. G. Redway, London (Reprinted 1985, J. Watson and B.A. Thomas, University of Manchester).

# References

Williamson, W.C. and Scott, D.H. (1894) Further observations on the organisation of the fossil plants of the Coal Measures. I. *Philosophical Transactions of the Royal Society of London*, **185**, 863-959.

Williamson, W.C. and Scott, D.H. (1895) Further observations on the organisation of the fossil plants of the Coal Measures. III. *Philosophical Transactions of the Royal Society of London*, **B 186**, 703-79.

Wilson, E. (1876) On the Permians of the north-east of England (at their southern margin) and their relations to the under- and overlying formations. *Quarterly Journal of the Geological Society of London*, **32**, 533-7.

Wilson, E. (1881) The Permian formation in the north-east of England, with special reference to the physical conditions under which these rocks were formed. *Midland Naturalist*, **4**, 97-101, 121-4, 187-91, 201-8.

Wilson, G.V., Edwards, W., Knox, J., Jones, R.C.B. and Stephens, J.V. (1935) The geology of the Orkneys. *Memoir of the Geological Survey UK*.

Witham, H.T.M. (1830) *Observations on fossil vegetables*. W. Blackwood, Edinburgh and T. Cadell, London.

Witham, H.T.M. (1831) Vegetation of the first period of an ancient world. *Philosophical Magazine*, **7**, 23-31.

Witham, H.T.M. (1833) *The internal structure of fossil vegetables*. A. and C. Black, Edinburgh.

Woodward, H. (1866-1878) A monograph on the British fossil crustaceae, belonging to the Merostomata. *Palaeontographical Society (Monographs)*, Pt. 1 (1866), 1-44; Pt. 2 (1869), 45-70; Pts 3-4 (1872), 71-180; Pt. 5 (1878), pp. 181-263.

Young, J. (1873) Notes on a section of strata containing beds of impure coal and plant remains showing structure at Glenarbuck, near Bowling. *Transactions of the Geological Society of Glasgow*, **4**, 123-8.

Yurina, A.L. (1969) Devonskaya flora tsentral'nogo Kazakhstana. *Materialy po Geologii Tsentral'nogo Kazakhstana, Moscow State University*, **8**, 1-143 [In Russian].

Zakharova, T.V. (1981) On the systematic position of the species 'Psilophyton' goldschmidtii from the Lower Devonian of Eurasia. *Palaeontologicheskii Zhurnal*, **23**, 111-8.

Zdebska, D. (1972) *Sawdonia ornata* (=*Psilophyton princeps* var. *ornatum*) from Poland. *Acta Paleobotanica*, **13**, 77-98.

Zeiller, R. (1886-1888) Flore fossile du terrain houiller de Valenciennes. *Études des Gîtes Minéraux de la France*, 1-731.

Ziegler, W. and Klapper, G. (1985) Stages of the Devonian System. *Episodes*, **8**, 104-9.

Zimmermann, W. (1926) Die Spaltöffnungen der Psilophyta und Psilotales. *Zeitschrift für Botanik*, **19**, 129-70.

# *Glossary*

These are brief explanations for some of the botanical terms used in this volume. For further explanations, the reader is advised to consult one of the many biological or botanical dictionaries now available, or one of the standard botanical text books.

**Abaxial** – The side of a leaf facing away from the stem or main axis. In most leaves, this is the lower surface.

**Abscission** – The controlled shedding of a leaf, branch, fructification or other organ.

**Actinostele** – Stele with star-shaped cross-section.

**Adaxial** – The side of a leaf facing the stem or main axis. In most leaves, this is the upper surface.

**Adventitious** – A structure that arises in an unusual way. Usually applied to masses of small roots that never secondarily thicken.

**Angiosperms** – Flowering plants.

**Annular thickenings** – Rings of secondary thickening in walls of vessels and tracheids.

**Annulus** – A cluster, band or ring of thick-walled cells in the sporangial walls of many ferns. The annulus, in drying out, causes tension in the sporangial wall, which eventually ruptures at the stomium.

**Antheridium** – The male sex-organ which produces the motile male gametes.

**Arborescent** – Tree-like.

**Archegonium** – The female sex-organ which produces the egg cell.

**Axil** – The upper angle between a stem and a lateral branch or leaf. Structures growing out of that angle (tubercles, branches, sporangia etc.) are said to be axillary.

**Bifurcate** – Division into two branches.

**Bipartite fronds** – Fronds that have a dichotomy of the main rachis near the base, dividing the distal part of the frond in two halves. Occurs typically in a range of Palaeozoic pteridosperm fronds.

**Bisporangiate** – Bearing both megasporangia and microsporangia.

**Bract** – A leaf-like structure.

**Bryophytes** – Non-vascular land plants known as mosses, liverworts and hornworts.

**Cambium** – Meristematic tissue in stems and roots which gives rise to secondary growth.

**Campylotropous** – Description of seeds and ovules where the chalaza and lagenostome are close together.

**Central plug** – A parenchymatous structure within the lagenostome/salpinx of an ovule.

**Centrarch** – Description of a protostele in which the protoxylem is central.

**Chalaza** – Base of the nucellus where the integuments are attached.

**Charcoal** – Carbonized remains of plant tissue burnt at very high temperatures, in which some internal structure of the plant may be still preserved.

**Circinate** – A type of leaf development in which the young leaf is inrolled with its apex central. When the leaf starts to unroll, it forms a crozier, as seen in many living ferns.

**Cortex** – Zone of tissue outside the stele.

# Glossary

**Cupule** – Cup-shaped protective structure containing one or more seeds or ovules.

**Cuticle** – Outer protective 'skin' covering the aerial parts of most land plants.

**Cyclocytic** – Stomata where the guard cells are surrounded by a ring of subsidiary cells.

**Decorticated** – A stem which has had the epidermis and all or part of the cortex removed prior to fossilization.

**Dehiscence** – Splitting. Here, generally used to refer to the splitting open of a sporangium to release the spores.

**Dichotomous** – A type of branching where an axis divides into two equal branches.

**Disseminule** – A part of the plant, such as a seed, that is released from the parent to achieve propagation.

**Distichous** – Leaves or branches arranged in two vertical rows on opposite sides of the stem. The leaves or branches may be alternate or as opposite pairs on the stem.

**Dorsiventral** – A flattened structure, such as a leaf, showing differences in structure between the upper and lower sides.

**Dwarf shoot** – A lateral branch, of limited growth, arising in a bract axil in the cones of early conifers.

**Eligulate** – Having no ligule.

**Embryo sac** – The megaspore in gymnosperms and angiosperms, containing the female gametophyte.

**Endarch** – Description of a vascular strand, in which the metaxylem develops to the outside of the protoxylem.

**Endosporal gametophyte** - A gametophyte that develops within the protective wall of the spore. The archegonia or antheridia are exposed by the opening of the spore wall.

**Epidermis** – Outermost cells of a plant, usually (but not always) in a single layer.

**Exarch** – Description of a vascular strand, in which the metaxylem develops to the inside of the protoxylem.

**Exine** – The protective outer layer of a spore or pollen grain.

**Frond** – A leaf, especially of ferns and some primitive gymnosperms.

**Funicle** – Stalk of an ovule.

**Fusain** – See Charcoal.

**Gametangiophore** – An upright, extension from the gametophyte, bearing archegonia or antheridia.

**Gametes** – Sexual reproductive cells, equivalent to the eggs and sperm of animals.

**Gametophyte** – The sexual, gamete-forming phase (or generation) of the life-cycle of a plant.

**Guard cells** – Usually a pair of cells surrounding the pore of a stomata, that controls the size of the aperture and thus the movement of moisture and gases in and out of the plant.

**Gymnosperms** – Plants that reproduce by 'naked' seeds (i.e. seeds not enclosed in a carpel).

**Heterospory** – The production of two types of spore by a plant - larger megaspores that each produce a female gametophyte, and smaller microspores that each produce a male gametophyte.

**Holotype** – When a taxon (species, genus, etc.) is first established, the holotype is the one fixed specimen of that taxon. It must always belong to that taxon, no matter what subsequent systematic revisions occur. If the holotype is unavailable, a replacement lectotype must be selected from the original collection. If the original collection is unavailable, then a neotype must be selected from another collection.

**Homospory** – The production of only one type of spore by a plant.

**Hydroid** – An elongate water-conducting cell found in the stems of bryophytes. It is similar in function to the xylem of vascular plants.

**Infrafoliar bladders** – Spongy tissue connected to the parichnos in certain lycopsids, which is thought to be part of an aerating system.

**Integuments** – Protective structures enclosing the nucellus in ovules.

**Intercellular flange** – A cutinized ridge on the inner surface of the cuticle, marking the junction between adjacent epidermal cells.

**Lagenostome** – An apical projection of the nucellus that assisted fertilization in primitive ovules. When it is very prominent, it is sometimes known as a salpinx.

**Leaf cushion** – The swollen basal part of the leaf, especially in lycopsids. It usually remains attached to the stem after the leaf lamina has become detached.

**Leaf scar** – An abscission mark left on the stem or leaf cushion after a leaf has become detached.

**Leaf trace** – The vascular strand that enters the base of the leaf.

**Lianescent** – Vine-like.

**Lignin** – A complex polymer deposited in the walls of vessels, tracheids and fibres to increase their strength.

**Ligule** – A small flap of tissue on the upper surface of the leaf in some lycopsids.

# Glossary

**Lumen** – Central cavity of a cell.

**Medulla** – See Pith.

**Megaphyll** – A large, usually planated, leaf with veins that may be parallel, radiating or meshed.

**Megasporangium** – A spore case producing megaspores.

**Megaspore** – A spore that produces a female gametophyte.

**Meristem** – A zone of actively dividing cells, producing new growth.

**Mesarch** – A description of a vascular strand, in which the older, smaller xylem cells (protoxylem) are in the middle of the strand.

**Mesophyll** – Internal photosynthetic tissue in leaves.

**Metaxylem** – Primary xylem formed after the protoxylem.

**Microphyll** – A small leaf with just a single, or in some cases a pair of veins running along its length.

**Micropyle** – A small pore remaining from the incomplete closure of the integuments in an ovule, through which a pollen grain or pollen tube has to pass to effect fertilization.

**Microsporangium** – A spore case producing microspores.

**Microspore** – A spore that produces a male gametophyte.

**Monopodial** – A style of branching where a main axis produces lateral, subsidiary branches.

**Mycorrhiza** – An association of fungi with the roots of a plant.

**Neighbour cells** – Cells surrounding the stomatal guard cells, which are morphologically indistinguishable from the other epidermal cells (contrast with subsidiary cells).

**Nucellus** – Tissue surrounding the embryo sac in an ovule. It is equivalent to a pteridophyte megasporangium.

**Ontogeny** – Growth and development through the life of an individual organism.

**Ovule** – A female reproductive structure in gymnosperms and angiosperms, which contains an embryo sac surrounded by the nucellus and integuments. It is known as a seed after fertilization.

**Papilla** – Small 'bump' on the plant surface.

**Paracytic** – Stomata where subsidiary cells lie with their long-axes parallel to the guard cells.

**Parenchyma** – Tissue of thin-walled, unspecialized cells that often make up a large part of non-woody plants and plant-organs.

**Parichnos** – A zone of loosely arranged parenchyma that extended from the cortex to the leaf in many lycopsids, which is thought to have had an aerating function. Can often be recognized in leaf scars as a pair of small marks on either side of the vascular trace.

**Pedicel** – A short stalk to which a flower or synangium is attached. Also sometimes used for the short stalk at the base of a pinnule in some fronds.

**Pericycle** – The outermost cells of the stele, consisting mainly of parenchyma. In roots, the pericycle initiates lateral root formation.

**Petiole** – The stalk of a leaf.

**Phloem** – Conducting tissue responsible for the movement of sugars and other nutrients throughout a plant.

**Photosynthesis** – The process whereby green plants trap light in chlorophyll and use it to synthesize carbohydrates from carbon dioxide and water.

**Phyllophore** – A leaf-bearing organ.

**Phyllotaxy** – The pattern of arrangement of leaves on a stem.

**Pinna** – A subdivision of a compound leaf or frond.

**Pinnule** – The ultimate division of a compound leaf or frond.

**Pith** – A zone of central parenchyma within the stele of a stem or root.

**Platyspermic** – Ovules and seeds with a flattened, bilateral symmetry.

**Plinth** – The apical part of the nucellus, below the salpinx.

**Pollen** – The microspores of angiosperms and certain groups of gymnosperm.

**Pollen drop** – Fluid exuded from the distal part of some gymnosperm ovules to capture pollen.

**Polystele** – Vascular system consisting of more than one stele.

**Pre-pollen** – Pteridophyte-like microspores of some primitive gymnosperms.

**Progymnosperm** – An extinct group of plants thought to be ancestral to the true gymnosperms. They had gymnosperm-like woody stems and pteridophyte-like spores.

**Propagule** – Any part of a plant capable of growing into a new individual, e.g. seeds and spores.

**Protostele** – The simplest type of stele consisting of a solid strand of xylem, surrounded by a cylinder of phloem and pericycle.

**Protoxylem** – The older, first-formed xylem in a vascular strand. In cross-section, it can usually be recognized by the smaller diameter of the cells.

**Pseudomonopodial** – A variation on dichotomous branching, where one branch is more prominent than the other.

# Glossary

**Pseudoparenchyma** – Tissue found in some fungi and algae, consisting of an interwoven mass of fine tubes.

**Pteridophytes** – A generalized term used for vascular plants, including ferns, horsetails and club mosses, that reproduce by spores.

**Pteridosperm** – A heterogeneous group of, mainly Palaeozoic, gymnosperms with large dissected leaves which superficially resemble fern fronds.

**Rachis** – The supporting axis of a compound leaf or frond, to which the leaflets or pinnules are attached.

**Radiospermic** – Seeds and ovules that are radially symmetrical.

**Rays** – Radially arranged lines of parenchyma cells in vascular tissue.

**Rhizome** – A horizontal stem, usually underground, that facilitates vegetative propagation.

**Rhizomorph** – Creeping 'stems' of certain algae and fungi.

**Salpinx** – A trumpet-shaped extension of the nucellus that aided pollen-capture.

**Saprophyte** – Plants and fungi that grow on the decaying remains of dead organisms.

**Scalariform thickenings** - Interlocking bands of secondary wall thickenings in vessels and tracheids, forming 'ladder-like' rows.

**Sclerenchyma** – Tissue with strengthened, usually lignified, cell walls.

**Sclerotic** – Thickened with lignin.

**Secondary growth** – The increase in girth of a plant by cell divisions in the cambium. Secondary wood in particular is an important means of increasing the girth of many plants, especially in gymnosperms and angiosperms.

**Seed** – A reproductive structure formed from a fertilized ovule.

**Sessile** – Unstalked.

**Sexine** – The outermost part of the protective coat of a spore.

**Siphonostele** – A stele consisting of a vascular cylinder with a central core or pith.

**Sparganum** – Cortex in which there are radiating bands of vertically-aligned fibrous cells.

**Sporangiophore** – A structure bearing one or more sporangia.

**Sporangium** – A spore case or capsule that produces spores.

**Spore** – A reproductive unit of one or more cells, produced by plants, protozoa and bacteria.

**Sporophyll** – A modified leaf, usually in a strobilus, on which a sporangium is borne.

**Sporophyte** – The spore-producing, non-sexual phase (or generation) in the life-cycle of a plant.

**Stele** – Sometimes known as the vascular cylinder, consisting of xylem and phloem.

**Sterome** – Outer zone of thick-walled cells in the stems of some non-vascular plants (e.g. certain mosses), which assisted with support of the plant.

**Stomata** – Small pores with guard cells in the epidermis, which facilitate the movement of moisture and gases in and out of the plant (singular – stoma).

**Stomium** – An area of thin-walled cells on a sporangium wall, where rupture takes place.

**Strobilus** – A well-defined, terminal, spike of fertile appendages with sporangia.

**Subsidiary cells** – Cells surrounding the stomatal guard cells, which are morphologically distinct from the other epidermal cells.

**Sympodium** – A discrete, axial vascular bundle, from which leaf traces are emitted at intervals.

**Synangium** – A fused cluster of elongate sporangia.

**Taphonomy** – The study of the process of fossilization.

**Terete** – Smooth, cylindrical and tapering.

**Thallus** – A plant body that is not differentiated into leaves, stems, root, etc.

**Tracheids** – Discrete, elongated, water-conducting xylem cells, joined by pits and open ends.

**Trichomes** – Epidermal hairs, that may have a protective function. In some cases, a gland occurs at the trichome tip from which an exudant may be produced.

**Trifurcate** – The production of three branches at one place.

**Trilete mark** – A 'Y'-shaped mark on a spore, formed through the development of the spores in tetrahedrally symmetrical groups.

**Tubercle** – A wart-like projection.

**Vascular plants** – Plants with conducting tissue (xylem and phloem) in the roots, stems and usually the foliage.

**Venation** – The pattern of veins on a leaf or pinnule.

**Vernation** – The way in which a young leaf or shoot is folded when in the bud.

**Vessel** – A series of open-ended cells, arranged end-to-end, to form an elongate tube, found in the xylem of many angiosperms, and in some ferns and gymnosperms.

**Xylem** – Woody conducting tissue responsible for the movement of water and solutes around a plant.

**Zygote** – The product of the sexual fusion of two gametes. In bryophytes and vascular plants, it forms the embryo from which develops the sporophyte.

# *Index*

Page numbers in **bold** type refer to figures and page numbers in *italic* type refer to *tables*.

Aberlemno Quarry 64
Acetate peel method 82, 142
*Achlamydocarpon* sp. 151
*Actinophyllum* sp. 43, 44
Adaptations for life 3, **4**
*Adiantites*
   *antiquus* (Ettingshausen) Kidston 179
   *hibernicus* Forbes 101-2
   *machanekii* Stur 179, **180**
Adpressions 10
*Aglaophyton major* (Kidston and Lang) D.S.
   Edwards 86, 89-90, **90**
*Alcicornopteris*
   *convoluta* Kidston 124
   *hallei* Walton 163
*Aldanophyton* 25-7
*Alethopteris* **201**
   *ambigua* Lesquereux 221
   *grandinioides* Kessler 221
   *lancifolia* Wagner 212
   *lonchitica* Sternberg 216
Algae 3, 28-9, 91, 172, 185, 233
*Algites virgatus* (Munster) Stoneley 233
Allen, J.R.L., and Dineley, D.L. 56
Allen, K.C. 169
   and Marshall, J.E.A. 97
Alvin, K.L. 131
American fossils *see* North America
*Amyelon* 149
   *bovius* Barnard 135
*Aneimites* acadica Dawson 120, 124
Angaran Palaeokingdom 107, 195, 203, 228

Angiosperms 126, 225
*Annularia* 216
   cf. *spicata* Gutbier 231
   *stellata* (Sternberg) Wood 231
Arachnids 92
Arber, E.A.N. 31
   and Goode, R.H. 101, 102
Arbuthnott Group 59, 62, 63-7
Archaeocalamitaceae 111, 155
*Archaeocalamites* 111, 196
   *radiatus* (Brongniart) Stur 102, 160, 174, 177, 186
Archaeopteridaceae 111
*Archaeopteridium tschermakii* (Stur) Kidston 171
*Archaeopteris* 6, 61, 97, 100
*Archaeosigillaria stobbsii* Lacey 172, 173-4, **174**
Archangelsky, S., and Arrondo, O.G. 161-2
Aridity 229
Arnsbergian-Chokierian ($E_2$-$H_1$) stage boundary 107
Arrondo, O.G., and Archangelsky, S. 161-2
Arthropods 91-2
Asbian 141-50, 150-2
*Asterophyllites equisetiformis* Brongniart 216, 231
Asteroxylaceae 89
*Asteroxylon*
   *elberfeldense* Kräusel and Weyland 91
   *mackei* Kidston and Lang 80, 88, **88**, 88-9, 92
   *setchellii* Read and Campbell 91
Atlantic Palaeoarea (Vakhrameev *et al.*) 235

# Index

Auchensail Quarry **78**, 78-80, **79**
Australia 25
*Autunia* 227
Autunian 225
Axes, spiny 46, **49**, 77
Aymestry Limestone Formation 37

Baggy Formation 61, **100**, 100-3, **101**
Ballanucater Farm 75-7
Banks, H.P. 43, 84
*Baragwanathia* 26-8, 57
Barber, C.A. 31, 34
Barinophytales 72, 74, 75
Barinophytopsida 57
Barker, W.R. 215, 216-17
Barnard, P.D.W. 126, 127, 136
    and Long, A.G. 126
*Barrandeina pectinata* Høeg 94
Bassett, M.G. (Edwards, D. *et al.*) 38-9
Bateman, R.M. 131, 137, 138, 145, 155, 160
    and Rothwell, G.W. 135, 136, 137
    *et al.* 131
Bay of Skaill 93-6
Beck, C.B. 161-2
Belgium 62
Bell, J., *et al.* 234
Bennettitales 181
Benson, M. 149
*Bensonites* 140
    *fusiformis* R. Scott 148
*Berwynia carruthersii* Hicks **32**, **34**, 34-5
Bhutta, A.A. 86, 88, 88-9
*Bilignea*
    *resinosa* Scott 162
    cf.*solida* Scott 135
Binney, E.W. 154
Bonamo, P.M., and Leclercq, S. 94
Boon, G. 208
Borders Region (Scotland) 113
*Bothrodendron wardiense* Crookall 181, 183
*Botrychiopsis* tundra 203
Botryopteridaceae 148-9
Botryopteridales 111, 148, 161, 199-200, 206
*Botryopteris*
    *antiqua* Kidston 148
    cf.*antiqua* Kidston 161
Boullard, B., and Lemoigne, Y. 91
Bower, F.O. 186
Bowmanitales 146, 199, **200**
*Bowmanites tenerrimus* (Ettingshausen) Hoskins
    and Cross 174-5
Brack-Hanes, S.D. 206
    and Thomas, B.A. 12, 145
Branch knots 91

Brindley, S.A. (Scott, A.C. *et al.*) 151, 152
Brithdir Beds 218-19
Brousmiche, C. 216
Brownstone Group 67
Bryophyta 28, 88, 168
Burdiehouse Limestone 113, 183
Burgess, N.D., and Edwards, D. 31, 44, 91
*Burnitheca pusilla* Meyer-Berthaud and Galtier
    161
*Buteoxylon* 126
    *gordonianum* Barnard and Long 135
    *Tantallosperma* correlation 134
Buteoxylonaceae 126, 135
*Bythotrepis* sp. 185

*Caia langii* Fanning, Edwards and Richardson 44,
    45
*Calamites* 96-7, **200**, 206, 218, 220
    *carinatus* Sternberg **212**, 220
    *pettycurensis* Scott 145
    *undulatus* Sternberg 221
Calamopityales 114, 116, 118, 119, 120, 124-6,
    160, 163, 179, 188
*Calamopitys radiata* Scott 163
Calamostachyales 199, **200**
*Calamostachys paniculata* Weiss 212
*Calathiops*
    *acicularis* Göppert 178
    *dyserthensis* Lacey 172, 175
    *glomerata* Walton 178-9
    *gothanii* Benson 179
    *renieri* Walton 179
    *trisperma* Smith 163
*Calathopteris heterophylla* Long 135
*Calathopteris-Calathospermum-Salpingostoma*
    134
*Calathospermum*
    *fimbriatum* Barnard 121, 136, 162
    *scoticum* Walton 162
Cf.*Calathospermum* 152
Calder, M.G. 155
Caldy Island (Dyfed) 62, 66
Callistophytales **202**, 202, 227
*Camptosperma berniciense* Long 122-4, **123**
Canonbie Coalfield 219, **220**
Capel Horeb Quarry 39-43, **40**
Carbonate beds 225
Carboniferous 8
    Lower 107-91
        palaeogeography 107, **109**
        plant fossils 112-14
        research history 20-1
        stratigraphy 107-9, **130**
        vegetation 110-12

# Index

Upper 8, 10, 193-221
   palaeogeography **195**, 195-6
   plant fossils 203-4
   research history 21-2
   stratigraphy **109, 110**, 196, **204**
   vegetation 196-203
Carboniferous-Permian boundary 225
*Cardiopteridium nanum* (Eicinwald) Walton 188
*Carpolithus* 179
   *puddlebrookense* Thomas and Purdy 171
Casts 10
Cathaysia Palaeokingdom 199, 227, 228
Cattybrook Claypit **210**, 210-14, **211**
*Caulopteris? peachii* Salter 94
Cementstone Group 112-27, **117**, 164
Chaloner, W.G. (Kevan, P.G. *et al.*) 85
Chaphekar, M. 145, 160, 183
Charales 6
Charophytes 28
Cheirostrobaceae 141, 146, 150
*Cheirostrobus pettycurensis* Scott **146**, 146
Chert *see* Rhynie chert
China 227-8, 236
Chlorophycophyta 6
   Ballanucater Farm 75
   Craig-y-Fro Quarry 72
   Freshwater East 46
   Llanover Quarry 67
   Perton Lane 43
   Rhynie 84
   Targrove Quarry 61
   Turin Hill 64
Cladoxyales 161
*Cladoxylon*
   *kidstonii* Solms-Laubach 119
   cf.*taeniatum* Bertrand 161
   *waltonii* Long 119
   cf.*waltonii* Long 134
Clayton, G. (Scott, A.C. *et al.*) 151, 152, 157-9, 161
Cleal, C.J., and Thomas, B.A. 213
*Clepsydropsis* 161
   *antiqua* Unger 119
Clocksbriggs Quarry (Wemyss Quarry) 64
Club-mosses *see* Lycopsida
*Clwydia decussata* Lacey 173-4
Clyde Plateau Volcanic Formation 157, **158, 159**, 164
Coal formation 186
Coal Measures 203, 204-7, 208-10, **209**, 210-14, **211**, 214-17
Coal-balls 21, 22, 199, 203-4, 219
Coalfields 195-6, 203, 217
Coenopteridales 199

Compressions 10
Conifers (Pinales) 3
   Permian 196, 227, **228**
   Triassic 225
   Upper Carboniferous 202-3
*Cooksonia* 38-9
   *caledonica* Edwards 63, 66, 67
   *cambrensis* Edwards 38, **45**, 46
   *downtonensis* Heard 42
   *hemisphaerica* Lang 42, 43, 46, 61, 62
   *pertoni* Lang 38, **39**, 42, 43, **44**, 44, 45
*Cordaitanthus*
   *flagellibracteatus* Barker 217
   *nostellensis* Barker 217
Cordaites 126, 196, **202**, 202-3, 214, 215, 216-17
'Corduroy plant' 96-7
Corynepteridaceae 161, 163
*Corynepteris* 161
Craig-y-Fro Quarry 70, 71-5, **72**
Crane, P.R. 227
Croft, W. 67
   and Lang, W.H. 68, 70, 71
Crookall, R. 21, 31, 126, 183, 206, 212, 218, 220
*Crossotheca* cf.*crepinii* Zeiller 216
Croyde Hoe Quarry 101, 103
*Cryptoxylon forfarense* Kidston 65
Cupules, ovulate, pteridosperm 136-7
Cuticles 6, 75, 76, 172, 175
Cwm Craig Ddu Quarry 38-9
Cyanochloronta 84
*Cyathocarpus* sp. 221
Cycadales (cycads) 3, **226**, 227, 228, 233, 234
Cycadopsida
   Cattybrook Claypit 211
   Jockie's Syke 220
   Middridge Quarry 233
   Nant Llech 206
   Nostell Priory Brickpit 214-15, 216
   Stairhill 231
   Teilia Quarry 176, 179
*Cyclostigma* 61
*Cyrtograptus murchisoni* Zone 31
*Cystosporites devonicus* Chaloner and Pettit 119
Czech Republic 62, 66

Daber, R. 25
*Dadoxylon* sp. 152, 218
Dasyclades 37
Davies, D. 21
*Dawsonites* 77, 80
   *arcuatus* Halle 69-70, **70**, 71
*Deheubarthia splendens* Edwards, Kenrick and Carluccio 68
Deltas, Upper Carboniferous 10, 195, 203

# Index

*Deltasperma fouldenense* Long 122, **123**, **133**, 138
Desiccation 25
Devonian 6-8, 53-103
   Lower **48**, **59**
   Middle 61
   palaeogeography 53-6, **55**
   plant fossils 59-60
   research history 19-20
   stratigraphy 56, **61**, **130**
   Upper 61
   vegetation 56-9
*Dichotangium quadrothecum* Rowe 170
*Didymosporites scottii* Chaloner 148
Dilcher, D.L., and Retallack, G.J. 120-1, 124
DiMichele, W.A. 144, 154
   (Bateman, R.M. *et al.*) 131
Dineley, D.L., and Allen, J.R.L. 56
*Diplolabis roemeri* (Solms-Laubach) Bertrand 147, **148**
*Diplopteridium* **113**
   *holdenii* Lele and Walton 167, 169-70, **171**, 172
   *teilianum* (Kidston) Walton 176, 178
Dispersal *see* Reproduction and dispersal
Ditton Group 59
Dix, E. 21, 204, 206-7, 213-14
*Dolichosperma*
   *pentagonum* Long 124
   *sexangulatum* Long 124, **125**
   cf.*sexangulatum* Long 137
Drepanophycopsida 89
*Drepanophycus spinaeformis* Göppert 68-9, 71, 76, **76**, 80
Drybrook Sandstone Formation 113, **167**, 167
Dryden Shale Formation 84
Dundee Formation 64
Dyfed **48**, 62, 66
Dyserth 172-5

Eames, A.J. 88
*Eccroustosperma langtonense* Long 122, **123**
Cf.*Eccroustosperma langtonense* Long 138
Edwards, D. 19, 25, 35, 42, 53, 66, 68, 73, 74
   and Burgess, N.D. 31, 44, 91
   and Edwards, D.S. 42
   and Fanning, U. 62
   and Lyon, A.G. 88
   and Rogerson, E.C.W. 42
   and Rose, V. 62
   *et al.* 38-9
      (Fanning, U. *et al.*) 44
Edwards, D.S. 85, 86, 89-90, 91
   and Edwards, D. 42

Eggert, D.A 86
El-Saadawy, W.E.L.-S., and Lacey, W.S. 86-7, 89
Emergences 6
Emsian 56, 61, 75-7, 78-80
*Endoxylon zonatum* (Kidston) Scott 166
Enville Formation 229
*Eosperma*
   *edromense* Long 122, **123**
   *oxroadense* Barnard 137, 138
Eospermaceae 122-4
Equatorial and low latitudes flora phytogeographic subunit
   American 77
   European 70
Equisetales 145
Equisetopsida 8
   Carboniferous
      Lower 107, **108**, 111
      Upper **197**, 199, **200**
   Cattybrook Claypit 211, **212**
   Devonian **54**, 59
   Glenarbuck 166
   Glencartholm 185, 186
   Jockie's Syke 220
   Kingswood End 150
   Laggan 153, 155
   Loch Humphrey Burn 159, 160
   Middridge Quarry 233
   Moel Hiraddug 173, 174
   Nant Llech 206
   Nostell Priory Brickpit 214, 216
   Oxroad Bay 130, 134
   Permian 225-7, **226**
   Pettycur 143, 145-6
   Plaistow Quarry 101
   Stairhill 231
   Teilia Quarry 176, 177
   Triassic **226**
   Wardie Shore 181
*Eristophyton*
   *beinertianum* Zalessky 135
   *fasciculare* Scott 162
   *waltonii* Lacey 162
*Eskdalia* 174, 183
   *fimbriophylla* Rowe 169
   *minuta* Kidston 185
   *variabilis* (Lele and Walton) Rowe 169, **170**
*Etapteris tubicaulis* Göppert 161
Etheridge, R. 29, 101-2
Etruria Formation 203, 217
Euramerian Palaeokingdom 107, 196, 225, 232
   Palaeophytic-Mesophytic transition 229
*Eurystoma*
   *angulare* Long **124**, 125, 171

# Index

*burnense* Long 124
Cf.*Eurystoma burnense* Long 137
Eurystomaceae 121
Evolutionary history 12
Extinctions 8

Fair Isle **96**, 96-100, **98**, **99**
Famennian 59
Fanning, U.
   and Edwards, D. 62
   *et al.* 44
Fern-like plants 58, 107, 111
Filicopsida (ferns) 8
   Carboniferous
      Lower 107, **108**, 111
      Upper 196, **197**, 199
   Cattybrook Claypit 211
   Devonian **54**
   Glenarbuck 166
   Glencartholm 185, 186
   Jockie's Syke 220
   Kingswood End 150, 152
   Loch Humphrey Burn 155, 159-60, 161
   Nant Llech 206
   Nostell Priory Brickpit 214, 216
   Oxroad Bay 130, 134
   Permian 225, **226**
   Pettycur 141, 142, 143, 146-8
   Stairhill 231
   Teilia Quarry 176, 177
   Triassic 225, **226**
   Weak Law 139
   Whiteadder 118, 119-20
Filzer, P. 85
Flemingitaceae 145
*Flemingites*
   *allantonensis* (Chaloner) Brack-Hanes and
      Thomas 119
   *scottii* (Jongmans) Brack-Hanes and Thomas
      144
Flett, J.S. (Kidston, R. *et al.*) 139
'Florensprung' 107
Florin, R. 236
Foel Formation 172-5
Foliar organs, pteridosperm 136
Forest of Dean 61
   *see also* Puddlebrook Quarry
Forests 8, 107, 229
   *see also* Tropical swamp-forests; Wadsley Fossil
      Forest
Fossil, definition 8
Fossilization processes 8-10, **9**
Freshwater East **45**, 45-9, **47**
Fructifications 74-5, 80, 176, 178, 186-8

Fungi 85, 91-2
Fusain **133**, 143, 150, 151-2, 169

Galtier, J.
   and Scott, A.C. 140
   (Scott, A.C. *et al.*) 151, 152, 157-9, 161
Gametophytes 3-4, 70, 86, **90**, 90
Garleton Hills Volcanic Formation, North Berwick
      Member 129-39, **130**, **132**, **133**
Garwood, E.J. (Kidston, R. *et al.*) 139
Gastaldo, R.A. 190, 209
Gedinnian 56, 59, 61-3
*Geminitheca scotica* Smith 162
*Genomosperma*
   *kidstonii* Long 116, 121, **122**, 127
   *latens* Long 116, 121, **122**, 122
Gensel, P.G. 42
Genus concept 11-12
Gigantopterids 228, 232
Ginkgophytes 227, 233
Glasgow
   area **190**
   'Fossil Grove' 188-91, **189**
Gleicheniaceae 199-200
Glenarbuck 149, **157**, 164-6, **165**
Glencartholm 184-8
Glencartholm Volcanic Group 184-8
Gondwana 25, 107, 195, 203
   Laurussia collision 195, 196, 225, 229, 236
Goode, R.H., and Arber, E.A.N. 101, 102
Gordon, W.T. 20, 127, 137, 139, 145, 149
   (Kidston, R. *et al.*) 139
Gorstian 35
*Gosslingia* 57
   *breconensis* Heard 68, 71, 73, **74**, 75
Gothan, W. 107
Graig Quarry 175
*Granulatisporites* 148
'Greenhouse effect' 229
Gronant Group 175-81
Gymnospermophyta 5, 8
   Carboniferous 107, 112, 200-1
   Devonian **54**, 59
   Loch Humphrey Burn 160
   Oxroad Bay 130-1
   Plaistow Quarry 101
   Triassic 225
   Whiteadder 120-4, 126

Hall, T.M. 101
Halle, T.G. 77
Hass, H. (Taylor, T.N. *et al.*) 91
Heard, A. 71, 72, 74-5
Heerlen classification 196

# Index

Hemsley, A.R. 58, 65
*Heterangium grievii* Williamson 149
Heterospory 5
*Heterotheca grievii* Benson 149
Hicks, H. 34, 35
*Hierogramma* 161
   *mysticum* Unger 119
Hilton Beck Plant Bed 235
*Hiltonia rivulii* Stoneley 235-6
*Holcospermum ellipsoideum* (Göppert) Walton 179
Holden, H.S. 148
Holden, J. (Bell, J. et al.) 234
Horne, J. (Kidston, R. et al.) 139
*Hornea lignieri* Kidston and Lang *see below* Horneophyton lignieri
*Horneophyton lignieri* (Kidston and Lang) Barghoorn and Darrah 86-8, **87**, 90
Horneophytopsida 85, 86-8
Horsetails *see* Equisetopsida
*Hostinella* 38, 39, 46
   *heardii* Edwards 71, 74
Hueber, F.M. 69
*Hydrasperma tenuis* Long 121, **122**, 122, **132**, 136-7, 137
Hydrasperman reproduction concept 135
*Hyenia* Zone 93, 94-5

Impressions 10, 211
*Inopinatella lawsonii* Elliot 36-8, **38**

Jockie's Syke **219**, 219-22

*Kaloxylon* sp. 155
*Kalymma* 152
   *tuediana* Calder 124
   cf.*tuediana* Long 163
*Karinopteris acuta* (Brongniart) Boersma 210, 212-13, **213**, 214
Keele Formation 203, 204, 229, 232
Keltie Water 80
Kerp, J.H.F. 227, 231
   and Poort, R.J. 236
Kerry Head (Irish Republic) 127
*Kerryia* 121
Kevan, P.G., *et al.* 85
Kidston, R. 20, 21, 65, 94, 107, 117, 170-1, 177, 178, 183-8, 199, 220
   and Lang, W.H. 19, 65, 81-2, 84, 85, 87, 88-9
   *et al.* 139
*Kidstonophyton discoides* Remy and Hess 90
Kilpatrick Hills *see* Glenarbuck; Loch Humphrey Burn
Kiltorcan assemblage 61

Kimberley Railway Cutting 236-8
King, Wickham 67
King, W.W. 46
Kingswood End 150-2, **151**
Kingswood Limestone 150
Kingwater 140-1
*Knorria* sp. 102
Kräusel, R., and Weyland, H. 68-9
*Krithodeophyton croftii* Edwards 71, 73-4, 75
Kuznetsk (Siberia) 62, 66

Lacey, W.S. 20, 126, 173-4
   and El-Saadawy, W.E.L.-S. 86-7, 89
   and Matten, L.C. 136, 162
Lagenostomales **112**, 112, **113**, 160, 162-3, 201-2
Lagenostomopsida
   Cattybrook Claypit 211-12
   Glenarbuck 166
   Glencartholm 185, 186-8
   Kingswood End 150, 152
   Laggan 153
   Lennel Braes 116
   Loch Humphrey Burn 162-3
   Moel Hiraddug 173
   Nant Llech 206
   Nostell Priory Brickpit 215, 216
   Pettycur 143, 148
   Plaistow Quarry 101
   Puddlebrook Quarry 168-9
   Teilia Quarry 176, 178-9
   Wardie Shore 181
   Weak Law 139-40
   Whiteadder 118, 119, 120-4
Laggan 152-5
Lang, W.H. 19, 34, 44, 45, 62, 76-8, 91, 94
   and Croft, W. 68, 70, 71
   *see also* Kidston, R., and Lang, W.H.
*Langiophyton mackei* Kidston and Lang 90
Langsettian 210-14
Langton Burn and Cove 126
Laurussia 25, 53, 70, 107
   *see also* Gondwana
*Laveineopteris loshii* (Brongniart) Cleal, Shute and Zodrow 212, 213, 216
*Laveineopteris loshii* Subzone 212, 213
*Laveineopteris rarinervis* Subzone 215
Leclercq, S., and Bonamo, P.M. 94
Leisman, G.A., and Schlanker, C.M. 155
Lele, K.M., and Walton, J. 169, 170-1, 172
Lemoigne, Y. 86, 88
   and Boullard, B. 91
   and Zdebska, D. 86

# Index

Lennel Braes 113-16
Lepidocarpaceae 111
*Lepidocarpon*
   *lomaxii* Scott 145
   *wildianum* Scott 145
*Lepidodendron* 65, 175, 176-7, **199**
   *aculeatum* Sternberg 206
   *calamopsoides* Long 119
   *mannebachensis* Presl 206
   *obovatum* 206
   *veltheimianum* Sternberg 183, 185
*Lepidodendron?*
   *pettycurense* Kidston 145
   *solenofolium* Smith 166
*Lepidodendropsis jonesii* Lacey 172, 174
*Lepidodendropsis-Rhacopteris-Triphyllopteris* 107
*Lepidophloios*
   *kilpatrickensis* Smith 160, 166
   *scottii* Gordon 145
   *wuenschianus* (Williamson) Walton 153-4, **154**
*Lepidostrobophyllum fimbriatum* (Kidston) Allen 169, 174, 185
*Lepidostrobus* 185
   *cylindrica* Gordon M.S. 145
   *ornatus* Brongniart 185, 206
   *variabilis* Lindley and Hutton 185, 206
*Lesleya* 235
Levée-banks 204-7, 210-14, 214-17
*Levicaulis arranensis* Beck 154
Libya 25
Lignin 10
Limestone Coal Group 188-91, **189**, **190**
Limonitization 10
Liverworts *see* Bryophyta
Llanbradach Quarry 217-19, **218**
Llangammarch Wells Quarry 35, **36**
Llanover Quarry 67-71
*Lobatopteris*
   *geinitzii* (Gutbier) Wagner 231-2, **232**
   *miltoni* (Artis) Wagner 216
*Lobatopteris micromiltoni* Zone 219-22
Loch Humphrey Burn 155-64, **156**, **157**, **158**, **159**
*Lonchopteris rugosa* Brongniart 212, 214
Long, A.G. 21, 117, 119, 120-1, 124, 125, 126, 127, 135, 136-7, 139-40
   and Barnard, P.D.W. 126
   and Walton, John 149
Long Quarry Formation **40**, **41**, 41-2
Low latitudes *see* Equatorial and low latitudes flora
Lower Brown Limestones 172-5

Lower Carboniferous *see* Carboniferous, Lower
Ludfordian Upper Perton Formation 43
Ludlow Series 26
*Lycopodites stockii* Kidston 186
Lycopsida 5, 8, 75, 153-5, 186, 229
   Auchensail Quarry 79
   Ballanucater Farm 76
   Bay of Skaill 94
   Carboniferous
      Lower 107, **108**, 111
      Upper 196-9, **197**, **199**
   Cattybrook Claypit 211
   Craig-y-Fro Quarry 72
   Devonian 53, **54**, 57, 58
   Glenarbuck 166
   Glencartholm 185-6
   Jockie's Syke 220
   Kingswood End 150
   Lennel Braes 116
   Llanover Quarry 68, 71
   Loch Humphrey Burn 160
   Moel Hiraddug 172
   Nant Llech 205-6
   Nostell Priory Brickpit 214, 215-16
   Oxroad Bay 130, 131-4
   Permian 225, **226**
   Pettycur 141, 143, 144-5
   Plaistow Quarry 101
   Puddlebrook Quarry 168, 169
   Rhynie 85
   Silurian 26-8
   Teilia Quarry 176-7
   Triassic **226**
   Victoria Park 188-91, **189**
   Wardie Shore 181
   Weak Law 139
   Whiteadder 118, 119-20
*Lycostachys protostelicus* Pant and Walton 154-5
Lydienne Formation (France) 127
*Lyginopteris bermudensiformis* (Sternberg) Patteisky 178
*Lyginopteris hoeninghausii* Zone 204, 206-7, 210-14
*Lyginorachis* 135, 136, 152, 162
   *arberi* Long 121
   *brownii* Calder 166
   *gordonii* Galtier and Gordon 140
   *kidstonii* Long 121
   *trinervis* Calder 162
   *waltonii* Calder 135, 155
Lyon, A.G. 88-9
   and Edwards, D. 88
*Lyonophyton*
   *rhyniensis* Remy and Remy **90**, 90
   *Sciadophyton* correlation 91

# Index

*Lyrasperma, scotica* (Calder) Long **114**, 124, 124-6, **125**

MacGregor, M., and Walton, J. 190
Majonicaceae 227, 233-6, **235**, 236
Marattiales 111, 163, 196, 200
*Margophyton? goldschmidtii* (Halle) Zacharova 77
Marine algae *see* Algae
*Mariopteris*
   *sauveurii* (Brongniart) Zeiller 216
   cf.*sphenopteroides* (Lesquereux) Zeiller 206-7
Marl Slate 229, 234, 236
Marshall, J.E.A., and Allen, K.C. 97
Matten, L.C. 97
   and Lacey, W.S. 136, 162
Mauchline Volcanic Group 229-33
*Mazocarpon pettycurense* Benson 119, 145
Medullosales *see* Trigonocarpales
Megacupules 162
Megaphylls 6
cf.*Melissiotheca* 138
*Melissiotheca longiana* Meyer-Berthaud 150, 152
Merker, H. 86
cf.*Mesoxylon* sp. 218
*Metaclepsydropsis* 161
   *duplex* (Williamson) Bertrand **147**, 147
*Metadineuron ellipticum* (Kidston) Galtier 147
Meyen, S.V. 88, 203, 235
Meyer-Berthaud, B. 152
   (Scott, A.C. *et al.*) 151, 152
Microcupules 162
Microphylls 6
Middle Rock Formation 208-10, **209**
Middridge Quarry 233-6
Midlands 217
Milford Haven Group 46, **47**
Miller, H. 19
*Milleria pinnata* Lang 94
Millstone grit 203
*Mitrospermum bulbosum* Long **126**, 126
Cf.*Mittagia seminiformis* Lignier 166
*Mixoneura* 232
Moel Hiraddug 172-5, **173**
*Monograptus riccartonensis* Zone 31
Moulds 10
*Musatea*
   *duplex* Chaphekar and Alvin 147
   *globata* Galtier 147
*Muscites plumatus* Thomas 167, 169
Mycelia 91
Mykura, W. 97
Myreton Quarry 66

Namurian 196, 203
Nant Llech 204-7, **205**
*Nematasketum* 91
   *diversiforme* Burgess and Edwards 44, 61, 62
Nematophytes 28, 90-1
*Nematophyton* 65
*Nematoplexus* 42
   *rhyniense* Lyon 91
*Nematothallus* 28, 41, 57-8, 71
   *pseudovasculosa* Lang 62
Namurian 113
*Neocalamites mansfeldicus* Weigelt 234
*Neuralethopteris jongmansii* Laveine 207, **208**
*Neuralethopteris jongmansii* Subzone 207, **208**
*Neuropteris antecedens* Stur 179
*Neuropteris antecedens* Zone 184, 188
*Neuropteris? huttoniana* King 234
New Albany Shales (USA) 127
New Red Sandstone 231
Newton Dingle (Shropshire) 62, 66
Niklas, K.J. 65, 121
*Noeggerathia* 200
Nomenclature 10-12, **11**
North America 199, 228, 232
North America Palaeokingdom 228, 232
North Gavel Formation 96, 97
Northern hemisphere 228
Nostell Priory Brickpit 214-17, **215**
*Nothia aphylla* Høeg 89-90

Oberste-Brink, K. 177
*Odontopteris cantabrica* Zone 204
*Odontopteris subcrenulata* var. *gallica* Doubinger and Remy 232, 233
Oil Shale Group of Scotland 113, 181-4
Old Red Sandstone 53, 67-71
Ovules 116, 120-4, **122**, **123**, 124-6, **125**, 137-8
Oxroad Bay 127-39, **128**, **129**, **130**
*Oxroadia*
   *conferta* Bateman 131, **133**
   *gracilis* Alvin **111**, **113**, 119, 131, **132**, **133**, 151
   *parvus* Long 135

*Pachytheca* 28, 34, 57-8, 65, 68, 72, 91
Palaeobotanical problems 8-12
Palaeoenvironmental analyses 12
*Palaeomyces* sp. 76-7, 91
*Palaeonitella cranii* (Kidston and Lang) Pia 91
Palaeophytic 3, 195
Palaeophytic-Mesophytic transition 225, 229
*Palaeostachya ettingshausenii* Kidston 216
Pangaea 107, 195, 196, 199, 225
Pant, D.D. 81, 86

# Index

*Paracalamites kutorgai* (Geinitz) Zalessky 234
Paralic Belt 195-6
*Paralycopodites brevifolius* (Williamson)
    DiMichele 119, **144**, 144-5
*Paripteris linguaefolia* Zone 214, 215
*Parka* 35, 61, 65
    *decipiens* Fleming 29, 58, 63, **64**, 64, 65, 67
Parrish, J.T., and Raymond, A. 107
*Paurodendron* 131
    *arranensis* Fry 155
cf.*Paurodendron* 134
Peach, B.N. (Kidston, R. *et al.*) 139
Peach, C.W. 65
Peat deposits 195-6, 196-9
*Pecopteris* sp. 206
Peel method *see* Acetate peel method
Peltaspermaceae 236-7
Peltaspermales 227, 234, 236
*Peltaspermum* 227
    *martinsii* (Germar) Poort and Kerp 234,
    235-7, 236
Pen-y-Glog Grit Formation 31, **32**
Pen-y-Glog Quarry 31-5, **32**, **33**
Pen-y-Glog Slate Formation 31, **32**
Penhallow, D.P. 65
Pennant Measures 218-19
Permian 196, 225-38
    palaeogeography 225, **227**
    plant fossils 229
    research history 22
    stratigraphy 225, **231**
    Upper 8, **230**, 232-6
    vegetation 225-8
Permineralization 113
Perton Lane 43-5
*Pertonella dactylethra* Fanning, Edwards and
    Richardson 44
Petrifactions 10, 113
Pettigrew, T.H. (Bell, J. *et al.*) 234
Pettitt, J.M. 138
Pettycur 141-50, **142**, **143**, 152
Pettycur Limestone 143, 144-5, 146, 149
*Phacelotheca pilosa* Meyer-Berthaud and Galtier
    150, 152
Phaeophycophyta
    Auchensail Quarry 79
    Capel Horeb Quarry 41
    Craig-y-Fro Quarry 71
    Freshwater East 46
    Llanover Quarry 67
    Perton Lane 43
    Rhynie 84
    Targrove Quarry 61
    Turin Hill 64

Phillips, J. 44
*Phlegmaria* 186
Photosynthesis 6
Phylogenetic radiation 112
Pilton Formation **101**
Pinales *see* Conifers (Pinales)
Pinopsida
    Cattybrook Claypit 212
    Kimberley Railway Cutting 236
    Middridge Quarry 233
    Nant Llech 206
    Nostell Priory Brickpit 216-17
    Permian and Triassic **226**
    Upper Carboniferous **197**
*Pitus*
    *antiqua* Witham 116
    *dayi* Gordon 139-40
    *primaeva* Witham 120, 139, **141**
*Plagiozamites middridgensis* Schweitzer 235
Plaistow Quarry 100-3
Plant architecture 5-6
Plant systematics 12-14
Polar ice-cap 203
Pollak Stollen Formation (Upper Silesia) 188
Pollen-organs, pteridosperm 138
Poort, R.J., and Kerp, 236
*Pothocites grantoni* Paterson 160, 181, 183, 184,
    **186**, 186
*Powysia bassettii* Edwards 35, **37**
Pre-adaptation hypothesis 3
Pre-Silurian vegetation, evidence 25-7
Preservation, modes 10, **11**
Přídolí 45-9, **48**
Progymnospermopsida 8
    Bay of Skaill 94-6
    Carboniferous **108**, 111, **197**, 200
    Devonian **54**, 58
    Glencartholm 185, 186
    Loch Humphrey Burn 155, 160, 161-2
    Moel Hiraddug 173, 175
    Permian and Triassic **226**
    Teilia Quarry 176, 177-8
*Protocalamites* 155
    *goeppertii* (Solms-Laubach) Bateman 155, 160
    *longii* Bateman 131, 134
    *pettycurensis* (Scott) Scott 145, 152
*Protocalamostachys*
    *arranensis* Walton 155, 160
    *farringtonii* Bateman **132**, 134
    *pettycurensis* Chaphekar 145, 160
*Protoclepsydropsis kidstonii* (Bertrand) Hirmer
    119-20
Protolepidodendrales 89
Protopityaceae 111

# Index

Protopityales 155, 161
*Protopitys*
  *buchiana* Solms-Laubach 161
  *scotica* Walton 161
*Protopteridium* 6, **58**, 58, 100
  *thomsonii* (Dawson) Kräusel and Weyland 93, 94, **95**, 96
*Prototaxites* 28, 57-8, 62, 65, 79, 80, 97
  cf.*calendonianus* (Lang) Kräusel and Weyland 68
  *hicksii* (Etheridge) Dawson 31, **32**, 34
  (*Nematophyton*) sp. 72
  *storrei* (Barber) Dawson 31
  *taitii* (Kidston and Lang) Pia 90-1
*Prototaxites-Pachytheca* assemblage 31-4
Provincialism 53, 56, 107, 196, 225
*Psalixochlaena berwickense* Long 120
*Psaronius* sp. 218
*Pseudoctenis middridgensis* Stoneley 234
*Pseudosporochnus* 58, 58
*Pseudovoltzia liebeana* (Geinitz) Florin 234, 236, **237**
Psilophytes 89
*Psilophytites* sp. 46, 49
*Psilophyton* 77
  *goldschmidtii* Halle 77
*Psilophyton* Zone 67, 70
*Psygmophyllum cuneifolium* (Kutorga) Schimper 234-5
Pteridophyta 3-4, 8, 225
Pteridosperms 115-16
  callamopityalean 196
  callistophytalean **202**, 202
  Carboniferous
    Lower **112**, 112, **113**, **114**, **118**
    Upper 196, **197**, 200-1, **201**, **202**
  lagenostomalean 201-2
  Laggan 155
  Loch Humphrey Burn 155, 162
  Moel Hiraddug 175
  Nant Llech 206
  Oxroad Bay 134-8
  Permian **226**, 227
  Puddlebrook Quarry 169
  Triassic **226**
  trigonocarpalean 196, **201**, 201
  Whiteadder 120-4
Pterinea Beds 38
Puddlebrook Quarry **167**, 167-72, **168**
*Pullaritheca* 121
  *longii* Rothwell and Wight **132**, 136-7
Purdy, H.M., and Thomas, B.A. 169, 171
Pyritization 10

Raudfjorden (Spitsbergen) 25
Raymond, A. 53-6, 107
  and Parrish, J.T. 107
Rayner, D.H. 76-7
Reconstructions, whole-plant 10, 134-8
Red Marls Formation 41
Red Sandstone Group 219
Red-beds 225
*Remia pinnatifida* (Gutbier) Knight 232
Remy, W. 86
  and Remy, R. 235
  (Taylor, T.N. *et al.*) 91
*Renalia* 86
*Renaultia* 206
*Renaultia?* *crepinii* (Stur) Kidston 206, **207**
Reproduction and dispersal 3-5
Research history 19-22
Retallack, G.J., and Dilcher, D.L. 120-1, 124
Rex, G.M.
  and Scott, A.C. 143, 149, 152
  (Scott, A.C. *et al.*) 151, 152
*Rhacophyton* Zone 100-3
*Rhacopteris* 111, 176
  (*Anisopteris*) 177
  *circularis* Walton 177
  *fertilis* Walton 178
  *geikiei* Kidston 184, 186
  *inaequilaterata* (Göppert) Stur 177
  *lindsaeformis* (Bunbury) Kidston 186
  *machanekii* Stur 177
  *petiolata* (Göppert) Schimper 177-8
  *robusta* Kidston 177
  *subcuneata* Kidston 175
  *weissii* (Walton) Hirmer 172, 175
*Rhetinangium arberi* Gordon 121, 149
*Rhodeopteridium*
  *machanekii* (Ettingshausen) Purkyňová 184, 186, 188
  *tenue* (syn. Rhodea tenuis Gothan) 177
*Rhynia* 6
  *Salopella* correlation 91
  *gwynnevaughanii* Kidston and Lang 80, **85**, 85-6, **87**, 92
  *major* Kidston and Lang 86
Rhyniaceae 73
Rhynie 19, 20, 80-92, **81**, **82-3**, **92**
Rhynie chert 8, 61, **81**, 81, **82-3**, **84**, 84
Rhyniophytina 86
Rhyniophytoids 26-9, **28**, 39, 41-6, 49, 56, 61-4, 67, 79
Rhyniopsida 72, 84-5
Rice, C.M., and Trewin, N.H. 84
Richardson, J.B., (Fanning, U. *et al.*) 44
Ridgeway Thrust Zone 210

# Index

Ritchie, A. 34
Rockhall Quarry 36-7
Rogerson, E.C.W.
 and Edwards, D 42
 (Edwards, D. *et al.*) 38-9
Roots and rooting structures 6, 135
Rose, V., and Edwards, D. 62
Rosettes 76
Rothwell, G.W.
 and Bateman, R.M. 135, 136, 137
 and Wight, D.C. 137
Rowe, N.P. 20, 167, 169-70, 174
Rufloriaceae 228
Rushall Formation 43

Saalfeld 127
Sahara 34
St Maughan's Group 61
*Salopella-Rhynia* correlation 91
*Salpingostoma dasu* Gordon 121, **122**, 122, 137
*Samaropsis*
 *bicaudata* Kidston 124
 *triangularis* (Geinitz) Seward 236, **237**
Sandwick Fish Bed **93**, 93
Sap-sucking aniamls 92
*Sartilmania jabachensis* (Kräusel and Weyland) 89
Savile, D.B.O. (Kevan, P.G. *et al.*) 85
*Sawdonia ornata* (Dawson) Hueber 76, 79-80
Saxonian 229
Saxony 28
Scheckler, S.E. 97
Schlanker, C.M., and Leisman, G.A. 155
Schweitzer, H.-J. 234, 236
*Sciadophyton*
 *Lyonophyton* correlation 91
 *steinmannii* Kräusel and Weyland **90**
 cf.*steinmannii* Kräusel and Weyland 70
Scotland 113, **115**
Scott, A.C. 21
 and Galtier, J. 140
 and Rex, G.M. 143, 149, 152
 *et al.* 151-2, 157-9, 161
*Scutelocladus variabilis* Lele and Walton 169
Secondary wood 6
Sedman, K.W. (Bell, J.*et al.*) 234
Seed plants *see* Gymnospermophyta
Selaginellaceae 169
*Selaginellites resimus* Rowe 169
Senni Beds 61, 67, 71-5, **72**
*Sennicaulis hippocrepiformis* Edwards 70, 71, 74
Sermeyaceae 200
*Setispora pannosa* (Alvin) Spinner 131, **133**
Siegenian 56, 57, 61, 67-71

*Sigillaria* sp. 177
Sigillariostrobaceae 119, 145
Silurian 6, 25-49
 palaeography 25, **27**
 plant fossils 29-31
 research history 19
 stratigraphy 25, **30**
 vegetation 26-9
Sites, choice *14*, 14-15, **15**
Skaill, Bay of *see* Bay of Skaill
Sloagar **96**, 96-100, **98**, **99**
Smith, D.L. 124, 136, 144, 162, 166
Smith, G.M. 88
Sorby, H.C. 208-9
Southern hemisphere 228
*Spathulopteris* 163
 *clavigera* (Kidston) Walton 179
 *decomposita* Kidston 184, 188
 *dunsii* Kidston 183
 *ettingshausenii* (Feistmantel) Kidston 179
 *obovata* (Lindley and Hutton) Kidston 184, 188
Specimen interpretation 10
Speck, T., and Vogelleher, D. 85-6
*Sphaerostoma ovale* (Williamson) Benson 149
cf.*Sphaerostoma* sp. nov 137-8
*Sphenobaiera* 227
 *digitata* (Brongniart) Florin 234
*Sphenophyllum* **200**
 *cuneifolium* (Sternberg) Zeiller 210, 212-13, 214
 *insigne* Williamson 146
*Sphenopteridium*
 *capillare* Walton 176, 178
 *crassum* (Göppert) Schimper 163, 178, 186
 *macconochiei* Kidston 184, 186, 188
 *pachyrrachis* (Göppert) Schimper 121, 124, 163, 178, 186, **187**
 *rigidium* (Ludwig) Potonié 101, **102**
*Sphenopteris* 101
 *affinis* Lindley and Hutton 120, 149, 183
 *bifida* Lindley and Hutton 137, 186-8
 *cuneolata* Lindley and Hutton 170
 *elegans* Brongniart 149
 cf.*filiformis* Kidston 179
 *obfalcata* Walton 170, 179
 *warei* Dix 206
Spicer, R.A., and Thomas, B.A. 12
Spitsbergen 66
Sporangia 62, 77
*Sporogonites exuberans* Halle 68, 80
Sporophytes 3-4
Stairhill 229-33, **232**

# Index

*Stamnostoma* **112**
   *huttonense* Long 120-1, **122**, 122, 137
   *oliveri* Rothwell and Scott 137, 138
*Staphylotheca kilpatrickensis* Smith 163
*Stauropteris*
   *berwickensis* Long 119, 134
   *burntislandica* Bertrand 147, 148
*Steganotheca striata* Edwards 42
Stele 6
   and leaves, evolution 86
Stems, pteridosperm 135-6
*Stenokoleos* 91
*Stenomyelon tuedianum* Kidston 124
Stephanian 204
Sterome 5
*Stigmaria* 190
   *ficoides* (Sternberg) Brongniart 144, 160, 164, 185
cf.*Stigmaria* 169
Stomata 6, 77
Stoneley, H.M.M. 22, 233, 234, 235, 236
Strathmore Group 61, 75-7, **78**, 78-80, **79**
Straw, S.H. 44
Strobili 152
Study techniques 10
Stur, D. 183
*Svalbardia scotica* Chaloner 96, 97-100
*Svalbardia* Zone 96
Swamp-forests *see* Tropical swamp-forests
Symbiosis 91
Synangia 138
*Syncardia* 161

*Taeniocrada* sp. 68
*Taeniopteris eckhardtii* Kurtze 235
*Taitia* 31
*Taitia*? 72-3
Tantallon area (East Lothian) **129**
*Tantallosperma setigera* Barnard and Long 137
*Tarella trouwenii* Edwards and Kenrick 71, 73, 75
Targrove Quarry 42-3, 61-3, 66
Taxonomic radiation 6-8
Taxonomy 11
Taylor, T.N. 126
   *et al.* 91
Tedeleaceae 163, 200, 206
Teilia Quarry 175-81
*Telangium* 101, 152
cf.*Telangium* 138
Terrestrial ecosystem, *in situ* 80, 91-2
*Tetrapterites visensis* Sullivan 169
*Tetrastichia*
   *bupatides* Gordon **132**, 135

*Eosperma* correlation 134
Thatto Heath 213
Thin sectioning method 20, 113, 116
Thomas, B.A. 12, 169, 183, 185, 206
   and Brack-Hanes, S.D. 12, 145
   and Cleal, C.J. 213
   and Purdy, H.M. 169, 171
   and Spicer, R.A. 12
*Thursophyton milleri* (Salter) Nathorst 94
*Tomiodendron* 169
*Tortilicaulis transwalliensis* Edwards 45, 46, **48**
Tournaisian 113, 116-27, **117**, 155
Townrow, J.A. **236**
Tracheids, *in situ* 60, 62
Tree stumps 8
   Kingwater 140-1, **141**
   Laggan 152, 153, **154**
   Nant Llech 205
   Victoria Park 188-91, **189**, **190**
   Wadsley Fossil Forest 208-10, **209**
Tree-ferns, Upper Carboniferous 196, 200, **201**
Trewin, N.H., and Rice, C.M. 84
Triassic 225
*Trichopherophyton teuchansii* Lyon and Edwards 88
*Trichopitys* 227
Trigonocarpales (Medullosales) 176, 179, 181, 233
   Permian 227
   Upper Carboniferous **201**, 201
Trimerophytes 57, 69-70, 71
Trimerophytopsida 68, 72, 76, 79
*Triradioxylon primaevum* Barnard and Long 126, 136
*Tristichia ovensii* Long 121
Tropical extra-basinal areas, Upper Carboniferous 195
Tropical swamp-forests
   Early Carboniferous 143, 144, 147-8, 149, 164-6
   Late Carboniferous 195, **198**, 202, 204-7, 210-14, 214-17, 221
Turin Hill 62, **63**, 63-7

*Ullmannia*
   *bronnii* Göppert 236
   *frumentaria* (Brongniart) Göppert 234, 235
Ullmanniaceae 227, 233-6, **234**, **235**, **236**
Ulodendroid-scars 183
*Ulodendron* 144-5, 183
Upper Black Limestone Group *see* Gronant Group
Upper Carboniferous *see* Carboniferous, Upper
Upper Roman Camp Formation **40**, 41-2
*Uskiella spargens* Shute and Edwards 68, **73**, 73

# Index

Vascular plants 6
  evolution 6, **7**
  fungi and arthropods interaction 91-2
Vegetational history 3
Vesicles 91
Victoria Park ('Fossil Grove') 188-91, **189**, **190**
Virginia 31-4
Visean 113-4, 141-50, 150-2, 155, 160, 164-6, 167-72, 172-5, 181-4
Vogelleher, D., and Speck, T. 85-6
Vojnovskyaceae 228
Volcanogenic sites 113

Wadsley Fossil Forest 208-10, **209**
Wagner, R.H. 221, 231-2
*Walchia* **228**
Walchiaceae 227
Wales **29**
  North **33**, 113
  South **48**, **59**, 217
  Welsh Borderland **29**, **48**, **59**
  *see also* Dyfed
Walton, John 20, 81, 153, 154, 155, 176-7, 178, 179, 188
  and Lele, K.M. 169, 170-1, 172
  and Long, A.G. 149
  and MacGregor, M. 190
Wardie Shales Formation 113, 181-4, **182**, 188
Wardie Shore 181-4, **182**
Water relations 6
Water transport 9
Weak Law 139-40
Welsh Borderland *see* Wales
Wenlock Series 28, 29, **33**, 34
Westphalian 195, 196, 199, 203, 214-17, 217-19

Weyland, H., and Kräusel, R. 68-9
Whiteadder 116-27, **117**, **118**
Whole-plant reconstructions 10, 134-8
Wight, D.C., and Rothwell, G.W. 137
Willard, D.A. (Bateman, R.M. *et al.*) 131
Williamson, W.C. 153
Wilsonia Shales Formation 38
Wind transport 9
Witham, H.T.M. 20, 114-6
Wood, thin sectioning method 20, 114, 116

*Xenotheca* 101
Xylem 5, 6

Zakharova, T.V. 77
Zdebska, D., and Lemoigne, Y. 86
Zechstein Sea 235
Zimmermann, W. 86
Zosterophyllales 75, 76
Zosterophyllopsida
  Auchensail Quarry 79-80
  Craig-y-Fro Quarry 72, 73, 75
  Devonian **56**, 56-7
  Llanover Quarry 67, 71
  Rhynie 85, 88
  Targrove Quarry 62
  Turin Hill 64-7
*Zosterophyllum* 46, **56**, 56-7
  *fertile* Leclercq 66
  *llanoveranum* Croft and Lang 67, 68, **69**, 73
  *myretonianum* Penhallow **65**, 65-6, **66**, 67
*Zosterophyllum* Zone 62, 63, 64, 65-7
Zygopterid Limestone 143, 149
Zygopteridaceae 146-8, 161